Making Medicines in Early Colonial Lima, Peru

Atlantic World

EUROPE, AFRICA AND THE AMERICAS, 1500–1830

Edited by

Benjamin Schmidt (*University of Washington*)
Wim Klooster (*Clark University*)

VOLUME 34

The titles published in this series are listed at *brill.com/aw*

Making Medicines in Early Colonial Lima, Peru

Apothecaries, Science and Society

By

Linda A. Newson

BRILL

LEIDEN | BOSTON

over illustration: Drawing of mechoacan in Antonio de Robles Cornejo's "Examen de los simples medici-
ales," 1617. Courtesy Spain. Ministerio de Cultura. Archivo del Real Jardín Botánico, Madrid. División 1, leg
fol. 204.

ibrary of Congress Cataloging-in-Publication Data

ames: Newson, Linda A., author.
itle: Making medicines in early colonial Lima, Peru : apothecaries, science
and society / by Linda A. Newson.
escription: Leiden ; Boston : Brill, 2017. | Series: Atlantic world :
Europe, Africa and the Americas, 1500-1830, ISSN 1570-0542 ; VOLUME 34 |
Includes bibliographical references and index.
lentifiers: LCCN 2017029635 (print) | LCCN 2017031403 (ebook) | ISBN
9789004351271 (E-book) | ISBN 9789004350632 (hardback : acid-free paper)
ubjects: LCSH: Pharmacists--Peru--Lima--History. |
Medicine--Peru--Lima--History. | Materia medica--Peru--Lima--History. |
Pharmacy--Social aspects--Peru--Lima--History. | Science--Social
aspects--Peru--Lima--History. | Lima (Peru)--Social conditions. | Lima
(Peru)--Intellectual life. | Lima (Peru)--Relations--Spain. |
Spain--Relations--Peru--Lima. | Peru--History--1548-1820.
lassification: LCC RS67.P42 (ebook) | LCC RS67.P42 L538 2017 (print) | DDC
615.10985/25--dc23
C record available at https://lccn.loc.gov/2017029635

ypeface for the Latin, Greek, and Cyrillic scripts: "Brill". See and download: brill.com/brill-typeface.

SN 1570-0542
SBN 978-90-04-35063-2 (hardback)
SBN 978-90-04-35127-1 (e-book)

For Keith

Contents

Appendices

Acknowledgments

This monograph had its genesis is a previous book I published with Susie Minchin entitled, *From Capture to Sale: The Portuguese Slave Trade to Spanish South America in the Early Seventeenth Century* (Leiden: Brill, 2007). In that book we examined the medicines and medical practices that were involved in the treatment of African slaves both during their transhipment from Senegambia to Peru as well as in Lima. Arising from this previous study, I speculated that a conjunction of political, cultural and environmental conditions in Peru favoured early experimentation with native *materia medica*. I started with the propositions that the Counter Reformation encouraged medical practitioners of a more "progressive" persuasion to migrate from Spain to escape scrutiny by the Inquisition and that, while the development of empirical methods was generally encouraged by Spain's encounter with the New World, this was especially so in Peru given its rich botanical and mineral resources. As such, I postulated that medical practice in Peru might have been more "progressive" than in Spain at the time. I chose to focus on apothecaries since more than other professional medical practitioners they were pivotal in supplying Lima with medicines and, unlike physicians, had to hand the raw materials with which to experiment and thus possessed the potential to be pioneers in the development of medicine. As this book reveals, many of these original ideas proved to have limited foundation, at least for the period studied. Nevertheless, the research yielded a number of interesting findings that challenge some existing understandings of the nature of early modern medical practice. At the same time it revealed some underexplored facets of Lima's early colonial social history.

Another stimulus behind this book lay in the very exceptional archival sources for the study of early colonial medicine that exist in Lima, which I had encountered during research for *From Capture to Sale*. Of particular importance is the Archivo de Beneficencia Pública, which contains hospital and *cofradía* records from the mid-sixteenth century. Records of this kind are quite exceptional for Europe, let alone for Spanish America. I felt they deserved more extensive research, and they still do, especially since studies of professional medical practice in early colonial Spanish America have generally been hampered by the lack of archival sources.

I owe scholarly debts to a number of researchers who have generously supported this project by bringing to my attention unfamiliar sources. First, I would like to thank the brilliant and generous young scholar, Samir Boumediene, who I met while we were researching in the Jesuit archive in Rome. It was

Samir who drew my attention to the unpublished manuscript by Antonio de Robles Cornejo in the Archivo del Real Jardín Botánico in Madrid that proved to be a vital source for this study and one which made me think through many of the ideas in this book. Second, I would like to thank Pablo Gómez for bringing to my notice a manuscript recently acquired by the John Carter Brown Library referring to the installation of the pharmacy of Diego de Tineo in Lima in 1555. Finally, I would like to thank Susie Minchin, who worked with me on *From Capture to Sale*. Without the groundwork that we undertook for that project, this book would not have been either conceived or possible.

In the course of this research I have worked in many archives and libraries in Peru, Colombia, Spain, and Italy. In particular I would like to acknowledge the service provided by Yolanda Auqui at the Archivo General de la Nacion and Laura Gutiérrez Arbulú at the Archivo Arzobispal in Lima. Finally, this research would not have been possible without the generous financial support provided by the British Academy, the Wellcome Trust, and the Leverhulme Trust.

List of Illustrations

Figures

Tables

Abbreviations

Archival Sources

AAL	Archivo Arzobispal de Lima
ABPL	Archivo de Beneficencia Pública, Lima
AGI	Archivo General de Indias, Seville
AGNC	Archivo General de la Nación, Bogotá, Colombia
AGNP	Archivo General de la Nación, Lima, Peru
	SO CO Santo Oficio Contencioso
AHIRA	Archivo Histórico Instituto Riva-Agüero, Lima
AHML	Archivo Histórico Municipal de Lima
AHNM	Archivo Histórico Nacional, Madrid
AMS	Archivo Municipal de Sevilla
ARJBM	Archivo del Real Jardín Botánico, Madrid
ARSI	Archivum Romanum Societatis Iesu, Rome
BNM	Biblioteca Nacional, Madrid
JCB	John Carter Brown Library, Providence, Rhode Island
RAHM	Real Academia de Historia, Madrid
UNMSM	Archivo Histórico "Domingo Angulo," Universidad Nacional Mayor de San Marcos

CHAPTER 1

Medicines: Empire, Science and Society

> … the medicines of people around the Atlantic rim were intermingled, affecting not only the history of places Europeans went, but the history of Europe itself. They did so because medicine was not only a set of ideas and practices but was also part of the fabric of the commercial and colonial enterprises of the day.[1]

∵

Harold Cook and Timothy Walker have observed that one of the great ironies of the study of the history of medicine is that very little attention has been paid to medicines.[2] This is not wholly true in the case of Spanish America since a few scholarly studies exist of a limited number of American plants, such as guaiacum, sarsaparilla, mechoacan, cinchona (quinine) and tobacco, which made it into European pharmacopoeias in the early colonial period.[3] The relative shortage of studies of New World medicines reflects Spain's

1 Harold J. Cook and Timothy Walker, "Circulation of Medicine in the Early Modern Atlantic World," *Social History of Medicine* 26 (3) (2013): 348.

2 Ibid., 345.

3 Simon Varey, Rafael Chabrán and Dora B. Weiner, eds., *Searching for the Secrets of Nature: The Life and Works of Dr. Francisco Hernández*, 2 vols.(Stanford: Stanford University Press, 2000); Londa Schiebinger, *Plants and Empire: Colonial Bioprospecting in the Atlantic World* (Cambridge, Mass.: Harvard University Press, 2004); Londa Schiebinger and Claudia Swan, eds., *Colonial Botany: Science, Commerce, and Politics in the Early Modern World* (Philadelphia: University of Pennsylvania Press, 2005); Antonio Barrera-Osorio, *Experiencing Nature: The Spanish American Empire and the Early Scientific Revolution* (Austin: University of Texas Press, 2006); Antonio Barrera-Osorio, "Knowledge and Empiricism in the Sixteenth-century Spanish Atlantic World," in *Science in the Spanish and Portuguese Empires: 1560–1800.*, eds. Daniela Bleichmar, Paula de Vos, Kristin Huffine and Kevin Sheehan (Stanford: Stanford University Press, 2009), 219–32. For cinchona see Fiammetta Rocco, *Miraculous Fever-Tree: Malaria, Medicine and the Cure That Changed the World* (New York: HarperCollins, 2003); Samir Boumediene, *La colonisation du savoir: une histoire des plantes medicinales de "Nuevo Mundo" (1492–1750)* (Vaulx-en-Velin: Éditions des mondes à faire, 2016), 187–292 ; Matthew J. Crawford, *The Andean Wonder Drug: Cinchona Bark and Imperial Science in the Spanish Atlantic, 1630–1800* (Pittsburg, PA: University of Pittsburgh Press, 2016).

 | DOI 10.1163/9789004351271_002

continued reliance in the sixteenth and early seventeenth centuries on Asia as its main source of both drugs and spices, and also that its interest in American medical products was restricted to finding substitutes for those which were in short supply in Europe or were expensive. As such, no large-scale transatlantic trade in American medicinal products developed in the early colonial period.[4] Meanwhile, very little is known about the medicines that were employed by professional medical practitioners in the Americas at that time.

In the New World the Spanish encountered a diversity of environments and cultures, including healing traditions, which should have encouraged intellectual curiosity and experimentation with native botanical and mineral products. However, focussing on Peru, and Lima in particular, this study reveals that contrary to what might be expected, professional medical practitioners in the period prior to 1650 adopted very few New World products and continued to rely largely on *materia medica* of Old World origin, whether they were imported or grown locally. It argues that the development of experimental methods was discouraged by state regulation, the activities of the Church and Inquisition, and by entrenched values and cultural traditions that governed not only medical practice but also colonial society as a whole.

The study focuses on the daily practices of apothecaries since they were primarily responsible for the acquisition of *materia medica* and the preparation of medicines. As recent historians of science and medicine have argued, scientific practices can only be understood within particular historical, cultural, and geographical contexts. Consistent with this judgment, the book considers how apothecaries were trained, ran their business, traded and prepared medicines, and established their place in society, activities that reflected a complex set of cultural values and beliefs that reflected both transatlantic connections and local circumstances. Apothecaries and the medicines they

---

4 Francisco Guerra, "Drugs from the Indies and the Political Economy of the Sixteenth Century," in *Analecta Medico-Historica: Materia Medica in the XVIth Century*, ed. Marcel Florkin (Oxford: Pergamon, 1966), 48–52; J. Worth Estes, "The European Reception of the First Drugs from the New World," *Pharmacy in History* 37(1)(1995): 11–21; Miguel Cordero del Campillo, *Crónicas de Indias: ganadería, medicina y veterinaria* (Salamanca: Junta de Castilla y León, 2001), 125–27. In fact, the interest in Asian plants led to early experiments with the large-scale production of ginger, cloves, and China root (probably ginseng) in Mexico and the Caribbean islands (Paula S. De Vos, "The Science of Spices: Empiricism and Economic Botany in the Early Spanish Empire," *Journal of World History* 17(4)(2006): 409–26; John Slater, "The Green Gold Fallacies: Myth and Reality in the Transatlantic Trade in Medicinal Plants (1493–1663)," in *Geografías médicas: orillas y fronteras culturales de la medicina hispanomericana (siglos XVI y XVII)*, eds. José Pardo-Tomás and Mauricio Sánchez Menchero (Mexico DF: Universidad Autónoma de México, 2014), 99–122.

prepared therefore open a window on colonial society in Lima as a whole in the early colonial period and in particular on the neglected role, organisation, and status of professional groups in general. The book goes beyond being a case study, however, and contributes to understanding of the relationship between medicine and empire, the nature of early modern medicine, and the role of different types of professional medical practitioners in the development of medicine, and indeed science in general. An overview of these wider processes and debates is therefore presented below in order that the significance of the book's empirical findings can be fully understood and their significance appreciated.

Medicine and Empire

Historians have often linked the spread of modern scientific medicine to the expansion of empire, seeing it as facilitating overseas exploration, settlement, and administrative control.[5] Most research in this field has focussed on the nineteenth century by which time most Spanish American regions had achieved political independence. Yet, recent research has demonstrated that the spread of western science was not necessarily dependent on imperial support or control. In Latin America, newly-independent states often adopted modernising agendas that incorporated public health initiatives based on Enlightenment ideas and scientific medicine, which in common with colonial powers served to legitimise state control through their perceived beneficial and civilisational effects.[6] Science in general is usually regarded as being beneficial to society

5 David Arnold, ed., *Imperial Medicine and Indigenous Societies* (Manchester: Manchester University Press, 1988), 11; Roy MacLeod, "Introduction," in *Disease, Medicine and Empire: Perspectives on Western Medicine and the Experience of European Expansion*, eds. Roy MacLeod and Milton Lewis (London: Routledge, 1988), 2–4; Andrew Cunningham and Bridie Andrews, eds., *Western Medicine as Contested Knowledge* (Manchester University Press: Manchester, 1997), 1–12.

6 For a review of recent work in this field see Diego Armus, ed., *Disease in the History of Modern Latin America: From Malaria to Aids* (Duke University Press: Durham, 2003) part translated as "Legados y tendencias en la historiografía sobre la enfermedad en América Latina moderna," in *Higienizar, medicar, gobernar: Historia, medicina y sociedad en Colombia*, eds. Jorge Márquez, Álvaro Casa, and Victoria Estrada (Medellín: Universidad Nacional de Colombia, 2004), 13–39. The Spanish version includes a more extensive bibliography. See also Marcus Cueto and Stephen Palmer, *Medicine and Public Health in Latin America: A History* (New York: Cambridge University Press, 2015), 58–105, and pp. 279–81 for a useful bibliography on the topic.

in bringing rationality and freedom from superstition, though has also been viewed by some as being detrimental in facilitating state control and repression.[7] Arguably the link between medicine and imperialism is more evident in the case of scientific medicine since its objective methods aim to produce the "truth," making state control and intervention in public health both easier to justify and more difficult to challenge. Scientific medicine it is argued was, and is, inherently imperialistic.[8]

In examining the relationship between medicine and empire, David Arnold has suggested that prior to the nineteenth century medicine was not employed to extend and consolidate state control,[9] while others have argued that in as much as there was a relationship, colonial powers were primarily concerned with the health of Europeans and made few formal attempts to transform indigenous medical practices.[10] This conclusion is generally reached by those conducting research on Asia and Africa, where the tropical fevers took a heavy toll of European lives.[11] In early colonial Spanish America circumstances were different. The disease environment was less threatening to the health of colonial servants, such as bureaucrats, soldiers and the clergy;[12] there it was the indigenous population that died in millions.[13] Even if the health of Europeans

---

7 Charles W.J. Withers and David N. Livingstone, "Introduction: On Geography and Enlightenment," in *Geography and Enlightenment*, eds. David N Livingstone and Charles W.J. Withers (Chicago: The University of Chicago Press, 1999), 8 and Charles W.J. Withers, *Placing the Enlightenment: Thinking Geographically about the Age of Reason* (Chicago: the University of Chicago Press, 2007), 4, referring to Max Horkheimer and Theodor Adorno's seminal work, *Dialectic of Enlightenment* (New York: Herder and Herder, 1972).

8 Cunningham and Andrews, *Western Medicine*, 6, 10.

9 Arnold, *Imperial Medicine*, 11–12, 18.

10 MacLeod, "Introduction," 6–9.

11 See for example, Philip D. Curtin, *Death by Migration: Europe's Encounter with the Tropical World in the Nineteenth Century* (Cambridge: Cambridge University Press, 1989).

12 In fact the disease environment became more threatening with the introduction of the tropical fevers, malaria and yellow fever, but their full impact was not felt until after the mid-seventeenth century (J.R. McNeill, *Mosquito Empires: Ecology and War in the Greater Caribbean, 1620–1914* (Cambridge: Cambridge University Press, 2010).

13 There is an extensive literature on this topic. For an introduction see Alfred W. Crosby, "Virgin Soil Epidemics as a Factor in the Aboriginal Depopulation in America," *The William and Mary Quarterly* 33 (2) (1976): 289–299; Noble David Cook, *Born to Die: Disease and New World Conquests, 1492–1650* (Cambridge: Cambridge University Press, 1998) and for Peru *Demographic Collapse: Indian Peru 1520–1620* (Cambridge: Cambridge University Press, 1981).

in the New World was not of prime concern to the Spanish state, two other motives encouraged state intervention in the medical field, which both had the potential to transform the nature of medical practice and the *materia medica* employed.

First, as Antonio Barrera-Osorio and Paula De Vos among others have shown, from earliest years of conquest Spain was interested in native medicinal products from the perspective finding *materia medica* that could be commercialised and generate profits to strengthen its financial position.[14] While early state-sponsored expeditions in search of new products may not have affected medical practice or the types of medicinal products used in the colonies directly, as will be demonstrated in this study, the process of acquiring knowledge about New World products indirectly fostered new empirical approaches to medicine, which had the potential to affect the nature of medical practice and indeed science more generally.

Second, state intervention into medical realm was also driven by the Catholic Spain's desire to discharge its Christian duty to provide charity and care for the sick and, through the vehicle of healing, fulfil its providential mission to "civilise" and convert native peoples to Christianity.[15] In order to achieve its colonial ambitions, the Habsburg state worked closely with the Church, and as will be shown in some circumstances, including in healthcare, even depended

14 Barrera-Osorio, *Experiencing Nature*, 5; De Vos, "The Science of Spices," 399–427; Schiebinger, *Plants and Empire*; Jorge Cañizares-Esguerra, *Nature, Empire, and Nation: Explorations of the History of Science in the Iberian World* (Stanford: Stanford University Press, 2006), 56–60; Daniela Bleichmar, Paula De Vos, Kristin Huffine, and Kevin Sheehan, eds. *Science in the Spanish and Portuguese Empires, 1500–1800* (Stanford University Press, 2009) *passim*; Pratik Chakrabarti, *Medicine and Empire 1600–1960* (Basingstoke: Palgrave Macmillan, 2014), 20–26. For a similar argument see Pamela H. Smith, *The Business of Alchemy: Science and Culture in the Holy Roman Empire* (Princeton: Princeton University Press, 1997), 5, and in the context of the Dutch empire Harold J. Cook, *Matters of Exchange: Commerce, Medicines, and Science in the Dutch Golden Age* (New Haven: Yale University Press, 2007).

15 John H. Elliott, *Empires of the Atlantic World: Britain and Spain in America 1492–1830* (New Haven: Yale University Press, 2006), 184–87; Francisco Guerra. "The Role of Religion in Spanish American Medicine," in *Medicine and Culture*, ed. F.N.L. Poynter (London: Wellcome Institute, 1969), 179–88; Francisco Guerra, *El hospital en hispanoamérica y filipinas 1492–1898* (Madrid: Ministerio de Sandiad y Consumo, 1994); David Sowell, *The Tale of Healer Miguel Perdomo Neira: Medicine, Ideologies, and Power in the Nineteenth-Century Andes* (Wilmington, Delaware: Scholarly Resources Inc., 2001), 14–18; Gabriela Ramos, "Indian Hospitals and Government in the Colonial Andes," *Medical History* 57 (2)(2013): 186–205.

upon it to deliver its colonial mission. It was only with the Bourbon reforms of the eighteenth century that the Crown sought to wrest control of medical practice from the Church and religious orders and promoted the professionalization of medicine.[16] This study shows how the Church's involvement in the medical sphere directly affected the philosophical basis of medical practice and even the particular medicines employed. It will argue that in some respects these influences, while different in kind, were no less interventionist than those later associated with scientific medicine sponsored by the secular state.

In the fifteenth and sixteenth centuries the Spanish state regulated the practice of medicine more than any other European country.[17] It did this through the founding of chairs in universities for the teaching of medicine and through the establishment of medical boards, known as *protomedicatos*, which licensed practitioners and inspected pharmacies and hospitals.[18] Crown regulation was strongest in Castile; in Valencia, Navarra and Aragón it did not exert the same degree of power.[19] It was the bureaucratic and regulatory framework that was being developed in Castile that served as the basic model for Spanish state control of medical practice in the New World. However in crossing the Atlantic, some modifications were made, notably to the jurisdiction and responsibilities of the *protomedicato*. These aimed to take account of American circumstances, particularly the general shortage of trained personnel and the presence of native *curanderos*.[20] It is worth noting that even in Castile state medical institutions were still in the process of formation and much medical practice remained unregulated; as in the rest of Europe, popular healers and priests played a vital role in caring for the sick.[21]

16 Adam Warren, *Medicine and Politics in Colonial Peru: Population Growth and the Bourbon Reforms* (Pittsburgh, PA: University of Pittsburgh Press, 2010), 20, 35–77.

17 Guenter B. Risse, "Medicine in New Spain," in *Medicine in the New World: New Spain, New France, and New England*, ed. Ronald L. Numbers (Knoxville: University of Tennessee Press, 1987), 15; Paula S. De Vos, "The Art of Pharmacy in Seventeenth- and Eighteenth-Century Mexico." PhD diss., University of California, Berkeley, 2001, 6.

18 Michele Clouse, *Medicine, Government and Public Health in Philip II's Spain: Shared Interests, Competing Authorities* (Farnham: Ashgate, 2011). For apothecaries in particular see pp. 111–41.

19 María Luz López Terrada, "Medical Pluralism in the Iberian Kingdoms: The Control of Extra-academic Practitioners in Valencia," *Medical History Supplement* 29 (2009): 9–13.

20 Pilar Gardeta Sabater, "El nuevo modelo del Real Tribunal del Protomedicato en la América española: trasformaciones sufridas ante las Leyes de Indias y el cuerpo legislativo posterior," *Dynamis* 16 (1996): 237–59.

21 Francisco Guerra, "Role of Religion," 179–81; Luis S. Granjel, *La medicina española renacentista* (Salamanca: Ediciones Universidad de Salamanca, 1980), 133–50; Nancy G. Siraisi, *Medieval and Early Renaissance Medicine: An Introduction to Knowledge and Practice* (Chicago: University of Chicago, 1990), 7–9; Enrique Perdiguero, "Protomedicato y

In Spanish America the establishment of this medical infrastructure was never fully achieved. Universities capable of awarding medical degrees were only established in Mexico and Lima in 1631 and 1634 respectively; in many other regions they did not appear until the eighteenth century.[22] The result was a severe shortage of licensed physicians. Although the Salamancan physician Hernando de Sepúlveda was appointed as *protomédico* in Lima in 1537, it was only in 1568 that Antonio Sánchez Renedo was appointed as royal *protomedico* with powers equivalent to *protomédicos* in Spain; similarly it was only in 1570 that the Crown established the *protomedicato* in New Spain with the appointment of the naturalist and court physician, Francisco Hernández.[23] The delay in the establishment of the *protomedicatos* in Spanish America has been attributed not to Philip II's lack of interest in medical care, but rather his over-ambitious health programme and the lack of funds to carry it through.[24] As will be demonstrated in this book, the medical infrastructure that was established was therefore more partial, fragmented, and hence often less effective than was envisaged in the metropolis, which potentially at least created space for the emergence of alternative forms of medical practice.

Practices of Medicine

The medical infrastructure that was being developed in Spain and its colonies occurred against a backdrop of changes in the nature of medical practice in

curanderismo," *Dynamis* 16 (1996): 101–102; Mary Lindemann, *Medicine and Society in Early Modern Europe* (Cambridge: Cambridge University Press, 1999), 123–29; Guenter B. Risse, *Mending Bodies, Saving Souls: A History of Hospitals* (New York: Oxford University Press, 1999), 73–109; David Sowell, *Tale of Healer*, 14–15.

22 John Tate Lanning, *The Royal Protomedicato: The Regulation of the Medical Professions in the Spanish Empire*, ed. John Jay TePaske (Durham: Duke University Press, 1985), 325–31.

23 Hernando de Sepúlveda was appointed as *protomédico* in 1537. He accompanied Pedro de Alvarado's expedition to Peru and he remained there as a supporter of Francisco Pizarro. Although he received the title from the Crown, his appointment was approved by the *cabildo* who exercised its power to license medical practitioners according to Spanish pragmatics (*Libros de cabildos de Lima*, 1: 141–142 27 Apr. 1537; Lanning, *Royal Protomedicato*, 29–30; Luis Deza Bringas, *Testimonios del linaje médico peruano en los libros del cabildo de Lima, siglo XVI* (Lima: Universidad de San Martín de Porres, Facultad de Medicina Humana, 2004), 31–35, 173).

24 Lanning, *Royal Protomedicato*, 11; David C. Goodman, *Power and Penury: Government, Technology and Science in Philip II's Spain* (Cambridge: Cambridge University Press, 1988), 261–64.

Europe and the growing power of the Inquisition. Humoral medicine dominated in late Renaissance times.[25] At that time, sickness was thought to derive from an imbalance in the humours that could be redressed through diet, bleeding, and the application of medicines that acted as purgatives or emetics. Ill health was regarded as a divine punishment for sin and therefore required no further explanation. Humoral medicine was based on a natural philosophy that was elaborated through the study of classical texts. However, new approaches to medical practice, associated with the names of the chemists (Theophrastus Bombastus von Hohenheim) (1493–1541) and Jan Baptist Van Helmont (1580–1644) among others, were emerging. These argued that illness was caused by some external factor that through observation and experiment could be identified and the ailment treated. For them the development of medical knowledge lay in understanding nature rather than in scholasticism.[26] The empirical methods they adopted challenged the existing natural philosophy that was based on Christian interpretations of classical texts, which had largely been undertaken by clerics or by scholars working for institutions allied to the Church. Any opposition to the prevailing natural philosophy might therefore be seen as an attack on Christianity itself.[27] Moreover the empirical methods advocated by Paracelsians often used chemicals and minerals and were regarded as dangerously close to witchcraft. For these reasons, what came to be known as Paracelsianism, was opposed by the Catholic Church particularly during the Counter Reformation when the Inquisition sought to suppress it.[28]

In an effort to maintain orthodox medical practice, in 1559 Philip II banned Spaniards from studying abroad thereby reducing their access to the main European centres of medical education at Bologna, Padua, Paris and Montpellier.[29] Like many European countries at the time, Castile sought to control the

25 George M. Foster, *Hippocrates' Latin American Legacy: Humoral Medicine in the New World* (Langhorne, PA: Gordon and Breach:, 1994), 4–7, 12–14; Rebecca Earle, *The Body of the Conquistador: Food, Race and the Colonial Experience in Spanish America, 1492–1700* (Cambridge: Cambridge University Press, 2012), 19–41.

26 Steven Shapin, *The Scientific Revolution* (Chicago: University of Chicago Press, 1996), 69.

27 Ibid., 126, 136; William B. Ashworth, "Catholicism and Early Modern Science," in *God and Nature: Historical Essays on the Encounter between Christianity and Religion*, eds. David C. Lindberg and Ronald L. Numbers (Berkeley: University of California Press, 1986), 148–53; Cueto and Palmer, *Medicine and Public Health*, 23.

28 José Maria López Piñero, "Paracelsus and His Work in Sixteenth and Seventeenth Century Spain," *Clio Medica* 8 (2) (1973): 119–31.

29 Allen G. Debus, "Paracelsus and the Delayed Scientific Revolution in Spain: A Legacy of Philip II," in *Reading the Book of Nature: the Other Side of the Scientific Revolution*, eds. Allen G. Debus and Michael T. Walton (Kirksville-Missouri: Sixteenth Century Journal

inflow of foreign literature. From 1558 the Crown banned foreign book imports and obliged printers to obtain licences to print. At the same time the Inquisition was entrusted with censoring the publication of books deemed to be heretical.[30] Some debate exists about the impact of the Counter Reformation and Inquisition on the development of medicine, and indeed science generally, in Spain.[31] Some scholars see these measures as having a detrimental effect, while others argue that their impact was limited because few Spaniards sought training abroad, the laws were often circumvented or ignored, and works by Paracelsus were not unknown. A tribunal of the Inquisition was established in Lima in 1570 and this book explores the possible impact of the Inquisition in both Spain and Peru on local medical practices. First, it evaluates the success of the Inquisition in restricting the introduction and circulation of what might be called "progressive" medical texts, here defined as those based on empirical and experimental methods that among other things might advocate the use of chemical and mineral *materia medica*. Second, it explores whether the creation of a conservative intellectual environment in Spain that opposed the development of new approaches to medicine might have encouraged certain medical practitioners to migrate to Spanish America in order to practise more freely;[32] they might include medical practitioners of Jewish or Muslim heritage, known as *Conversos*, who although they had converted to Christianity, were particularly vulnerable to scrutiny by the Inquisition and being charged with heresy. Research for this book began with the proposition that for both groups the distant geographical location of Peru would have made it a particularly

Publishers, Inc., 1998), 148. All those studying overseas were ordered home except those enrolled in the theologically more orthodox colleges at Bologna, Rome, Naples, and Coimbra.

30 Henry Kamen, *The Spanish Inquisition: An Historical Revision* (London: Weidenfeld and Nicolson, 1997), 103–34.

31 José Maria López Piñero, *Ciencia y técnica en la sociedad española de los siglos XVI y XVII* (Barcelona: Labor Universitaria 1979), 141–44; David C. Goodman, "Philip II's Patronage of Science and Engineering," *The British Journal for the History of Science* 16(1) (1983): 50–53 and *Power and Penury*, 220–21; Kamen, *Spanish Inquisition*, 104–108. For recent evaluations of the impact of Paracelsus in Spain see: José Rodríguez Guerrero, "Censura y Paracelsismo durante el Reinado de Felipe II," *Azogue* 4 (2001) URL: http://www.revistaazogue.com [Accessed 19 May 2016]; Miguel López Pérez, "Spanish Paracelsus Revisited and Decontaminated," *Azogue* 7 (2013) URL: http://www.revistaazogue.com [Accessed 19 May 2016]; Mar Rey Bueno, "Los paracelsistas españoles: medicina química en la España moderna," in *Más allá de la Leyenda Negra: España y la revolución científica*, ed. Victor Navarro Brotons and William Eamon (Valencia: Instituto de Historia de la Ciencia y Documentación López Piñero, 2007), 41–55.

32 See for example, the Sevillian surgeon Pedro López de León who is discussed in Chapter 5.

attractive destination for those groups, but as will be shown in Chapter 5 it has found little evidence to support it.

While the regulation of medical practice and the control of the literature entering the New World by the Inquisition may have promoted a conservative intellectual environment for the practice of medicine in early colonial Peru, several factors promoted innovation. As in Spain, the medical infrastructure in the New World had its limitations, leaving space for the emergence of alternative methods of healing. Equally important was the rich diversity of plants, minerals, and other natural products found in the Andean region that aroused curiosity and encouraged experimentation. Of particular importance was the abundance of minerals, including mercury, which was a common ingredient in medicines promoted by Paracelsians.

Medicine and Science

Spain's encounter with New World peoples and environments generated new knowledge that challenged existing practices and beliefs. In fact, scholars such as Jorge Cañizares-Esguerra, Antonio Barrera-Osorio, and Mauricio Nieto have demonstrated that Spain's experience in the Americas, particularly during the early years of exploration had a profound effect on intellectual developments in Europe, including the Scientific Revolution.[33] They have argued that Spanish scientific achievements, notably in the fields of cartography, cosmology and natural history, were fundamental to the development of modern science. More specifically, others have suggested that the abundance of minerals found in the New World furthered an interest in the development of chemical methods associated with modern scientific medicine.[34] Despite the putative contributions of Spanish science to the development of modern science, Cañizares-Esguerra argues that they have generally been neglected in

33 Jorge Cañizares-Esguerra, "Iberian Science in the Renaissance: Ignored How Much Longer?," *Perspectives on Science* 12(1)(2004): 86–124, "Iberian Colonial Science," *Isis* 96 (2005): 64–70, and *Nature, Empire, and Nation*, 14–45; Mauricio Nieto Olarte, "Scientific Practices in the Sixteenth-Century Iberian Atlantic," in *Theorising the Ibero-American Atlantic*, eds. Harald E Braun and Lisa Vollendorf (Leiden: Brill, 2013), 149; Barrera-Osorio, *Experiencing Nature* and "Knowledge and Empiricism." For the importance of the encounter with foreign lands in contributing to the Enlightenment see Withers and Livingstone, "Introduction," 12.

34 Alix Cooper, *Inventing the Indigenous: Local Knowledge and Natural History in Early Modern Europe* (Cambridge: Cambridge University Press, 2007), 87–92; Chakrabarti, *Medicine and Empire*, 9.

the discourse around the Scientific Revolution and Enlightenment. This he attributes to the focus of historians of science on the contributions of astronomy, physics and mathematics to the exclusion of other scientific fields and to the prejudice of Protestant Northern Europeans against Catholic Spain. As such, the usefulness of the concept of the Scientific Revolution itself has been questioned especially by scholars of Spanish science.[35] It is not the aim of this book to demonstrate the intellectual contributions of Peruvian colonial science to modern science in general, but rather through a study of pharmacy practice in Lima contribute to this discourse by providing an enhanced understanding of the complex nature of science in Peru in the early modern period revealing the importance of both local circumstances and Atlantic connections.

The book's focus on medical *practice* follows recent research in the history of science. The history of medicine and indeed science in general has until recently been written around the achievements of a few prominent individuals or small bodies of practitioners and their role in the progress and triumph of scientific medicine. In 1967 George Basalla developed a seminal model about the spread of western science, in which he envisaged non-Western societies to be recipients of scientific knowledge and practices that had diffused from Europe.[36] His model has been criticised on many grounds. Of particular relevance for this study, is the consensus that western science was not the product of a one-way process, but that colonial practices were forged through interactions and exchanges with local scientists within both the colonies and Europe, and that they varied according to different cultural, social, political, and geographical settings.[37] Most scholars now see the history of science as a product of social processes that recognise the plurality and coexistence of modes of enquiry.[38] As such, in discussing the Scientific Revolution, Steven Shapin has argued that there is a need to move beyond such "free-floating concepts" to investigate the human practices that contributed to knowledge and that these can only be understood in their particular historical and geographical

35 See Victor Navarro Brotóns and William Eamon, "Spain and the Scientific Revolution: Questions and Conjectures," in *Más allá de la Leyenda Negra: España y la revolución científica*, eds. Victor Navarro Brotons and William Eamon (Valencia: Instituto de Historia de la Ciencia y Documentación López Piñero, 2007), 27–38.

36 George Basalla, "The Spread of Western Science," *Science* 156 (1967): 611–22. Basalla's model describes the diffusion of western science as a whole but it has relevance to medicine.

37 MacLeod, "Introduction," 2–7. Many of the criticisms are summarised in Dhruv Raina, "From West to Non-West Basalla's Three-stage Model Revisited," *Science as Culture* 8 (4) (1999): 497–516.

38 Pamela H. Smith, "Science on the Move: Recent Trends in the History of Early Modern Science." *Renaissance Quarterly* 62 (2) (2009): 358–61.

contexts.[39] The importance of geography in the development of science, specifically of the Enlightenment, has also been stressed by a number of scholars.[40] Some twenty years ago, Roy Macleod reported that there were increasing attempts to "localize [sic]" science.[41] Inasmuch as this has occurred in studies of Latin American science and medicine, they have focussed on the period of the Bourbon reforms and especially the Enlightenment;[42] while it remains the case that fewer studies have been conducted on professional medical practices in the late sixteenth and seventeenth centuries.[43]

Consistent with this perspective, this book aims to situate medical practice and pharmacy in particular within the specific cultural, historical and geographical context of Lima during the early colonial period. Very little has been written on pharmacy in colonial Spanish America. Paula De Vos has pioneered work in this field through her research on late colonial Mexico; her insights have been the touchstone for this study.[44] Luz María Hernández Sáenz has also studied apothecaries as a distinct professional group in late eighteenth and early nineteenth century Mexico.[45] In many respects this book complements these studies in its focus on the early colonial period and on Peru, a region that from a colonial perspective was geographically more distant from Spain and where the natural environment and cultures of the Andes presented different opportunities and challenges for the practice of pharmacy.

Apothecaries have been selected as the focus of this book since they were the medical practitioners directly involved in the practical tasks of acquiring

39 Shapin, *Scientific Revolution*, 4.

40 See David N. Livingstone and Charles W.J. Withers, eds. *Geography and Enlightenment* (Chicago: The University of Chicago Press, 1999); David N. Livingstone, *Putting Science in its Place: Geographies of Scientific Knowledge* (Chicago: The University of Chicago Press, 2003); Charles W.J. Withers, *Placing the Enlightenment: Thinking Geographically about the Age of Reason* (Chicago: The University of Chicago Press, 2007).

41 MacLeod, "Introduction," 3.

42 Antonio Lafuente, "Enlightenment Science in an Imperial Context: Local Science in the Late-Eighteenth-Century Hispanic World," *Osiris* 15 (2001): 155–73; Juan José Saldaña, "Science and Happiness during the Latin American Enlightenment," in *Science in Latin America: A History*, ed. Juan José Saldaña (Austin: University of Texas, 2006), 51–92. See also the chapters in the same volume by Luis Carlos Abboleda and Diana Soto Arango, and by Antonio Lafuente and Leoncio Lopez-Ocón.

43 James E. McClellan III, *Colonialism and Science: Saint Domingue in the Old Regime* (Baltimore: Johns Hopkins University Press, 1992), 6.

44 De Vos, "Art of Pharmacy," and "From Herbs to Alchemy: the Introduction of Chemical Medicine to Mexican Pharmacies in the Seventeenth and Eighteenth Centuries," *Journal of Spanish Cultural Studies* 8(2)(2007): 135–68.

45 Luz María Hernández Sáenz, *Learning to Heal: The Medical Profession in Colonial Mexico, 1767–1831* (New York: Peter Lang, 1997).

materia medica as well as preparing medicines, and were therefore more likely than physicians to be innovators in medical practice. They were uniquely qualified to judge the qualities of products and in a position to experiment with different doses and combinations and assess their efficacy. Such empirical methods were intrinsic to the art of pharmacy and for this reason apothecaries have been regarded as pioneers in the development of science.[46] Indeed, it has been argued that it was from apothecaries that physicians acquired knowledge of the natural world that enabled them to challenge the prevailing natural philosophy necessary for the development of modern science.[47] Scholars recognise that it may not have been the initial intention of apothecaries to provide alternative explanations for the causes of illness, but rather provide more effective treatments, which they continued to employ within in a traditional Galenic framework.[48] Nevertheless, they argue that it was through practice rather than theoretical reasoning that medical science progressed.[49] This book will examine the extent to which apothecaries in Lima were involved in empirical and experimental methods and will demonstrate that inasmuch as they used these procedures they did not represent a change in the natural philosophy underlying medical practice.

Practitioners of Medicine

This study is not a comprehensive account of medical practice in early colonial Lima. It focuses on professional medicine and does not explore in detail what might be called the popular sector or what Lawrence Brockliss and

46 The same argument has been made for alchemists who engaged with the natural world of minerals. See: Smith, "Science on the Move," 355–56.

47 Paula Findlen, *Possessing Nature: Museums, Collecting, and Scientific Culture in Early Modern Italy* (Berkeley and Los Angeles: University of California Press, 1994), 9–10, 241–56. For the role of apothecaries in undertaking early experiments in Italy and the adoption of recipes by physicians see: William Eamon, *Science and the Secrets of Nature: Books of Secrets in Medieval and Early Modern Culture* (Princeton: Princeton University Press, 1991), 7–11, 147–51, 357–59.

48 Lorraine Daston and Katharine Park, *Wonders and the Order of Nature 1150–1750* (New York: Zone Books, 1998), 136–37.

49 Smith, *Business of Alchemy*, 4. The edited volume by Pamela H. Smith, Amy R. Meyers and Harold J. Cook, *Ways of Making and Knowing: The Material Culture of Empirical Knowledge* (Ann Arbor: University of Michigan Press, 2014)) also argues that history of science was not built on a history of concepts alone, but on a history of making and using objects to understand the world.

Colin Jones have termed “the medical penumbra.”[50] While this may be regarded as a limitation, paradoxically we probably know more about the character o popular medicine in terms of methods of healing and medicinal products than we do about the *daily practices* of professional practitioners.[51] This is because popular healers were sometimes charged with witchcraft and brought before the ecclesiastical courts or Inquisition. It is worth noting, however, that the accused that appear in these studies may not be representative of the bulk of popular healers, whose practices were generally tolerated unless they involved in obviously superstitious or idolatrous methods, such as magic or divination.[52] Other scholars have found evidence for the practices of popular healers in legal cases brought by slave owners claiming that they had been sold slaves who were sick or had an undisclosed ailment that popular healers had been unable to cure.[53] This evidence is relatively abundant for Lima since African slaves comprised over half of the city’s population of some 27,000 in 1636.[54]

---

50 Lawrence Brockliss and Colin Jones, *The Medical World of Early Modern France* (Oxford: Clarendon Press, 1997), 8–20, 228–29.

51 See for example María Emma *Mannarelli*, “Inquisición y mujeres: las hechiceras en el Perú durante el siglo XVII,” *Revista Andina* 3(1) (1985): 141–56 and *Hechiceras, beatas y expósitas: mujeres y poder inquisitorial en Lima* (Lima: Fondo Editorial del Congreso del *Perú*, 1998); Ana Sánchez, *Amancebados, hechiceros y rebeldes* (*Chancay, siglo XVII*) (Cusco: Centro de Estudios Regionales Andinos, Bartolomé de las Casas, 1991); Alejandra B. Osorio, “‘El Callejón de la Soledad’: Vectors of Cultural Hybridity in Seventeenth-century Lima,” in *Spiritual Encounters: Interactions between Christianity and Native Religions in Colonial America*, eds. Nicholas Griffiths and Fernando Cervantes (Birmingham: University of Birmingham Press, 1999), 198–229, and “Hechicerías y curanderías en la Lima del siglo XVII: Formas femeninas de control y acción social,” in *Mujeres y género en la historia del Perú*, ed. Margarita Zegarra (Lima: CENDOC-Mujer, 1999), 59–75; Leo Garofalo, “Conjuring with Coca and the Inca: the Andeanisation of Lima’s Afro-Peruvian Ritual Specialists, 1580–1690,” *The Americas* 63 (1)(2006): 53–80; Linda A. Newson and Susie Minchin, *From Capture to Sale: The Portuguese Slave Trade to Spanish South America in the Early Seventeenth Century* (Leiden: Brill, 2007), 248, 251–52, 264–65; Claudia Brosseder, *The Power of Huacas: Change and Resistance in the Andean World of Colonial Peru* (Austin: University of Texas Press, 2014).

52 Nicholas Griffiths, “Andean *Curanderos* and their Repressors: The Persecution of Native Healing in Late Seventeenth- and Early Eighteenth-Century Peru,” in *Spiritual Encounters: Interactions Between Christianity and Native Religions in Colonial America*, eds. Nicholas Griffiths and Fernando Cervantes (Birmingham: University of Birmingham Press, 1999), 185–97.

53 Newson and Minchin, *From Capture to Sale*, 252, 264. The legal cases are to be found in the section “Causas de Negros” of the Archivo Arzobispal, Lima.

54 Frederick P. Bowser, *The African Slave in Colonial Peru, 1524–1650* (Stanford: Stanford University Press, 1974), 341.

Due to scholarly interest in popular healers, not only in Peru, but also other parts of the Spanish empire, there is an emergent view that in early modern times a form of medical pluralism existed and that medical practices were a fusion of indigenous, African, and European medical traditions.[55] The focus here on professional medical practitioners suggests that this view might need to be modified to some extent, at least in the case of early colonial Lima. This study suggests that there was not a unified free market in which medical practitioners competed for clients, but that it was segmented, with the professional sector being relatively distinct from the popular sector with which there was little exchange and from which few medicinal products were adopted. The study therefore questions the appropriateness of the concept of a "medical marketplace," as proposed by Roy Porter for early modern England, to Spanish colonial cities.[56] It also urges caution in believing that there was a free flow of medical ideas, practices and products.

Studies of the professional sector of medicine in early colonial Peru have focussed on institutional aspects of medical practice, including the *protomedicato*, the teaching of medicine in universities, and the establishment of hospitals.[57] Very little has been written about the daily practice of medicine and especially the role of apothecaries. Despite the book's ostensibly narrow focus on one group within the medical profession, it is clear that the role and status of apothecaries cannot be fully understood without reference to physicians

---

55 For example, Gonzalo Aguirre Beltrán, *Medicina y magia: el proceso de aculturación en la estructura colonial*, rev. ed. (México: Universidad Veracruzana, Instituto Nacional Indigenista, 1992); Noemí Quezada, *Enfermedad y maleficio: el curandero en el México colonia*, 2nd ed. (México: Universidad Nacional Autónoma de México, Instituto de Investigaciones Antropológicas, 2000); Luz Adriana Maya Restrepo, "Botánica y medicina africanas en la Nueva Granada, siglo XVII," *Historia crítica* 19 (2000): 27–47; Martha Few, *Women Who Live Evil Lives: Gender, Religion and the Politics of Power in Colonial Guatemala* (Austin: University of Texas Press, 2002); Laura Lewis, *Hall of Mirrors: Power, Witchcraft, and Caste in Colonial Mexico* (Durham, NC: Duke University Press, 2003); Pablo F. Gómez, "The Circulation of Bodily Knowledge in the Seventeenth-century Black Spanish Caribbean," *Social History of Medicine* 26:3 (2013), 383–402.

56 Roy Porter, "The Patient's View," *Theory and Society* 14(2)(1985): 175–98.

57 Key texts are: Hermilio Valdizán and Angel Maldonado, *La medicina popular peruana*, 3 vols. (Lima: Torres Aguirre, 1922); Juan B. Lastres, *Historia de la medicina peruana*. 3 vols. (Lima: Impr. Santa María, 1951); Guerra, *El hospital*, 429–75; Lanning, *Royal Protomedicato*; Miguel Rabí Chara, *Del Hospital de Santa Ana (1549 a 1924) al Hospital Nacional Arzobispo Loayza (1925 a 1999): 450 años de protección de la salud de las personas* (Lima: INDECOPI, 1999) and *El hospital de San Bartolomé de Lima: la protección y asistencia de la gente de color de 1646 a 1821 y evolución posterior: examen de un proceso histórico, social y sanitaria* (Lima: Grahuer, 2001); Warren, *Medicine and Politics*.

and surgeons, who are thus drawn into the discussion rather than excluded from it. Apothecaries warrant particular attention, however, for more than other professional medical practitioners they were pivotal in supplying Lima with medicines, interacting on the one hand with transatlantic merchants and on the other with local producers and traders of *materia medica*. They, unlike physicians, had to hand the raw materials with which to experiment and therefore had the potential to be pioneers in the development of medicine.

While the study focuses on the role of apothecaries in the practice of medicine, through concentrating on their daily practices, it also provides insight into their social standing within the city and into the manner in which professional groups in general organised their businesses and were regulated. Lima's social structure differed in some respects from other colonial cities. As capital of the Viceroyalty of Peru the city symbolised and promoted the authority of the Spanish state throughout South America and as such its social hierarchy reflected the traditional values that underpinned society in Spain.[58] Its viceregal role also meant that the state's bureaucratic and ecclesiastical presence was more substantial than elsewhere in South America potentially providing greater opportunities for regulation and social control. However, due in part to Peru's geographical distance from Spain, in reality state control was still limited, as shown in attempts to regulate medical practice and to define social groups in law as envisaged in the *sistema de castas*.[59] In the period under study, Peru was emerging from a bloody conquest and civil wars and the colony's social hierarchy was still in the process of consolidation. It faced the challenges of incorporating diverse ethnic groups and maintaining a hierarchy when opportunities for wealth creation in trade and mining provided opportunities for social advancement, even if ultimately the parameters within which people forged their lives were defined by the colonial state.[60] In such circumstances,

58 For the way in which the authority of the king was promoted through ceremonies see Alejandra B. Osorio, *Inventing Lima: Baroque Modernity in Peru's South Sea Metropolis* (New York: Palgrave Macmillan, 2008).

59 For succinct accounts of the *sistema de castas* see: Magnus Mörner, *Race Mixture in the History of Latin America* (Boston: Little, Brown and Company, 1967), 53–73; R. Douglas Cope, *The Limits of Racial Domination: Plebeian Society in Colonial Mexico City, 1660–1720* (Madison: The University of Wisconsin Press, 1994), 9–26; and Ilona Katzew, *Casta Paintings* (New Haven: Yale University Press, 2004), 39–61.

60 For the development of social hierarchies and their limitations see: Lewis, *Hall of Mirrors*, 15–41; Irene Silverblatt, *Modern Inquisitions: Peru and the Colonial Origins of the Civilized World* (Durham: Duke University Press, 2004), 4–27; María Elena Martínez, *Genealogical Fictions: Limpieza de Sangre, Religion, and Gender in Colonial Mexico* (Stanford: Stanford University Press, 2008), 271–72; Andrew B. Fisher and Matthew D. O'Hara, "Introduction:

individuals resorted to different ways of signifying their social status, for example, by dress, charitable works, or group membership.[61] This study demonstrates that apothecaries sought to indicate their status by distancing themselves from popular healers and by adopting the cultural attributes of elites. The impact of their social positioning, the study argues, was generally to weaken the innovative potential of their profession, already discouraged by the Church, Inquisition, and regulation.

It is hoped that this book will be a small response to James Lockhart's call for scholars to "simply to find out what went on in the Spanish colonies"[62] and contribute to knowledge of the social history of Lima at a time when its social and political structures were still in the process of formation and idealised views of how the colonies should be governed were adapting to the reality of the New World. It illustrates how individuals and groups forged their economic and social lives in a context where Spanish laws and regulations had to adapt to local circumstances and were only partially effective.[63]

Prospectus

Through an examination of the way in which apothecaries were trained, established their businesses, acquired and prepared medicines, this book argues that apothecaries had the potential to be innovators in science, especially in the New World where they encountered new environments and diverse healing traditions. However, it shows that despite experimental tendencies among some *boticarios* in Peru, especially in the very earliest years of conquest before regulatory measures were introduced, apothecaries in Lima generally adhered to traditional humoral practices, continuing to import *materia medica* from Spain and raise Old World plants locally, rather than adopt native botanical materials or take advantage of the region's rich mineral resources. It suggests that this adherence to orthodox practices did not derive primarily from

Racial Identities and Their Interpreters in Colonial Latin America," in *Imperial Subjects: Race and Identity in Colonial Latin America*, eds. Andrew B. Fisher and Matthew D. O'Hara (Durham: Duke University Press, 2009), 12–15: Rachel S. O'Toole, *Bound Lives: Africans, Indians, and the Making of Race in Colonial Peru*. Pittsburgh PA: University of Pittsburgh Press, 2012), 4.

61 Fisher and O'Hara, "Introduction," 15–23.

62 James Lockhart, *Spanish Peru 1532–1560: A Colonial Society* (Madison: The University of Wisconsin Press, 1968), 233.

63 Tamar Herzog, *Upholding Justice: Society, State, and the Penal System in Quito (1650–1750)* (Ann Arbor: University of Michigan Press, 2004), 5–8.

effective state regulation, which was applied flexibly and inconsistently according to local circumstances, but rather reflected the extent to which humoral beliefs were entrenched in Spanish thought and had permeated popular beliefs with the support of the Church and Inquisition. It responds to Roy MacLeod's appeal for scholars of colonial science to conduct more research "from the perspective of colonial life."[64]

---

64 Roy Macleod, "Nature and Empire: Science and the Colonial Enterprise," *Osiris* 15 (2000): 6.

CHAPTER 2

Learning to Make Medicines

> Study with learned men with many years' experience in the pharmacy, who are not immoral but virtuous.[1]

∵

Any understanding of pharmacy methods must necessarily take account of the training and licensing of apothecaries, which was specified in law. It was the intention that laws governing medical practice in Spain, specifically Castile, should be mirrored in its colonies.[2] These laws laid down the type of training that physicians, apothecaries and surgeons were to receive, which varied according to the tasks that each was expected to perform. Whereas in theory physicians were to be university educated, for the most part surgeons learned their skills through practical training. Apothecaries fell in between. On the margins of professional medicine and of lower social status were empirics, such as bonesetters, midwives and barbers, who acquired specialist technical skills through experiential learning.

Differences in training effectively created a social hierarchy of medical professionals that was based primarily on the level of university education each had received. Since university education was only open to those who could demonstrate purity of blood or *limpieza de sangre*, the legal training requirements in theory excluded persons from certain ethnic and social groups from aspiring to become physicians or apothecaries. Even though apothecaries generally had a lower status than physicians, Spanish laws reserved for them the exclusive right to prepare medicines. Despite the existence of laws regarding the training and licensing of apothecaries, as this chapter demonstrates, the *protomédicos* and *cabildos*, which were encharged with their compliance, were

1 Archivo del Real Jardín Botánico, Madrid, [hereafter ARJBM] División 1, legajo 17 Prohemiales 3 Libro de examen de los simples medicinales Antonio de Robles Cornejo 1617. The Spanish reads "aprenda con hombres doctos y experimentados de muchos años en la botica que no sea vicioso sino virtuoso."

2 *Recopilación de las leyes de los reynos de las indias* (Madrid: Gráficas Ultra, 1943), 2: lib. 5 tit. 6 ley 2: 140.

 | DOI 10.1163/9789004351271_003

forced to interpret them flexibly in the face of a lack of funding for medical education and a shortage of personnel.

Makers of Medicines

According to Spanish law, physicians were expected to attend the sick, diagnose illnesses, and prescribe medicines, but in theory at least were not permitted to prepare medicines, which remained the preserve of apothecaries.[3] A proposal to change Castilian ordinances to permit physicians to prepare medicines where no apothecary was available had been mooted in the late fifteenth century, but it was not incorporated into legislation.[4] The separation of tasks between physicians and apothecaries appears to have been driven less by an anxiety about the physical harm that physicians might inflict on patients and more by a concern about the fraud and excessive prices that might follow, if they both prescribed and prepared their own medicines. For the same reason the Castilian ordinances banned physicians from entering into business partnerships with particular apothecaries, stipulating that patients were to be free to send their prescriptions to any *boticario* they wished.[5] A particular worry was possible collusion between practitioners who were relatives.[6] However, physicians and surgeons who contravened the ban on preparing medicines received only modest fines for the first and second offences; in theory at least the third brought a two-year ban on practising in the place where they worked.[7] While these laws had been enacted in Spain and in theory were to apply to its colonies, they were often contravened. This was particularly true during the civil wars in Peru that immediately followed Spanish arrival, when in the absence of licensed practitioners and limited regulation, medicine was practised

3 *Recopilación de las leyes, pragmáticas reales, decretos, y acuerdos del Real Proto-Medicato: hecha por encargo y dirección del mismo Real Tribunal por Don Miguel Eugenio Muñoz* (Valencia: Imprenta de la Viuda de Antonio Bordazar, 1751), cap. XIII ordinance 3: 159–60 1537.

4 José Damián González Arce, "Los proyectos de ordenanzas generales de médicos, cirujanos y boticarios de Castilla (ca.1491–1513)," *Dynamis* 31(1)(2011): 222. It was the intention that the ordinances would apply to all Castilian territories, but there were disputes between physicians, opposition from the local authorities in other cities which had jurisdiction over medical practice, and the Royal Court, which meant that it was never adopted.

5 González Arce, "Proyectos de ordenanzas," 222.

6 *Recopilación de las leyes, pragmáticas reales* (Muñoz), cap. XIV ordinance 2: 170.

7 *Recopilación de las leyes, pragmátics reales* (Muñoz) cap. XIII ordinance 2: 159. Fines for the first and second offences were 10,000 and 20,000 *maravedís* respectively, which at 272 *maravedís* for one peso, were about 37 and 74 pesos respectively.

by a multiplicity of individuals from diverse backgrounds with differing levels of skill. Even after relative political stability had been achieved, the appointment of a royal *protomédico* encharged with the regulation of medical practice did not occur in until 1568; thereafter there were often gaps in appointments, when the *cabildo* rather than the state assumed responsibility for healthcare, as it had done since the city was founded in 1535.

The earliest medical practitioners to arrive in Peru accompanied armies of conquest. Some were licensed physicians, such as Hernando de Sepúlveda who accompanied Pedro de Alvarado to Peru in 1534 and was appointed as the first *protomédico* by the *cabildo* in Lima.[8] Another was Alvaro de Torres who provided medical support for Governor Cristóbal Vaca de Castro in the Peruvian civil wars of the 1540s.[9] Both these practitioners had graduated from the University of Salamanca and operated as both physicians and surgeons. However, many others, probably the majority, went as surgeons on armies of conquest when they performed the multiple roles of diagnosing sickness, treating wounds, and prescribing and preparing medicines. A military manual published by seasoned soldier Captain Bernardo Vargas de Machuca in 1599 specified the basic medicines that surgeons should carry with them in the field and indicated the types of remedies they should apply.[10]

In the early stages of colonial rule, therefore, it was surgeons rather than *boticarios* who prepared medicines; it was only after political stability had been achieved that trained *boticarios* moved to the New World. Meanwhile some surgeons who had arrived in Lima at an early date stayed on and sought acceptance as physicians or permission to prepare medicines, sometimes claiming they had lost their titles to practise or by demonstrating they had exceptional skill. Andrés Gonzales, who was of Portuguese descent and became a wealthy *boticario* and trader of medicines in Cartagena, does not appear to have had any medical training. When he was investigated as a foreigner in 1620, he testified that he had arrived in the Indies as a soldier, but finding a shortage of *boticarios* in the city had taken over the *botica* of one who had recently

8 *Libros de cabildos de Lima*, eds. Bertram T. Lee *et al.* (Lima: Torres Aguirre, 1935–1963), 1: 141–42 Licencia a dotor [sic] Sepúlveda 27 Apr. 1537; Deza Bringas, *Testimonios del linaje médico*, 31–35.

9 Gustavo Delgado Matallana, *Historia de la enseñanza médica en el Perú* (Lima: Universidad Nacional Mayor de San Marcos, 2008), 1–6.

10 Bernardo de Vargas Machuca, *The Indian Militia and Description of the Indies*, ed. Kris Lane, trans. Timothy F. Johnson (Durham, NC: Duke University Press, 2008), 60–69. See also Chapter 6.

died.[11] The *protomédico* and *cabildo* in Lima generally resisted moves by those without formal training to become licensed,[12] but it did allow some surgeons to practise surgery without licence.

Despite efforts by the municipal authorities and the *protomédico* to draw distinctions between physicians, surgeons and *boticarios*, they were blurred in the eyes of many and even in official sources practitioners were referred to inconsistently. Hence, when Francisco de Alva was made *boticario* to the Inquisition in 1574, he was described as being a surgeon and *boticario*.[13] Similarly, at the end of the century, in a legal case pursued by Pedro de Vergara for the recovery of debts for medicines he had dispensed, he was described as both a licensed *médico* and *boticario*.[14] Any local concern there may have been about physicians and surgeons preparing medicines focussed not on the illegality of the practice or the types of medicines they dispensed, but rather on their poor quality.[15] Hence in 1538, the *cabildo's* reaction to complaints that physicians and surgeons were selling poor quality medicines and charging excessive prices was not to impose fines and ban them from dispensing drugs, but rather to send inspectors to destroy the offending medicines and regulate their prices.[16]

Education and Practical Training

Despite the fact that many medicines were prepared illegally by persons other than *boticarios*, the educational and training requirements of different types of licensed medical practitioners were clearly distinguished. Physicians first had to obtain a Bachelor of Arts degree from a recognized university, which took four years, and then study for four years in a Faculty of Medicine in order to obtain a bachelor's degree in medicine. Finally, they had to work under the

11 Archivo General de Indias (AGI) Escribanía de Cámara 589B Denuncia contra Andrés Gonzales, boticario, Portugués 1620.

12 *Libros de cabildos de Lima*, 1: 161–62 Mandamiento de cirugía 30 Aug. 1537.

13 *Libros de cabildos de Lima*, 7: 549–550 Título de familiar al bachiller Alba [sic] 21 Jan. 1574. This position was later taken over by Juan de Bilbao (*Libros de cabildos de Lima*, 8: 201 Familiar del santo oficio a Juan de Bilbao 24 Feb. 1576).

14 Archivo General de la Nación, Peru (hereafter AGNP) Santo Oficio Contencioso (hereafter SO CO) 8–81 Pedro de Vergara, médico, licenciado, contra Nuño Rodríguez de Acevedo 23 Dec. 1599.

15 No detailed accounts exist of inspections of private pharmacies, but those that exist for hospital pharmacies (discussed in Chapter 4) and for those conducted in Spain, suggest that some inspections may have also specified that certain products should be acquired.

16 *Libros de cabildos de Lima*, 1: 181 Sobre las medicinas 23 Jan. 1538.

supervision of recognised doctors for two years before they were allowed to practise. Despite the last requirement, physicians, unlike *boticarios* and surgeons, actually received very little practical training.[17] However, they held the highest status due to their university education. In contrast, according to Spanish regulations in 1528 and 1563, surgeons and *boticarios* were not required to attend to university before being examined, but only show that they had practised under a recognised *boticario* or surgeon for four years.[18] From 1588 it was stipulated that licensed *boticarios* had to be at least 25 years old,[19] although training often began somewhat earlier. Writing in Peru, the Salamanca-trained physician Antonio de Robles Cornejo argued that training should not begin before the age of 15, since younger scholars did not have the strength or capacity to work and learn.[20] This was echoed by Antonio de Aguilera in his guidance to *boticarios* in his *Exposición sobre las preparaciones de Mesue* (1569) where he advised that a good apothecary should have "prudence and knowledge" (prudencia y ciencia), the acquisition of which required time. As such, he judged that although training might start at age 12 or 14 to 16 or 18, it took until 22, 24, or 26 for a *boticario* to acquire the necessary skills by working alongside an expert pharmacist.[21]

All apothecaries received on-the-job training, but not all undertook academic study. Nevertheless, the latter was considered desirable, since *boticarios* had to have knowledge of Latin in order to interpret physicians' prescriptions and to read manuals written on the preparation of medicines, which in the early sixteenth century were generally published in Latin.[22] This meant that in Spain at least apothecaries, unlike most surgeons, generally had a university education.[23] In the mid-sixteenth century it was proposed that physicians

17 *Recopilación de las leyes, pragmáticas reales* (Muñoz), cap. X ordinance 1: 137–38.

18 *Novísima recopilación de las leyes de España* (Madrid, 1805) lib. 8 tit. 13 ley 1: 106–107; *Recopilación de las leyes pragmáticas reales* (Muñoz), cap. XI ordinances 1–4: 150–52, cap. XII ordinances 1 and 2: 155–56.

19 *Novísima recopilación de las leyes de España*, lib. 8 tit. 13 ley 1: 106 1588.

20 ARJBM División 1, leg. 17 Prohemiales 3 Libro de examen de los simples medicinales Antonio de Robles Cornejo 1617. It reads that at age 15 years "tendra de esta edad fuerzas para trabajar en lo que fuere necesario y entendimiento libre para los secretos del [arte de farmacia]."

21 Antonio de Aguilera, *Exposición sobre las preparaciones de Mesue* (Alcalá: Juan de Villanueva, 1569), fols.17v–18.

22 Aguilera, *Exposición*, fol. 17; Christóval Suárez de Figueroa, *Plaza universal de todas ciencias y artes* (Madrid: Luis Sánchez, 1615), 301.

23 In fact some surgeons did attend university and were examined in Latin, in which case they were known as *cirujanos latinos*. Other surgeons without a university education, who were examined in Spanish, were known as *cirujanos romancistas*.

should write prescriptions in the vernacular, obviating the need for apothecaries to have knowledge of Latin. However, this suggestion was consistently rejected by the Crown which was anxious to restrict knowledge of potential harmful medicines to literate experts.[24] Latin therefore remained a legal prerequisite for obtaining a license to practise as an apothecary.[25] It also supported the status interests of *boticarios* since it distinguished them from surgeons and other healers and sellers of drugs, and allied them with higher status physicians. While those *boticarios* trained in Spain may have possessed Latin, as will be shown below this was not always the case in Lima.[26] It is worth noting that apart from Latin, *boticarios* were also required to have knowledge of arithmetic in order to work out weights and measures.[27]

From 1593 Spanish laws forbade women from practising pharmacy or owning *boticas* even if they did not practise pharmacy themselves but employed a licensed practitioner or salaried *oficial*.[28] However, widows often inherited *boticas* from their deceased husbands and in Seville the *cabildo* often permitted them to retain ownership of a pharmacy and either employ a trained *boticario* or pass its administration to a male relative.[29] This may have also been the case in Lima, where before eventually selling her deceased husband's *botica* Marina de Alva employed a Spanish *boticario* to administer it.[30] In Mexico City in the late eighteenth century a number of *boticas* were registered as being owned by women.[31] These may have been owned without licence since the law remained in force throughout the colonial period. Nowhere is any justification provided

24 Clouse, *Medicine,* 132.

25 *Recopilación de las leyes, pragmáticas reales* (Muñoz), cap. XII ordinance 1: 155, cap. XIV ordinance 9: 179.

26 ARJBM División 1, leg. 17 Prologo. Libro de examen de los simples medicinales Antonio de Robles Cornejo 1617.

27 ARJBM División 1, legajo 17 Prohemiales 2 Libro de examen de los simples medicinales Antonio de Robles Cornejo 1617.

28 *Novísima recopilación de las leyes de España,* lib 8. tit. 13 ley 3: 107 2 Aug. 1593. This stipulation was contrary to Roman law which allowed it (Lanning, *Royal Protomedicato,* 231). This restriction was not raised in the drafting of laws to be followed by the *protomédico* in Spain in the early sixteenth century (González Arce, "Proyectos de ordenanzas reales," 220–226).

29 Mercedes Fernández-Carrión and José Luis Valverde, *Farmacia y sociedad en Sevilla en el siglo XVI* (Sevilla: Servicio de Publicaciones del Ayuntamiento de Sevilla, 1985), 20.

30 AGNP Real Audiencia. Causas Civiles leg. 16 cuad. 81 fol. 422 Pleito de ejecución promovido por Baltasar de Zamora y Pedro Tenorio, mercaderes, vecinos de la ciudad de Trujillo y por otros acreedores, contra las bienes y herederos del bachiller Francisco de Alva, boticario morador en la Ciudad de los Reyes. 1576.

31 Hernández Sáenz, *Learning to Heal,* 151; De Vos, "Art of Pharmacy," 51–52.

for the ban on women practising pharmacy,[32] though moral attitudes of the time generally militated against the participation of women in the public sphere.[33] Even though women may not have been licensed to practise, and could not legally own *boticas*, they appear to have undertaken auxiliary activities in the pharmacy. In fact, Antonio de Robles Cornejo advised *boticarios* to employ their wives in keeping clean its utensils and equipment, and suggested they might even sell medicines when the *boticario* was absent.[34] Of course, the activities that wives actually undertook in the pharmacy may have extended beyond those specified. However, it remained the case that formal medical training was not available to them.

Apothecaries from Spain

As noted above, a few unlicensed practitioners with experience of preparing medicines in the field may have stayed on to practise and sought licences to do so, but in the earliest years of the colony most *boticarios* in Lima came from Spain with the intention to setting up businesses. Commonly these apothecaries had been licensed in Spain where they had possessed a *botica* or had worked as a salaried employee; some were the sons of *boticarios* who were following in the family tradition and once established in Lima continued to trade medicines with their relatives in Spain.[35] There was a lack of employment opportunities in Spain at this time and *boticarios* may have seen greater career prospects overseas.[36] However, migrating to the Americas required a licence from the *Casa de Contratación* and acquiring one was not a simple process. Initially the

---

32 Lanning, *Royal Protomedicato*, 231.

33 Nancy E. Van Deusen, *Between the Sacred and the Worldly: The Institutional and Cultural Practice of Recogimiento in Colonial Lima* (Stanford: Stanford University Press, 2001), 20–24.

34 ARJBM División 1, leg. 17 Prohemiales 6 Libro de examen de los simples medicinales Antonio de Robles Cornejo 1617.

35 Fernández-Carrión and Valverde, *Farmacia y sociedad*, 64. For *boticarios* in Spain who sent their sons to the Americas see: Diego, son of Antonio Bernal (Seville), AGI Contratación 5537 L1 fol. 210v. (Image 420) Licence granted to Diego Bernal 1557; Pedro, son of Francisco García (Plasencia) AGI Contratación 5255 N1 R68 Licence granted to Pedro García 9 May 1598; Juan, son of Pedro García de Poveda (Calahorra) AGI Contratación 5222 N1 R9 Licence granted to Juan García de la Poveda 1572.

36 For the lack of opportunities for trained physicians in Spain see Kristy Wilson Bowers, *Plague and Public Health in Early Modern Seville* (Rochester: University of Rochester Press, 2013), 29.

Crown encouraged immigration, but in the wake of the turbulent years of conquest and the rebellion of Gonzalo Pizarro, it decided to control the number of migrants going to Peru, especially of single men, for whom it was judged there were insufficient rewards in the form of *encomiendas* and limited employment opportunities. Immigration therefore became restricted to those who were deemed to be productive members of society and the migration of families was favoured.[37] Hence, the licences granted to individuals were generally given for a specified number of years only, after which they were to return to Spain or, if they were married, send for their wives.[38] The rule was difficult to enforce, but prosecutions were conducted if the opportunity arose. Hence, at the end of the sixteenth century the *boticario* Cristóbal Gómez, who had left his wife in Seville, was imprisoned in Lima and ordered to pay a fine of 300 pesos before being sent back to Spain as a prisoner. There he complained that all his wealth was in Peru and he successfully petitioned the *Casa de Contratación* to allow him to return to Peru with his wife, children and two servants.[39]

Licences to travel to the Americas also stipulated where individuals were to reside and for how long.[40] They might also require them to practise their profession and no other; migrants also had to provide bonds as security, especially if there was some doubt about their intentions.[41] Hence, the *boticario* Yuste de León who received a licence to travel to Tierra Firme had to provide a bond of 200,000 *maravedís* that he would practise his profession,[42] while Luis de León, travelling to the Tierra Firme as a servant to the *boticario*, Gerónimo Luis, was similarly required to post a bond for 200,000 *maravedís*, though in this case it was to ensure that he would remain in the region and not move to another for

37 Woodrow Borah, *Early Trade and Navigation between Mexico and Peru, Ibero-Americana* 38 (Berkeley and Los Angeles: University of California, 1954), 98–101, 107; Peter Boyd -Bowman, "Spanish Patterns of Emigration to the Indies to 1600," *Hispanic American Historical Review* 56 (4)(1976): 583.

38 *Recopilación de las leyes de las Indias*, lib. 1 tit. 7 ley 14 19 Oct 1544 and lib. 7 tit. 3 ley 8 2 Dec. 1578.

39 AGI Indiferente General 2104 N135 Licence granted to Cristóbal Gómez, boticario, 1599. For another case of single men being required to return to Spain for their wives see AGI Justicia 356 no 2 ramo 1 El fiscal contra Alonso Bernal, boticario vecino de Sevilla 1562–1569.

40 *Recopilación de las leyes de las Indias*, lib. 9 tit. 26 ley 41 19 Oct. 1566, 6 Oct 1578.

41 Ibid, lib. 9 tit. 26 ley 43 17 Apr. 1553, 7 Feb 1567; Ida Altman, *Emigrants and Society: Extremadura and Spanish America in the Sixteenth Century* (Berkeley: University of California, 1989), 208, 253.

42 AGI Contratación 5537 L3 fol.303 (Image 605) Licence granted to Yuste de León 5 Jan. 1569.

eight years.[43] It seems that greater restrictions were placed on those who were unmarried or who went as servants.

Licensed apothecaries were slow to arrive. In 1538 it was said there was only one *boticario* in the city, probably Alonso de Alemán.[44] At time, he and a number of physicians and surgeons were said to be selling expensive and poor quality out-of-date medicines, while others were preparing them illegally.[45] Another early arrival was *bachiller* Francisco de Alva, who became one of the most prominent apothecaries in Lima.[46] He like many other *boticarios* travelling to Lima in the sixteenth century migrated with his family. Another immigrant *boticario* was Cristóbal Carrasco, who had been examined in Madrid by the *protomédico* and chamber physician, Doctor Juan Gutiérrez de Santander, and had been practising in Spain for fifteen years. He had owned a *botica* in Plasencia and in 1581 was licensed to travel to Peru with his wife, two children and four servants.[47] Those who owned *boticas* in Spain often took drugs and equipment from their pharmacies with them. One such apothecary was Mançio Gasco who was a licensed *boticario* and *vecino* of Seville, who accompanied the physician Doctor Castellanos to Trujillo.[48] Similarly in 1591 when Juan Sánchez sailed to Peru with his wife and three children, he took with him fifteen medical books and nearly one hundred types of medicines worth 585 reals.[49]

Not all *boticarios* travelled to Peru with the intention of setting up new businesses. Some accompanied eminent officials, particularly the Viceroy. Unlike those who established businesses, these were mainly single men. Francisco

---

43 AGI Contratación 5538 L1 fol.405 (Image 809) Licence granted to Luis de León 14 Mar. 1582.

44 *Libros de cabildos de Lima*, 2: 115 El bachiller Pacheco malquerencia Dec. 1545. He had received a licence to travel to Tierra Firme on 3 Apr. 1538 (AGI Contratación 5536 L5 fol. 103r. Image 191).

45 *Libros de cabildos de Lima*, 1: 181 Sobre las medicinas 23 Jan. 1538, 1: 254 5 [Precios excesivos] Dec. 1538, 1: 256 [visita de la botica desta ciudad] 6 Dec. 1538, 1: 387 [Curas con e[n] salmos] 13 Oct. 1539.

46 In the sources consulted for this study, Francisco de Alva is the only *boticario* referred to as a *bachiller* (AGNP Real Audiencia leg. 16 cuad. 81 Inventario de los bienes del bachiller Francisco de Alva 9 Feb. 1576). He left for Lima with his wife about 1550 (AGI Lima 566 L6 N420), so his university education would have been in Spain.

47 AGI Indiferente General 2092 N86 Licence granted to Cristóbal Carrasco, boticario, 1581.

48 AGI Indiferente General 2090 N53 Licence granted to Mançio Gasco, boticario, 1578.

49 AGI Contratación 1097 N 5 197–202 Registro del navío Santa Catalina 1592. He appears in the ship's passenger list with his wife Francisca de Ortega and their children Ana, Gabriel and Mariana (Images 905–906). He was travelling with his father Cristóbal de Salcedo, aged 52, and his mother Catalina Rodríguez, who was a former resident of Nombre de Dios but had been living in Seville for 15 years. See also: AGI Contratación 5234B N1 R36 Licence granted to Juan Sánchez 1591 and Appendix 1.

Gutiérrez accompanied Diego López de Zúñiga y Velasco, the fourth Conde de Nieva in 1560,[50] and in 1569 Gerónimo Rodríguez travelled to Peru with Francisco de Toledo.[51] Unlike most other single migrants, no restrictions were placed on them in terms of their destination, length of stay, or the occupation they were to practise. The Spanish physician Antonio de Robles Cornejo who lived in Peru for many years observed that it had excellent medical practitioners, but was critical of those who arrived in the early years of Spanish settlement. He suggested that it was the less educated and skilled in all professions that migrated at that time, such that Peru had some apothecaries and physicians who were "ignorant and lazy."[52]

University Medical Education

The professional apothecaries who migrated from Spain had generally been formally educated and trained, but over time the majority were Creoles for whom the absence of the title 'bachiller' suggests that in the early colonial period at least most *boticarios* did not attend university. Aspirant *boticarios* did not have to attend courses in medicine, but in any case this was difficult since medical education in Peru was slow to become established. The University of San Marcos was founded in 1551 and in 1569 some medical instruction started with the arrival of the royal *protomédico*, Antonio Sánchez Renedo, who taught natural philosophy and astrology.[53] However, despite various proposals to found medical chairs that would enable the University to award medical degrees, this was delayed because of the lack of funds. This reflected in part the general lack of student interest in studying medicine because of the low status of the profession compared to the law or the Church.[54] The situation

50 AGI Contratación 5537 Lib. 2 fol. 97 (Image 195) Licence granted to Francisco Gutiérrez 7 Mar. 1560.

51 AGI Contratación 5537 Lib. 3 fol. 340 (Image 679) Licence granted to Gerónimo Rodríguez 1 Feb. 1569.

52 ARJBM División 1, leg. 17 Prologo Libro de examen de los simples medicinales Antonio de Robles Cornejo 1617.

53 Luis Antonio Eguiguren, *Alma mater: origines de la Universidad de San Marcos (1551–1579)*(Lima, 1939), 257–58, 337; Lastres, *Historia de la medicina peruana*, 49–55, 87–94; Lanning, *Royal Protomedicato*, 29–30; Matallana, *Historia de la enseñanza médica*, 76.

54 Susan E. Ramírez, *Provincial Patriarchs: Land Tenure and the Economics of Power in Colonial Peru* (Albuquerque: University of New Mexico Press, 1985), 176–77, 226; Eduardo Zarate, *Inicios de la escuela de medicina de Lima* (Lima: No Publisher, 2005), 27–28, 30. In Spain students showed an equal lack of interest in studying medicine (Javier Alejo

deteriorated after 1599 when the salaries paid to the incumbents of all university chairs were significantly reduced and support for the chair of medicine and philosophy, which had commanded a salary of 800 pesos, was withdrawn. In 1619 Viceroy Esquilache considered the lack of medical education to be a pressing issue since the general lack of physicians meant that barbers and surgeons were carrying out harmful treatments and were commanding large incomes without, it was said, being able to read or write.[55] He therefore argued that two chairs of medicine should be established. However, these did not materialise until 1634, when under growing pressure from the *cabildo* and the then *protomédico*, Doctor Juan de la Vega, the governing viceroy, the Conde de Chinchón, assigned income from the sale of *solimán* (sublimate of mercury) to support two medical chairs of *prima* and *visperas* with salaries of 600 pesos and 400 pesos respectively;[56] a move that was approved by the Crown in 1638.[57] Even so, salaries were often not provided and the chairs frequently remained vacant.[58]

Inasmuch as aspirant *boticarios* attended university, they would have pursued a general degree in the arts that enabled them to learn Latin, before undertaking practical training alongside licensed *boticarios*. In reality, those

---

Montes and Maria Carmen Rodríguez García, "Los estudios de la facultad de medicina en la Universidad de Salamanca de finales del siglo XVI," *Espacio, tiempo y forma, Serie IV, Historia moderna* 7 (1994): 38–39.

55 Archivo Histórico 'Domingo Angulo', Universidad Nacional Mayor de San Marcos (hereafter UNMSM), Archivo Histórico Tesorería General, Libro de Caja, ano 1619 (1577–1617] no. 28 Relación del estado, rentas y cátedras de la universidad de los reyes, Virrey Príncipe de Esquilache 10 Apr. 1619. It was suggested that the two chairs could be could be funded by the Crown supplementing the salaries paid to physicians to treat patients in the hospitals of Santa Ana and San Andrés by 300 *pesos ensayados* each.

56 AGI Lima 45 N4 fols. 146–147 El Conde de Chinchón 21 Apr. 1634, Márquez de Mancera 29 May 1640; Luis Antonio Eguiguren, *Diccionario histórico cronológico de la universidad real y pontificia de San Marcos y sus colegios* (Lima: Torres Aguirre, 1940), 1: 1016–1024; Lastres, *Historia de la medicina peruana*, 2: 51, 87–92; Lanning, *Royal Protomedicato*, 327–28, 331; Luis Martín, *The Intellectual Conquest of Peru: The Jesuit College of San Pablo, 1568–1767* (New York: Fordham University Press, 1968), 98; Delgado Matallana, *Historia de la enseñanza médica* 10–25, 104–112. These salaries were under half of those paid to professors of theology, law and the arts (Lanning, *Royal Protomedicato*, 327). The first chairs were filled by the *protomédico* Juan de la Vega (*prima*) and Andrés de la Rocha (*vísperas*) (José Antonio Suardo, *Diario de Lima (1629–1639)*, ed. Rubén Vargas Ugarte (Lima: "Lumen," 1936), 2: 76–77). The latter had only graduated with a doctorate in medicine in 1630 when he was examined by Juan de la Vega (Suardo, *Diario*, 107).

57 *Recopilación de leyes de las Indias*, lib. 1 tit. 22 ley 33 7 Mar. 1638.

58 AGI Lima 337 Real cédula 19 Jun. 1689.

working in *boticas* might acquire some knowledge of Latin without attending classes. They did, however, need to be literate.

Preparatory Schooling

The educational opportunities for students in Lima varied according to their class and ethnic background. Elite parents paid for their sons to be educated at primary level by private tutors, who could be either professional teachers or members of the religious orders. Tutors were usually contracted to teach reading, writing and arithmetic on an annual basis for a specified salary. Commonly they were paid 100 to 125 pesos for this instruction, though it could be less if the tutor resided in the pupil's home or more expensive if more advanced forms of instruction were required. For example, a merchant, Juan de Cardenas, paid a tutor 300 pesos over a year and a half for instructing his son in accounting practices.[59] Private tuition was only a practical proposition for the wealthiest families. Overtime, as the number of children increased, primary schools were established. Here fees were about half the cost of private tuition. Some of the first steps in education were taken by the religious orders that often ran classes attached to their convents, but other private schools also sprang up. By the beginning of the seventeenth century there were said to be ten or twelve approved primary schools, three of which were supported by the government to teach the poor free of charge. There were said to be about one thousand pupils attending these schools.[60] With the expansion of schools, in 1594 guidelines were drawn up that specified how pupils should be taught reading, writing, and simple arithmetic, and how they should conduct themselves.[61] About 1630 there were said to be twenty-eight *escuelas públicas* run by professional teachers for children.[62]

The desirability of founding secondary schools, which were residential and would prepare students for a university education by teaching students in Latin, was recognised at an early date. In 1575 Viceroy Toledo attempted to found

---

59 For examples of teaching contracts see: AGNP Protocolos Siglo XVI Ramiro Bote 14 fols. 1370-1370v. Francisco Díaz y Juan de Arriaça celebran concierto 14 Oct. 1597; Eguiguren, *Diccionario histórico cronológico*, 1: CCXCII, 2: 294–300; Luis Martín and Jo Ann Geurin Pettas, *Scholars and Schools in Colonial Peru* (Dallas, TX: Southern Methodist University, 1973): 115–17.

60 Bernabé Cobo, *Obras* (Madrid: Ediciones Atlas, 1956), 2: 414; Carlos Daniel Valcarcel, *Historia de la educación colonial* (Lima: Editorial Universo, 1968), 2: 31–32.

61 The guidelines are reproduced in Martín and Pettas, *Scholars and Schools*, 118–22.

62 Fray Buenaventura de Salinas y Córdova, *Memorial de las historias del Nuevo Mundo Pirú* [sic]. Vol. 1 (Lima: Universidad Nacional Mayor de San Marcos, 1957), 257.

the secular college or *colegio* of San Felipe y San Marcos for the sons of conquistadors and *beneméritos*, but it did not materialise until 1592 and then only admitted 20 students. Its ordinances banned "gente baja" (the lower classes), Mulatos and Zambahigos from attending as well as those condemned by the Inquisition.[63] Like many other institutions whose financial support derived in part from *encomiendas*, the decline in Peru's native population severely affected the College's income.[64] By the beginning of the seventeenth century, therefore, students had to pay their own subsistence costs.[65] Meanwhile, in 1582, the Jesuits had founded the *colegio real* of San Martín for the sons of conquistadors, *vecinos* and other "gente noble" (elite persons).[66] In the early seventeenth century and in contrast to San Felipe and San Marcos, the *colegio real* of San Martín was said to be flourishing and providing a good education, specialising in teaching philosophy and theology.[67] In 1612 it had 220 students who each paid 150 to 200 *pesos de nueve reales* a year for instruction, food, lodging, and medical care.[68] Despite supposedly being a school for the elite, unlike the *colegio* of San Felipe y San Marcos, initially it seems to have admitted persons from all ethnic backgrounds for by the mid-seventeenth century the Rector had to

63 Cobo, *Obras*, 2: 438–40. For the ordinances of the colegio of San Felipe y San Marcos by the Marqués de Cañete in 1592 see: AGI Lima 323A Ordenanzas del colegio de San Felipe y San Marcos 8 Aug. 1592; Pedro Manuel Alonso Marañón, "Constituciones del colegio real de San Felipe y San Marcos de Lima," *Estudios de historia social y económica de América*, 15 (1997): 419–34 which contains a transcription of the original manuscript in the Biblioteca Nacional Madrid, Manuscript 3043.

64 Cook, *Demographic Collapse*.

65 AGI Lima 337 La universidad de San Marcos 1 May 1601; UNMSM, Archivo Histórico Tesorería General, Libro de Caja, ano 1619 (1577–1617) no. 28 Relación del estado, rentas y cátedras de la universidad de los reyes, Virrey Príncipe de Esquilache 10 Apr. 1619.

66 ARSI Perú 4 fol.7 Catalogo común de las cosas y personas de la provincia del Perú 1583; Cobo, *Obras* 2: 436–38; AGI Lima 208 N11 2 May 1586. Its foundation was approved by Viceroy Martín Enríquez in 1586.

67 AGI Lima 322 Colegio de San Felipe y San Marcos 28 May 1604; AGI Lima 35 Número 5 Libro v fol. 30v Relación de las provincias, conventos … que tiene las ordenes en el Perú 162; UNMSM, Archivo Histórico Tesorería General, Libro de Caja, ano 1619 (1577–1617] no. 28 Relación del estado, rentas y cátedras de la universidad de los reyes, Virrey Príncipe de Esquilache 10 Apr. 1619.

68 AGI Lima 275 fol. 521v. Relación de lo que se paga de vino, aceite i medicinas … 1612. See also Reginaldo de Lizárraga, *Descripción breve de toda la tierra del Perú, Tucumán, Río de la Plata y Chile.* Biblioteca de autores españoles 216 (Madrid: Ediciones Atlas, 1968), cap. 46: 35; Antonio Vázquez de Espinosa, *Compendio y descripción de las Indias occidentales*. Biblioteca de autores españoles 231 (Madrid: Ediciones Atlas, 1969), 306; Salinas y Córdova, *Memorial*, 182; Eguiguren, *Diccionario histórico cronológico*, 2: 302.

be reminded not to admit Mulattoes and Mestizos.[69] Finally, the *colegio* of San Toribio was founded in 1594 for those seeking ordination as secular priests and was said to have "good and virtuous students."[70] Educational provision was dogged by jurisdictional disputes between the secular authorities and the religious orders. The latter, especially the Jesuits at the College of San Pablo, ran popular classes for lay students and thereby competed with providers elsewhere, and especially the University of San Marcos. This dispute was partially resolved in 1580 after which San Pablo became the preparatory school for students wishing to study the humanities at the University. At the same time, San Pablo received the exclusive right to teach Latin and Greek, but it could not award its own degrees.[71] Henceforth the University was the only institution that could award bachelors' degrees, masters' degrees, and doctorates.

Educational Opportunities for Non-Elites

The educational opportunities for non-elites varied by ethnic group. The Crown was of two minds about the education of Indians. On the one hand it considered the education of native elites to be fundamental to its mission to Christianise and 'civilise' native peoples and to be means of facilitating their political control; on the other it was concerned that literacy and knowledge of the law might enable indigenous people to challenge the privileged position of Spaniards and Creoles.[72] Others argued that schools for native children would divert much needed financial support from the Spanish sector to Indians, who as conquered and 'uncivilised' people were less worthy of education.[73] In as much as it was thought desirable to educate Indians, state support focussed on provision for native elites. It was late in coming, however. A royal school for the sons of indigenous leaders was not established in Lima until 1618 when the Crown founded the Colegio del Príncipe in El Cercado.[74] Prior to this, the

69 Cobo, *Obras*, 2: 425; Martín, *Intellectual Conquest*, 36–38.

70 Cobo, *Obras*, 2: 440–41; UNMSM, Archivo histórico Tesorería General, Libro de Caja, ano 1619 (1577–1617) no. 28 Relación del estado, rentas y cátedras de la universidad de los reyes, Virrey Príncipe de Esquilache 10 Apr. 1619.

71 Martín, *Intellectual Conquest*, 16–33; Eguiguren, *Alma mater*, 495–98.

72 Monica Alperrine-Bouyer, *La educación de las élites indígenas en el Perú colonial* (Lima: Instituto de Estudios Peruanos, 2007), 31–33.

73 Ibid., 84–85.

74 Alcira Dueñas, *Indians and Mestizos in the "Lettered City"* (Boulder: University Press of Colorado, 2010), 16–18; Alperrine-Bouyer, *Educación de las élites indígenas*, 75, 89; John Charles, "Trained by Jesuits: Indigenous letrados in Sixteenth-Century Peru," in *Indigenous*

religious orders had offered classes to Indians, with the Jesuits establishing a school in El Cercado in 1568.[75]

Unlike Indians, who were regarded as free subjects of the Crown and therefore did not have to demonstrate *limpieza de sangre*, this requirement represented a potential obstacle to the education and later to the acquisition of a license to practise for non-elites.[76] As María Elena Martínez has shown, the concept of *limpieza de sangre* was originally framed in Spain on the basis of genealogy and religious adherence, but even there it was never clearly defined.[77] In Spanish America the presence and intermarriage of persons from different ethnic groups, made the demonstration of *limpieza de sangre* even more difficult, such that the granting or denying of the status of *limpieza de sangre* came to rely on oral testimony, reputation, or skin colour, thereby rendering it open to interpretation.[78] The University of San Marcos was modelled on the universities of Salamanca and Valladolid in Spain, where university attendance required students, at least in theory, to be legitimate and demonstrate *limpieza de sangre*,[79] but even in Spain the requirement to demonstrate purity of blood for entry into university and obtain a license to practice medicine was often overlooked.[80] Similarly, in Peru the adherence of the authorities to this legal requirement was not constant over time and was often interpreted flexibly.[81]

In Lima, initially at least, the concept of *limpeza de sangre* did not have racial connotations and there does not appear to have been any formal barrier to Indians, Blacks or persons of mixed race from attending university there.

Intellectuals: Knowledge, Power and Colonial Culture in Mexico and the Andes, eds. Gabriela Ramos and Yanna Yannakakis (Durham: Duke University Press, 2014), 61–62, 72–74.

75 Eguiguren, *Diccionario histórico cronológico*, 1: 773–74; Martín and Pettas, *Scholars and Schools*, 18.

76 John Tate Lanning, *Academic Culture in the Spanish Colonies* (Oxford: Oxford University Press, 1940), 40.

77 For the origin of the concept of *limpieza de sangre* and its incorporation into Spanish legislation see Martínez, *Genealogical Fictions*, 25–60.

78 Ibid., 168–70.

79 Stafford Poole, "The Politics of *Limpieza de Sangre*: Juan de Ovando and His Circle in the Reign of Philip II," *The Americas* 55 (1999): 363, 388.

80 José Luis Valverde and Guillermina López Andújar, "Los disposiciones legales sobre el certificado de 'limpieza de sangre' en relación al ejército profesional farmacéutico," *Boletín de la sociedad española de la historia de la farmacia* 31 (1981): 259–69; Kamen, *Spanish Inquisition*, 233–34, 241–43.

81 Lanning, John Tate, "Legitimacy and *Limpieza de Sangre* in the Practice of Medicine in the Spanish Empire," *Jahrbuch für Geschicte von Staat, Wirtschaft und Gesellschaft Lateinamerikas* 4 (1967), 40–41, 46–52, 58–60.

Indeed, in 1576 Viceroy Francisco Toledo saw the university education of Mestizos as rendering them useful citizens, who through their knowledge of native languages could be instrumental in converting the Indians to Christianity and integrating them into colonial society.[82] In the 1602 constitution of the University of San Marcos, the only social group specifically excluded under statute 238 were those who had been sentenced by the Inquisition.[83] The rather liberal stance taken by the University in the early colonial period allowed people from diverse ethnic backgrounds to attend. However, an individual's lack of purity of blood always be could be raised at any stage, perhaps by those seeking to eliminate rivals or damage reputations.[84] In the seventeenth century, there were growing concerns, often from white professionals seeking to protect their privileged position, that the influx of non-whites into the University was damaging its reputation and its degrees. This led to a harder stance being taken on admissions.[85] In 1678 a *cédula* was issued reinforcing statute 238, but this time specifying that Mestizos, Mulattoes, *Zambos* and *Quarterones* should be excluded.[86] Even so, the edict was not consistently applied.

For those wishing to progress to the University of San Marcos another significant obstacle was cost. Its 1602 constitution specified that those wishing to graduate with a bachelor's degree in law or the arts had to pay 200 reals, or in theology and medicine 300 reals. A *licenciatura* cost 310 reals a year. This did not include the chickens, sweetmeats, candles,[87] and other goods and tips paid to various university personnel, as well as the cost of a dinner for the rector,

---

82 Cobo, *Obras*, II 406 Viceroy Toledo 3 Oct. 1576; Van Deusen, *Between the Sacred and the Worldly*, 55, 217n120.

83 Constituciones y ordenanças de la Vniuersidad, y Studio General de la Ciudad de los Reyes del Piru, 1602 at: https://archive.org/details/constitucionesyoo1univ [Accessed 10 Mar. 2016] Statute 238 p. 41. This included those whose parents or grandparents had been sentenced by the Inquisition.

84 See for example Juan Méndez Nieto, who despite calling himself licenicado, actually failed to get complete his medical studies in Spain, it is suggested due to the lack of a certificate of *limpieza de sangre* (Juan Méndez Nieto, *Discursos medicinales* (Salamanca: Universidad de Salamanca, 1989), XI–XII). See also the ridicule of the noble birth of Dr. Francisco Machuca in the poetry of Juan del Valle y Caviedes, in Juan del Valle y Caviedes, *Obra completa*, ed. Daniel R Reedy (Caracas: Biblioteca Ayacucho, 1984), 50–52.

85 Lanning, *Royal Protomedicato*, 182–87 and "Legitimacy and *Limpieza de Sangre*", 46–48; José R. Jouve Martín, *The Black Doctors of Colonial Lima: Science, Race, and Writing in Colonial and Early Republican Peru* (Montreal: McGill-Queen's University Press, 2014), 21–24.

86 *Recopilación de las leyes de las Indias*, lib. 1 tit. 22 ley 57 10 Dec. 1678.

87 Actually 'hachas de cera', which were large candles composed of four tapers with four wicks.

doctors and officials of the University on the day of the examination.[88] While this represented a considerable outlay, the University statutes did allow fees to be reduced for those of modest means.[89] Nevertheless, the cost must have been an obstacle for many since in 1601 the University itself petitioned the Crown to allow students to graduate up to the level of *licenciatura* free of charge.[90]

As will be shown in the next chapter, establishing a pharmacy was an expensive proposition, so that those who aspired to be licensed *boticarios* and own their own business invariably came from elite families whose sons were probably taught by private tutors or else attended one of the city's *colegios*. Some may have gone on to university, but others probably acquired some knowledge of Latin at one of the *colegios* where classes were often conducted in Latin.[91] Others who were employed in *boticas* or worked in hospitals or convents would have learned their trade through working alongside licensed practitioners from whom they might learn sufficient Latin to be able to read physicians' prescriptions. Nevertheless, it is clear that not all *boticarios* had knowledge of Latin, for Antonio de Robles Cornejo, after 25 years' experience in the Americas including Peru, felt compelled to write his 'Examen de los simples medicinales' in Spanish in order to address the fact that many *boticarios* were unfamiliar Latin. He argued that this lack of knowledge of how to make compound medicines, which depended on reading manuals written in Latin, resulted in them relying on imports from Spain which were often of poor quality and had lost their potency by the time they arrived.[92]

On the Job Training

The opportunities for learning the art of pharmacy through on the job training were rather limited in Lima; even in 1630 there were only twelve recognised

88 Constituciones y ordenanças de la Vniuersidad, y Studio General de la Ciudad de los Reyes del Piru, 1602 at: http://www.archive.org/details/constitucionesy001univ [Accessed 28 Oct. 2015] Statute 191 p. 32–32v., Statute 212 fol. 36v.-37. Cobo. *Obras*, 2: 413 gives slightly different charges: bachelor of arts or law 25 pesos; bachelor of theology 35 pesos; and master of arts 1,000 pesos. Doctors of theology and law were charged 2,500 pesos and doctors of medicine 3,000 pesos, all pesos of 9 reals.

89 Constituciones y ordenanças de la Vniuersidad, Statute 242, p. 42; Lanning, "Academic Culture," 47.

90 AGI Lima 337 Universidad de San Marcos 1 May 1601.

91 Martín, *Intellectual Conquest*, 40–41; Alperrine-Bouyer, *Educación de las élites indígenas*, 194.

92 ARJBM División 1, leg. 17 Prologo, Libro de examen de los simples medicinales 1617.

boticarios in the city.[93] In some cases knowledge of pharmacy was passed from father to son. Pedro de Bilbao was undoubtedly inducted into the art by his father, Juan de Bilbao, who was an eminent *boticario* in Lima in the late sixteenth century and from whom he inherited his *botica*.[94] However, as in Spain, others sought formal apprenticeships.[95] The contractual obligations of apprentices in general were specified by the guild associated with their trade.[96] Initially, the *cabildo* in Lima tried to regulate the city's economic activities indirectly through setting prices; later it did so more directly through encouraging the establishment of guilds or *gremios* under its jurisdiction.[97] In 1552 the *cabildo* ordered different trades to nominate four candidates from their professions to serve as *alcaldes* from which the *cabildo* would then select two. Tradesmen generally resisted any form of regulation, but with the arrival of larger numbers of immigrants from Spain and the emergence of artisans from other ethnic groups, they were persuaded of the merits of forming guilds. Even though the establishment a guild signified greater scrutiny by the *cabildo*, it could function to protect the privileged position of its members.[98] Comprised mainly of *maestros*, among other things the guilds examined and licensed artisans, supervised the quality of goods they produced, and settled disputes among their members.[99] However, guild ordinances, which often excluded certain ethnic groups, had to be approved by the *cabildo*, and the elections of guild officials, the examination of apprentices, and inspections of

93 Salinas y Córdova, *Memorial*, 257. At the same there were nine *médicos*, ten surgeons and three Spanish barbers' shops with many *oficiales*.

94 AGNP Protocolos Siglo XVI 14 Ramiro Bote fols. 1629–1631 Juan de Bilbao dona a Pedro de Bilbao, su hijo, de una botica con todas sus botes y drogas 3 Aug. 1596. Similarly in Mexico City in the eighteenth century, pharmacy was often a family profession (Hernández Saenz, *Learning to Heal*, 147–48).

95 For fourteenth century apothecary apprenticeships in Spain see Javier Sorní and José María Suñe, "La farmacia en Barcelona desde *Alfons el Benigne* a *Pere el Ceremonios* (1327–1387)," *Boletín de la sociedad española de historia de la farmacia* 36 (1985): 70–71. At that tiem they were known as spicers rather than *boticarios*. For apprenticeships associated with various trades in Lima see: Lyn Brandon Lowry, "Forging an Indian Nation: Urban Indians Under Spanish Colonial Control, Peru 1535–1765." PhD Diss. University of California, Berkeley, 1991, 209–213; Bianca Premo, *Children of the Father King: Youth, Authority and Legal Minority in Colonial Lima* (Chapel Hill: University of North Carolina, 2005), 55.

96 Francisco Quiroz Chueca, *Gremios, razas y libertad de industria: Lima colonial* (Lima: Universidad Nacional Mayor de San Marcos, 1995), 19–20.

97 The trades to which this applied included tailors, shoemakers, silversmiths, carpenters and hosiers (Libros de cabildos 3: 55 Oficios mecánicos 21 Jan. 1549).

98 Francisco Quiroz Chueca, "Gremios en Lima," in *Lima en el siglo XVI*, ed. Laura Gutiérrez (Lima: Instituto Riva-Agüero, 2005), 495–510.

99 Lowry, "Forging an Indian Nation," 152.

work premises, while they were conducted by guild officials were overseen by an *alcalde* and the *escribano* of the *cabildo*.[100]

A guild for surgeons and barbers existed in Lima in the sixteenth century,[101] but for the sixteenth and early seventeenth centuries at least no evidence has been found of a guild or *gremio* being established for *boticarios*. Similarly, one does not appear to have existed in Mexico City.[102] Paula Ronderos has suggested that a guild for *boticarios* was established in Bogotá, but the source she cites does not specifically mention a *gremio*, but only refers to the "leyes reales" that governed the profession, which would seem to refer to Spanish laws rather than the ordinances of a guild.[103] In Lima, the licensing of *boticarios* and the inspection of pharmacies fell under the jurisdiction of the royal *protomédico*, or in his absence the *cabildo*. In many respects this paralleled institutional arrangements in Spain, where in common with other Mediterranean countries, *gremios* for *boticarios* existed in Aragon, Cataluña and Valencia, but in Castile, where the *protomedicato* was established it was the body responsible for regulating the profession.[104] Hence, the existence of the *protomedicato* in Lima coupled with the relatively small number of practitioners in the city probably rendered a guild unnecessary.

Even though there was no guild for apothecaries, apprenticeships took the same form as for other trades. A two-year apprenticeship contract between one Gerónimo de Espino and the maestro *boticario*, Graviel de España, in 1619, specified that the apprentice would receive no salary, but would be provided with a bed, food, and clothes and be looked after when he was sick.[105] This

---

100 Quiroz Chueca, *Gremios, razas y libertad*, 25–29, 42, 55–56 and "Gremios en Lima," 512, 519–23.

101 Lastres, *Historia de la medicina peruana*, 2: 79; Francisco Quiroz Chueca, *Gremios coloniales peruanos* (Lima: Universidad Nacional Mayor de San Marcos, 1991), 14–16.

102 De Vos, "Art of Pharmacy," 61n2 also doubts that *boticarios* in Mexico City formed a professional organisation or guild.

103 Paula Ronderos, "El arte de boticarios durante la primera mitad del siglo XVII en el Nuevo Reino de Granada," *Fronteras de historia* 12 (2007): 176.

104 Francisco Javier Puerto Sarmiento, "La farmacia renacentista española y la botica de El Escorial," in *La ciencia en el Monasterio del Escorial*, ed. Francisco Javier Campos and Fernández de Sevilla (Madrid: Real Universitario Escorial-María Cristina, 1993), 77–78, 83–84; Goodman, *Power and Penury*, 225–27; Clouse, *Medicine*, 114.

105 AGNP Protocolos Siglo XVII 233 Bustamente fols. 520v–521v. Gerónimo Espino natural del pueblo de Callao… entro a aprehender el arte de boticario con Graviel de España 2 Feb. 1619. For an apprenticeship for a surgeon see: AGNP Protocolos Siglo XVI 154 fols. 284v–285v. Alonso Luzero pone apprendiz a su hijo Diego Luzero con los cirujanos Juan Rabuelo y Martín Sánchez 30 Aug. 1547. He was apprenticed at only 12 years old. For apprenticeship contracts for other trades see Teresa C. Vergara, "Growing Up Indian: Migration, Labor, and Life in Lima (1570–1640)," in *Raising an Empire: Children in Early Modern*

contract was similar to those drawn up for apprentice apothecaries in Spain in the sixteenth century.[106] In some cases, the apprentices came from Spain. In these cases they appear to have been paid a salary and at the end of their apprenticeship would be provided with clothing that reflected their Spanish heritage. Hence, in 1591 one Francisco Salguero, a citizen of Jerez, contracted to serve the *boticario* Blas de Medina for three years as an apprentice. He was to be provided with lodging, bed, board and medical care and at the end of the contract was to be given an "outfit of Castilian cloth of a colour of his choosing, which should include a jerkin, cape and wide trousers, stockings, breeches and shoes, and a hat, two shirts and two doublets …".[107] It is not clear whether Francisco Salguero was paid a salary, but one Andrés Vázquez, who came from Villaviciosa in Spain, who also contracted to work for Blas de Medina for a year, received 120 *pesos de a nueve* as a salary.[108] The 1613 census of Lima does not record the presence of any apprentice pharmacists, whereas a number of Indians were being trained as apprentice surgeons and barbers.[109]

Apart from formal apprenticeships, other individuals might learn the art informally by working alongside licensed *boticarios*, including those practising in hospitals where most workers were African slaves or less commonly Indians. Some of them, such as Juan Mandinga a slave in the Hospital of San Andrés, even ascended to the position of *boticario*, but very few who learned their skills informally appear to have been examined and licensed.[110] This included the Dominican priest, Martín de Porres, who was the illegitimate offspring of a

Iberia and Colonial Latin America, eds. Ondina E, González and Bianca Premo (Albuquerque: University of New Mexico Press, 2007), 86, 92–95.

106 Pilar Herrera Hinojo, "Contratos de aprendizaje en farmacia en el siglo XVI," *Boletín de la sociedad española de historia de farmacia* year 19, 73 (1968): 73–79; Clouse, *Medicine*, 128–30.

107 AGNP Protocolos Siglo XVI 51 Rodrigo Gómez Baeza fols. 1113–1113v. Blas de Medina concierto con Francisco Salguero 25 Sep. 1591. It reads "un vestido de paño de castilla del color que lo quisiere el dicho Francisco Salguero que ese entren de sayo y capote y saragueles [from the Arabic word sarawīl meaning wide trousers] e medias calzas e zapatos y sombrero e dos camisas e dos jubones …"

108 AGNP Protocolos Siglo XVI 52 Rodrigo Gómez Baeza fols. 1003–1005v. Blas de Medina concierto con Andrés Vázquez 14 Oct. 1592.

109 *Padrón de los Indios de Lima en 1613*, comp. Miguel de Contreras, and ed. Noble David Cook (Lima: Universidad Nacional de San Marcos, 1968), 132, 222.

110 Archivo Histórico del Instituto Riva-Agüero (hereafter AHIRA) Maldonado A-III-306 fol. 115 Libros de egresos e ingresos del hospital de San Andrés 1612.

Spanish father and a former African slave mother. He became renowned for his healing powers and was canonised as a saint in 1962.[111]

Examinations and Licences

After working alongside licensed *boticarios*, apprentices submitted themselves to examination, during which they were questioned on both theoretical and practical aspects of pharmacy.[112] The process is exemplified by the title of *boticario* granted to the Limeño-born Bernardo Gil, who was examined by the *protomédico*, Iñigo de Hormero, and the licensed *boticario*, Luis Nieto Maldonado in 1609.[113] Gil was described as a small man about 25 years old, with a wound on the right hand and three moles on the face, who witnesses testified had been practising for six years. The examiners then asked him many and repeat questions in Latin and Spanish about the preparation of simple and complex medicines. They also required him to read a manual in Latin by Mesuë on how to make and preserve complex medicines. In theory the examination was supposed to take place in a hospital,[114] but this source provides no information on where the examination was held.

When Gil was declared to have satisfied the examiners and given the license to practise, he made the sign of the cross and swore that he would practise the art of pharmacy ("arte de boticario") skilfully and competently ("[h]abil y suficiente"). He was instructed to give alms to the poor and to pledge that he would not to put poison in medicines. The latter assurance might appear unnecessary given the commitment of the profession to healing the sick, but it reflected Spaniards' constant fear that they might be killed by non-Christians. In Spain this fear had led to Jews, Muslims and later *Conversos* and *Moriscos* being forbidden from practising pharmacy, such that *boticarios* had to present certificates of *limpieza de sangre*, as well as legitimacy before they could receive a licence to practise.[115] In Seville ordinances relating to the practice

111 See Celia Cussen, *Black Saint of the Americas: The Life and Afterlife of Martín de Porres* (Cambridge: Cambridge University Press, 2014).

112 *Recopilación de las leyes, pragmáticas reales* (Muñoz), cap. 12 art. 3: 157.

113 Archivo Histórico Municipal de Lima (hereafter AHML) Libro 4 de cédulas y provisiones, fol. 161 Título de Bernardo Gil, boticario 7 Nov. 1609.

114 *Recopilación de las* leyes, *pragmáticas reales* (Muñoz), cap. 12 art. 3: 157.

115 *Recopilación de las leyes, pragmáticas reales* (Muñoz), cap. 6 art. 2: 73–74; Goodman, *Power and Penury*, 219–20. For obvious reasons, Indian candidates were not required to demonstrate *limpieza de sangre* (Hernández Sáenz, *Learning to Heal*, 148), but there is no evidence that any sought to be licensed in Lima at this time.

of pharmacy specified that an apothecary wishing to establish a shop had to be an "Old Christian, of pure blood and not a Moor or Jew nor those recently converted to our holy Catholic faith or Morisco, Negro or Mulatto."[116] In the particular case of Bernardo Gil it was only noted that he was the legitimate son of Lope and Maria Gil and no certificate of *limpieza de sangre* appears to have been presented or required. However, it was not until five years later in 1614 that his license to practise as a *boticario* was registered by the *cabildo*.[117]

In 1588 Spanish ordinances stipulated that the cost of entry to the examination for *boticarios* was four *escudos* of gold, equivalent to eight pesos. While this was half of that paid by aspiring physicians and surgeons,[118] it was equivalent to about two to three months wages of a native labourer and therefore an obstacle for those of limited means.[119] In fact, in 1667 the fee was tripled to 12 *escudos*, though it remained half of that paid by physicians.[120] Fees were not paid directly to the examiners, who received an annual salary for conducting examinations. In 1593 the salary paid to the *protomédico* for this task was set at 100,000 *maravedís* (about 370 pesos), while other examiners were paid 60,000 *maravedís* and assessors 20,000 *maravedís*.[121] These salaries were paid out of fees paid by students, but sometimes the number of candidates was too small to generate sufficient income to pay them.[122]

The licensing of physicians, surgeons, barbers, and *boticarios*, as well as the inspection of *boticas* was supposed to be done by a *protomédico*. *Protomédicos* were normally appointed by the Crown, but in Peru there were often long gaps between appointments. This not only led to many people practising without license, but also provided opportunities for the *cabildo* to exert its control over medical practice. In 1581 in the absence of a *protomédico*, the *cabildo* appointed

116 Archivo Municipal de Sevilla (hereafter AMS) Varios Antiguos 370 Ordenanzas de boticarios 1 Mar 1591. This stipulation was not always followed for during an inspection of boticas in Seville in 1631, one belonging to Diego Gómez was continuing to operate even though he had been imprisoned by the Inquisition for two years and had received a licence to practise (AMS Varios Antiguos 370 Visita de boticas 1631).

117 *Libros de cabildos de Lima*, 17: 631 Bernardo Gil boticario presenta su título de tal boticario 18 Jul. 1614.

118 *Recopilación de las leyes, pragmáticas reales* (Muñoz), cap. 7 art.1: 91–92.

119 One *escudo* of gold was equivalent to 16 reals, while forced labourers were paid one real a day.

120 AHML Libros de cédulas y provisiones 12, fols. 284v–285 Ordenanzas de médicos, cirujanos, barberos, oculistas, arbolarios y boticarios y visitas dellos 9 Jul 1667.

121 *Recopilación de las leyes, pragmáticas reales* (Muñoz), cap. 7 art.3: 95.

122 Lanning, *Royal Protomedicato*, 77.

Alvaro de Torres, a knowledgeable and experienced doctor, to issue licences;[123] his appointment as *protomédico* was subsequently confirmed by the Viceroy.[124] The *cabildo* also asserted its control over the practice of pharmacy through periodically ordering all *boticarios* to present their titles.[125] If the examination of the titles of surgeons is anything to go by, not all those submitting their titles were automatically approved despite the fact that there was a shortage of licensed practitioners.[126] In 1552 six surgeons presented their titles to the *cabildo* for approval, but two were regarded as insufficient. One Francisco Sánchez was deemed not to have demonstrated his competence ("suficiencia"). This may have been the same Francisco Sánchez who in 1537 had attempted to obtain a licence claiming he had lost his title. On that occasion he had been allowed to practise, because he had shown himself to be "hábil y de espiriencia [skilled and experienced]," but he was not permitted to prepare medicines.[127] The other, Salvador de Figueroa, was only permitted to treat minor cases, whether or not he was accompanied by a physician. However, the shortage of surgeons meant that the *cabildo* often turned a blind eye to the existence of legal titles, in addition to which there were many people who practised without license. In 1616 the *protomédico*, Doctor Melchor de Amusco, observed that,

> many people with little fear of god and less regard for the law are practising medicine giving themselves titles and calling themselves *licenciados*, doctors and [there are] others who use the art of barbers, *boticarios*, those who treat hernias, herbalists and charmers and everything related to and concerning the said arts and trades without possessing title, license or authority to do so and without understanding.[128]

These circumstances reflected in part the shortage of licensed practitioners, but also that *boticarios* were often consulted, particularly by the popular classes, including slaves, because their services were cheaper than those of

---

123 *Libros de cabildos de Lima*, 9: 452–53 Por protomédico al Licenciado Torres 6 Nov. 1581.

124 Lanning, *Royal Protomedicato*, 29–30.

125 *Libros de cabildos de Lima*, 13: 185 Que ysiban [exhiban] los títulos de cirujanos y barberos y boticarios Lima 6 Nov. 1598.

126 *Libros de cabildos de Lima*, 3: 616 Cirujanos los que son suficientes al parecer del Licenciado Torres 11 Nov. 1552.

127 *Libros de cabildos de Lima*, 1: 161–62 Mandamiento de cirugía 30 Aug. 1537.

128 *Libros de cabildos de Lima*, 18: 157 Alonso Alférez de Jodar fiscal y alguacil del protomédico 30 Oct. 1616.

physicians.[129] Moreover, it demonstrates that while there was knowledge of the law, it was not effectively upheld by the *protomedicato* or *cabildo* and indeed was frequently ignored.

Female Medical Learning

As noted above, women were forbidden by law from owning *boticas* or practising pharmacy, yet it is clear that some women did acquire medical knowledge and dispense medicines. Native women, Africans, and those of mixed ethnicity were most prominent in the popular sector, but in 1616 the *cabildo* noted that many Spanish women and *Mestizas* were also practising some form of medicine, including acting as midwives.[130] Where then did women acquire such knowledge? Inasmuch as women were educated, they were taught by private tutors, later in convent schools by *beatas*, or else in lay schools. When formal schools were founded at the end of the sixteenth century they were exclusively for boys, such that girls had to be educated separately.[131] Elites esteemed schooling in convents since it was seen to enhance a girl's chances of making a good marriage, because it imparted Christian moral and spiritual values, and protected a girl's virginity. Also, the higher cost of this type of education might reflect positively on the economic and social standing of the family.[132]

At first convent education for girls not intending to take holy orders was limited. In fact only in 1575 did a papal bull permit the education of women in convents in the Americas, though it seems that in Lima some instruction had been occurring there informally prior to that time.[133] Initially students had to have a papal licence in order to study in convents, so the process of admission was often protracted and their numbers limited. However, after 1618 another papal bull enabled the local bishop to determine their admissibility.[134] In 1600 there were only three convents in Lima in 1600 – La Encarnación, Santa Clara, and

129 See the redhibitoria case in Archivo Arzobispal, Lima (hereafter AAL) Causas de Negros leg. 4 doc. 7 Hernando de Valdés, contra el licenciado don Diego de Morales 1620.

130 *Libros de cabildos de Lima*, 18: 157 Alonso Alférez de Jodar fiscal y alguacil del protomédico 30 Oct. 1617.

131 Luis Martín, *Daughters of the Conquistadores: Women of the Viceroyalty of Peru* (Albuquerque: University of New Mexico Press, 1983), 73–74.

132 Van Deusen, *Between the Sacred and the Worldly*, 102–111.

133 Martín, *Daughters*, 74; Van Deusen, *Between the Sacred and the Worldly*, 214n78.

134 Martín, *Daughters*, 77; Van Deusen, *Between the Sacred and the Worldly*, 107–108, 240n42.

La Concepción – the first two charging 100 pesos and 200 pesos a year respectively for food and lodging. Although by the mid-seventeenth century there were eight convents, together they educated only about 250 secular girls.[135] Secular students, known as *seglares* or *educandas*, were taught separately from novices. The nuns instructed them in them the basic skills of reading, writing, arithmetic, sewing, cooking, music, drama and sometimes Latin.[136] It seems likely that given the Christian responsibility to care for the sick, as in European convents, women might also informally acquire healing knowledge and skills, including the use of herbs and preparation of medicines.[137]

Convents were large institutions since in addition to the nuns, novices, and secular students they housed, they accommodated a large number of resident servants and slaves who provided services for both individuals and the institution as a whole.[138] According to Bernabé Cobo, the convent of La Encarnación had 700 residents consisting of 300 nuns and novices and 400 servants, slaves and secular students.[139] Such large convents resembled miniature towns and offered spaces for socialising and the exchange of ideas and skills between women from different social and ethnic backgrounds.[140] While those seeking holy orders sought primarily to develop their spirituality, they had access to libraries and often pursued literary and musical interests.[141] Far from being totally closed off from the outside world, the nuns also communicated with the

135 Van Deusen, *Between the Sacred and the Worldly*, 107–109, 240n47. According to Bernabé Cobo (Obras, 2: 429), La Encarnación was only licensed to admit 24 *seglares* though it is not clear what date this observation refers to. Van Deusen (pp. 172–75) provides a table of the numbers of *seglares* in convents in the seventeenth century.

136 Martín, *Daughters*, 79.

137 Monica Green, "Bodies, Gender, Health, Disease: Recent Work on Medieval Women's Medicine," *Studies in Medieval and Renaissance History*, ser. 3, 5 (2005): 13–14; Sharon T. Strochhia, "The Nun Apothecaries of Renaissance Florence: Marketing Medicines in the Convent," *Renaissance Studies* 25(5)(2011): 627–47; Gianna Pomata, "Practising Between Heaven and Earth: Women Healers in Seventeenth-century Bologna," *Dynamis* 19 (1999): 131–43.

138 Martín, *Daughters*, 174–77, 185–91, 199, 221–22.

139 Cobo, *Obras*, 2: 429. At the same time in the convent of La Concepción there were 250 nuns and novices and 250 servants and slaves (p. 431). For the numbers of residents in convents in Lima in the seventeenth century see: Van Deusen, *Between the Sacred and the Worldly*, 172–75. By 1700 it had over 1,000 residents (Asunción Lavrin, "Female Religious," in *Cities and Society in Colonial Latin America*, ed. Louisa Schell Hoberman and Susan Migden Socolow (Albuquerque: University of New Mexico Press, 1986), 175.

140 Martín, *Daughters*, 199; Susan Migden Socolow, *The Women of Colonial Latin America.* (Cambridge: Cambridge University Press, 2000), 99–102.

141 Lavrin, "Female Religious," 185–87.

convent's servants and slaves and they were visited, sometimes illicitly, by relatives living outside.[142] There were considerable opportunities for the acquisition of medical knowledge of different kinds.

The dowry for entry into the convent of La Encarnación was 6,000 pesos and double if a separate apartment and a slave servant were desired.[143] However, in certain circumstances the dowry required might be waived or reduced. In 1628 the abbess of the convent of La Concepción petitioned the bishop to allow the admission of a poor widow to the black veil without paying a dowry, because she was regarded as an asset for the convent

> ... because she has the special gift of treating every kind of illness and is knowledgeable about women's illnesses [...] for she has cured many nuns in this convent with licence of the precentor [...] and has restored to complete health those suffering from diseases which physicians and surgeons have not managed to cure.[144]

Those who entered convents as secular students generally went on to elite marriages.[145] Others took vows of chastity, took the black veil, and lived in closed orders. Yet others sought more solitude away from the large convents or preferred to live in the world as *beatas* rather than under the stricter supervision of closed orders.[146] Some of those who became *beatas* may have been unable

---

142 Martín, *Daughters*, 219–30.

143 Boleslao Lewin, ed., *Descripción del Virreinato del Perú: crónica inédita de comienzos del siglo XVII* (Rosario, Universidad Nacional del Litoral, Facultad de Filosofía, Letras y Ciencias de la Educación, 1958), 58–59; Bowser, *African Slave*, 104–105. Writing of Spanish America in general, Lavrin ("Female Religious," 177) comments that in the sixteenth and seventeenth centuries dowries for nuns taking the black veil were generally between 1,000 and 2,000 pesos, but by the end of the seventeenth century had risen to 3,000 pesos. The figure of 6,000 pesos is quite high since the average dowry for convents in colonial Spanish America was normally between 3,000 and 4,000 pesos (Socolow, *Women*, 95–96).

144 AAL Monasterio de la Concepción leg. 4 exp. 1 Bernardina del Castillo Orihuela, abadesa 6 Sep. 1628. It reads "... porque tiene gracia particular en curar cualquiera enfermedad y sabe mucho en materia de las de mujeres [...] por las muchas religiosas que ha curado a este convento con licencia del señor chantre [...] le ha dado entera sanidad en enfermedades que los médicos y cirujanos no han acertado a curar ..." Such waivers might also apply to women with musical abilities (Martín, *Daughters*, 181–82).

145 See Kathryn Burns, *Colonial Habits: Convents and the Spiritual Economy of Cuzco, Peru* (Durham: Duke University Press, 1999), 218–24 for the destination of those educated in the convent of Santa Clara in Cuzco in the sixteenth century.

146 Van Deusen, *Between the Sacred and the Worldly*, 119, 188; Socolow, *Women of Latin America*, 106–107.

to afford the high cost of becoming full members of closed orders. These nuns followed the rules of the Third Orders of the various religious orders, took vows of chastity, and wore habits, but lived independently or later in the seventeenth century in *beaterios*. These women taught, preached, and undertook charitable activities that included collecting alms and caring for the sick.[147] In Lima their spirituality was commonly recognised and they were widely consulted by the population at large on both religious and medical matters.[148] One example, was Isabel Medel Cansino, a *beata* of the Third Order of San Francisco, who treated as many as ten to twenty African slaves in her own house and employed two slaves, one of whom was responsible for collecting herbs and medicines for the pharmacy.[149] Her treatments were said to include:

> meat, bread, wood, wine, eggs, piedra lipes [copper sulphate], oil, cardenillo [verdigris], fish, tallow, virgin oil, alum, lavender, espingo [ishpink], pingo pingo,[150] vilcatongo,[151] mastic, chochos,[152] honey, aniseed, fat, sulphur, candles and salt.

Conclusion

The ordinances that governed the training, examination and licensing of apothecaries in Peru were the same as those that operated in Spain. However, local circumstances dictated that state control of the practise of pharmacy was more limited. These included the failure of the state to fund medical education in the universities and to support high level *protomédicos* who might

147 Van Deusen, *Between the Sacred and the Worldly*, 24–25 and "Circuits of Knowledge among Women in Early Seventeenth-Century Lima, in *Gender, Race and Religion in the Colonization of the Americas*, ed. Nora E. Jaffary (Aldershot: Ashgate, 2007), 138, 146. On *beatas* in Cuzco see: Kathryn Burns, "Andean Women in Religion: Beatas, 'Decency', and the Defence of Honour in Colonial Cuzco," in *Gender, Race and Religion in the Colonization of the Americas*, ed. Nora E. Jaffary (Aldershot: Ashgate, 2007), 81–91.

148 Van Deusen, "Circuits of Knowledge," 141.

149 AGNP SO CO Ca 40 doc 383 fols. 461–480 Doña Isabel Medel Cansino 1636; AGNP SO CO Ca 2 doc 8 fols. 621–22 Fragment of an invoice drawn up by Isabel Medel Cansino, no date.

150 Pingo pingo (*Ephedra andina*) is a native Andean herb that was used as a diuretic and depurative.

151 This was probably a native herb. It was put in chicha and used as a purgative by one, María Ynes, in Chancay, who during an ecclesiastical *visita* in 1662 was accused of being a *hechicera* (Sánchez, *Amancebados, hechiceros y rebeldes*, 27).

152 Lupin *Lupinus mutabilis* Sweet.

bring stricter compliance with the law. This left the administration of healthcare largely in the hands of the *cabildo*, which faced with a shortage of medical expertise, due in part to limited opportunities for training, often licensed individuals who did not have the requisite certificates, but displayed particular knowledge and skills that they had acquired informally. Despite Antonio de Robles Cornejo's judgement that early migrants from Spain were the least educated, some did have a university education and formal training, whereas most of those practising later had little recourse other than to acquire knowledge and skills informally.

Probably the greatest difference between Peru and Spain was the greater diversity of ethnic backgrounds from which *boticarios* were drawn in Lima. Despite legislation that restricted pharmacy practice to those who could demonstrate *limpieza de sangre*, in Lima few serious attempts were made to prevent Blacks or those of a mixed ethnic background from practising the art, reflecting in part the demographic dominance of those of African heritage in the city. Likewise there is no evidence that women were prevented by law from practising pharmacy, though in many cases they were probably restricted to work in ancillary activities rather than preparing medicines themselves. The presence of such an ethnically diverse population, where the regulation of medical practice was only partially effective, where distinctions between licensed medical practitioners were blurred, and where only a small proportion of the population had been formally trained, created an environment that should have fostered innovation. Yet, as Chapters 5 and 6 will demonstrate through an analysis of the types of medicines prepared, *boticarios* adhered to traditional humoral beliefs and practices and depended on Old World *materia medica*. How can this apparent paradox be explained? It seems clear that the prevalence of humoral medicine did not depend on state regulation or formal training. Knowledge could be acquired through working alongside licensed *boticarios* or in hospitals. It might also be gleaned through private study, where most medical books available for study in Lima were orthodox texts that had been sanctioned by the Inquisition. Perhaps more significantly, as Chapter 5 will show, humoral beliefs were entrenched in popular thought to the extent that lay people often applied treatments according to its principles.

CHAPTER 3

The Medicines Business

This profession is not for the poor.[1]

Once apothecaries had been trained, the opportunities open to them to practise their art varied largely according to the financial resources they possessed. Contemporary observations and advice given to those aspiring to be independent *boticarios* converge in the judgement that in order to establish a pharmacy one had to be a person of moderate means. Christóval Suárez de Figueroa's encyclopaedia of the qualities required of those practising different professions suggested that the pharmacy profession was not one for the poor. This view was echoed by Antonio de Robles Cornejo who with experience in Peru advised that no one should attempt to open up a *botica* "without being rich or at least with substantial funds."[2] The need for *boticarios* to have significant financial resources was underlined in the city ordinances of Seville in 1591 which specified that in order to obtain a licence, a *boticario* not only had to pass the requisite examinations, but also show that he possessed 500 *ducados* (690 pesos) in order to purchase simples for the preparation of medicines.[3] There is no evidence that a similar local ordinance was drawn up in Lima, though there was no lesser need for *boticarios* to have significant financial backing. The reason given was not so much the cost of the premises or labour, but the high cost of medicines. It was generally considered that those

1 Suárez de Figueroa, *Plaza universal*, 302. Paula De Vos, "The Apothecary in Seventeenth- and Eighteenth-Century Mexico: Historiography and Case Studies in Medical Regulation, Charity, and Science," *Colonial Latin American Historical Review* 13(3)(2004): 253 notes that this guide was based on Tomaso Garzoni's, *La piazza universale di tutte le professioni del mondo* [1587] 2 vols. eds. Paolo Cherchi and Beatrice Collina (Turin: Giulio Einaldi, 1996), but suggests that the role of apothecaries is likely to have been similar in Spain and Italy.

2 He says, "sin ser rico o al menos con gruas riquezas mucha cantidad de pesos." ARJBM División 1, leg. 17 Libro de examen de los simples medicinales 1617. These comments echo those of Antonio de Aguilera, who in his *Exposición sobre las preparaciones de Mesue*, fols. 20v.-21v. advised that *boticarios* should be "rich or at least be of sufficient means."

3 AMS Varios Antiguos 370 Ordenanzas de boticarios 1 Mar 1591; Fernández-Carrión and Valverde, *Farmacia y sociedad*, 15.

 | DOI 10.1163/9789004351271_004

of moderate means could buy good quality medicines at the time they were needed, whereas the less affluent could only afford poor quality medicines or had to buy them on credit, which meant they actually paid more for items of lesser quality. Given the high cost of establishing and maintaining a *botica*, there were relatively few in the city. Even in 1630 there were only twelve to eighteen recognised *boticas* in Lima, four of which were attached to hospitals.[4]

For many *boticarios* who acquired licences, the expense of setting up an independent business was not a realistic economic proposition. Moreover, established *boticarios* often discouraged those they had trained or who were in their employment from setting up independent businesses for fear of competition and that they would take their customers with them. In the case of surgeons and barbers, the guild persuaded the municipal authority in Lima to introduce a regulation stipulating that apprentices and newly trained journeymen known as *oficiales* could not set up shops within four blocks of their master's practice for a year and a day.[5] A similar ordinance, does not seem to have applied to *boticarios*, but there were private efforts to restrict the competition. For example, in 1618 Luis Nieto Maldonado rented a *botica* in Cuzco from one Fernando de Cartagena and his four-year contract specified that for two years following the end of the contract he could not set up a new *botica* in the city, either personally or through an intermediary, though he could purchase one of the two that already existed.[6] In any case, the establishment of any new *botica* required municipal approval.

In common with other trades, many trained *boticarios* who lacked the resources to set up independent businesses continued to work in the salaried employment of a *maestro* until they had secured sufficient capital to open up their own pharmacies.[7] In the 1630s, the apothecary to the Inquisition, Mateo Pastor, had at least two *oficiales* working in his *botica*.[8] *Licenciado* Bartolomé Díaz Cabeza de Vaca appears to have aspired to be an independent *boticario*, but he could only realise this ambition in 1605 by entering into a *compañía* or partnership with Francisco Martín Reyna to purchase a pharmacy that had formerly belonged to Diego de Tineo. In this *compañía* Díaz Cabeza de Vaca

4 Salinas y Córdova, *Memorial*, 257. At the same there were said to be nine *médicos*, ten surgeons and three Spanish barbers' shops with many *oficiales*. Juan Bromley, however, suggests there were 18 *boticas* ("La ciudad de Lima en el año 1630," *Revista histórica* 24 (1959): 286).

5 Quiroz Chueca, *Gremios coloniales peruanos*, 14–16.

6 "Un botica colonial," *Revista del archivo histórico del Cuzco* 4 (1953): 279.

7 Quiroz Chueca, *Gremios, razas y libertad*, 19–23.

8 AGNP SO CO 44–394 fols. 191–210 Prisión, secuestro e inventario de bienes de Tome Cuaresma 20 Nov. 1635.

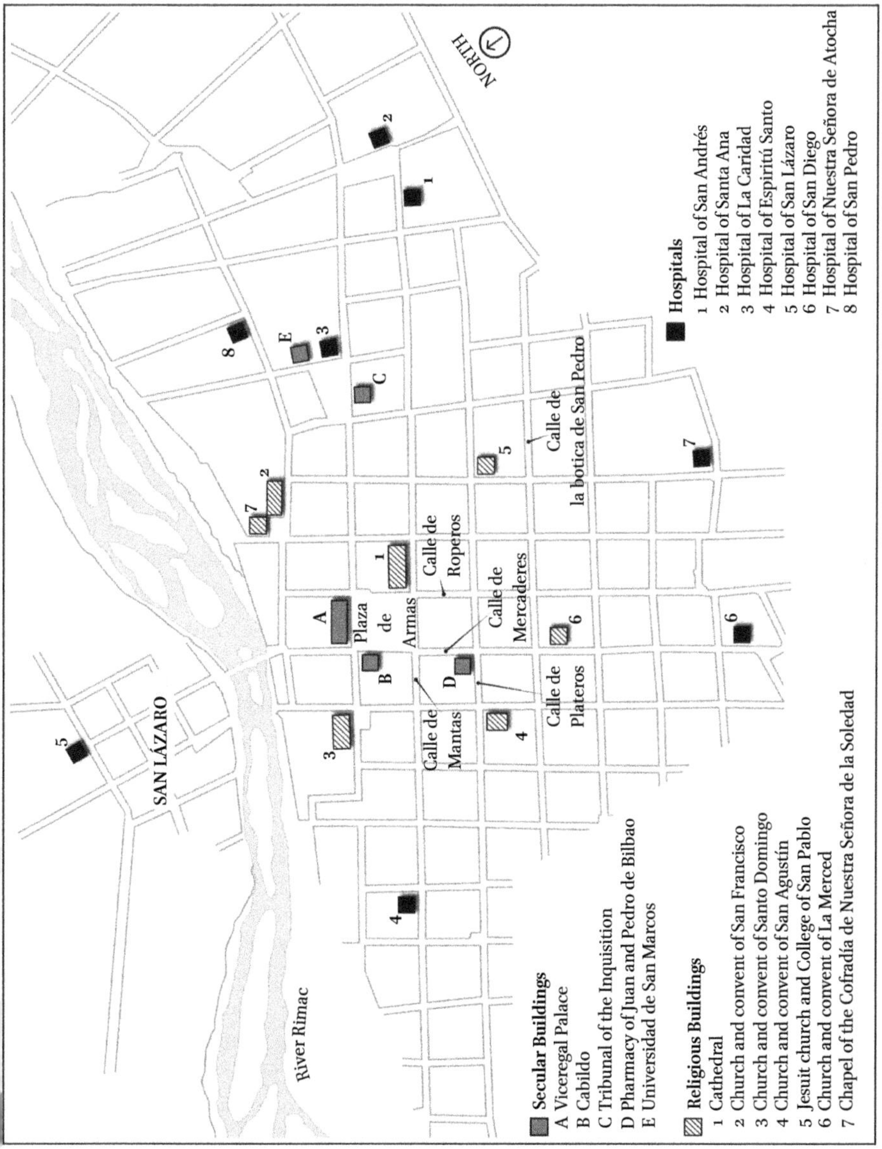

FIGURE 1 *Lima in the early seventeenth century.*

was only able to afford a one-third share.[9] Prior to this time, in 1596 he had worked under contract for Blas de Medina for the annual salary of 500 pesos.[10] This high salary probably reflected Díaz Cabeza de Vaca's experience and status as a licensed *boticario* since the annual contracts of *oficiales* working in Lima's *boticas* generally specified salaries of 200 to 250 *pesos de a nueve*.[11] These salaries were consistent with those of *oficiales* in other trades such as tailors, masons, carpenters or silkworkers, where most earned less than 250 pesos a year but skilled workers might receive as much as 500 to 600 pesos.[12] Unlike other trades, however, where *oficiales* did not normally reside in the house of the *maestro*,[13] those working in pharmacies seem to have lived on the premises and received food, lodging and clothing.

Other trained *boticarios* sought paid employment in public institutions such as hospitals, convents, or prisons.[14] (see Figure 1) Apothecaries employed in the Hospitals of Santa Ana and San Andrés were paid between 300 and 400 pesos a year, part of which might be paid in kind in the form of accommodation and food. For example, Rodrigo de Vargas, who held the post of *boticario* of the Hospital of Santa Ana for over thirty years was given a house in the hospital grounds and a ration of three pounds of beef and mutton, three loaves of bread of more than one pound, six bottles of local wine, and twelve blocks of soap from Spain.[15] This salary was only slightly less than that paid to physicians in the same hospitals and considerably more than the 200 pesos and 50 pesos paid to surgeons and barbers respectively.[16]

---

9 AAL Testamentos leg. 5 exp.1 fols. 7v–8 Testamento de Bartolomé Díaz Cabeza de Vaca 10 Apr. 1608.

10 AGNP Protocolos Siglo XVI 55 Rodrigo Gómez Baeza fols. 187–187v. Contract between Blas de Medina and Bartolomé Díaz Cabeza de Vaca 13 Aug. 1596.

11 For contracts between *boticarios* and *oficiales* see: AGNP Protocolos Siglo XVI 54 Rodrigo Gómez Baeza (1594) fols. 641–641v. Contract between Luis Nieto Maldonado and Juan de Arce 7 Jun. 1595; AGNP Protocolos Siglo XVI 55 Rodrigo Gómez Baeza fols. 980–980v. Contract between Luis Nieto Maldonado and Juan de Horosco 12 Jun. 1596.

12 Francisco Quiroz Chueca, "Artisans and Journeymen in Colonial Lima." Unpublished manuscript, 2014.

13 Quiroz Chueca, *Gremios, razas y libertad*, 22.

14 Unlike hospitals and convents, Lima's four public prisons did not seem to have salaried *boticarios* (AGI Lima 134 Cofradía de los pobres de las carceles de la ciudad de los reyes 1598; AGI Lima 131 Los hermanos de los pobres de las carceles de los reyes 27 Feb. 1593).

15 Archivo de Beneficencia Pública, Lima (hereafter ABPL) 9086 fols. 104–105 Visita al Hospital de Santa Ana fols. 104–105 no date [1588]. He was appointed to the hospital in 1567 and was still working there in 1606.

16 ABPL 9086 fols. 272–73 Visita al hospital de Santa Ana 26 May 1588; ABPL 9083 fol. 90 Libro mayor de rentas del hospital de Santa Ana desde 1593 a 1629; ABPL 9084 fols. 132, 145,

Acquiring a *Botica*

Residents of Lima who became independent *boticarios* acquired pharmacies in different ways. In the earliest years of colonial rule those apothecaries who had been licensed and practised in Spain often brought medicines and equipment with them to set up a pharmacy business, generally migrating with their families, servants or assistants. When the *boticario* Pedro de la Fuente moved to Lima in 1534, he took with him 500,000 *maravedís* (about 1,800 pesos) worth of goods, though it is not known what proportion was destined for the establishment of a *botica*.[17] Pharmacy was often a family profession such that some *boticarios* who settled in the Americas were sons of licensed practitioners in Spain.[18] Once established in the New World, pharmacies might pass from father to son, as was the case with Pedro de Bilbao in Lima who inherited his *botica* from his father, Juan de Bilbao, for whom he had worked for five years.[19] However, more evidence exists for the purchase of *boticas* either individually or as part of a joint venture.

---

162 Libro de cuentas de gastos del hospital de Santa Ana año 1598; ABPL 9086 fols. 6–7, 124 Visita al hospital de Santa Ana 1606; AGI Lima 122 fol. 38 Información del hospital de Santa Ana 1570. In 1582 the salary of the *boticario* at the hospital of San Andrés was only 150 pesos (AGI Lima 126 fols. 9v., 16v. El hospital real de los españoles de advocación San Andrés 1582), but in the early seventeenth century was 400 pesos, equivalent to that paid in the hospital of Santa Ana (Cobo, *Obras*, 2: 444). However, the ordinances for the hospital of San Andrés in 1577 made no provision for a salaried *boticario*, rather a priest took care of the *botica* which was supposed to be inspected by a physician or *protomédico* every three months (Guillermo Lohmann Villena and Maria Justina Sarabia Viejo, *Francisco de Toledo: Disposiciones gubernativas para el virreinato del Perú 1575–1580* (Seville: Escuela de Estudios Hispano-Americanos, 1989), 318. For salaries for *boticarios* in Mexico City in the eighteenth century see Hernández Sáenz, *Learning to Heal*, 162, which indicates similar differentials between different types of medical practitioner.

17 AGI Contratación 5536 L2 N167 Licence granted to Pedro de San Martín and Pedro de Fuente, boticario 21 Mar. 1534.

18 Joaquín Herrera Dávila, "Las boticas sevillanas de 1631," *Boletín de la sociedad española de historia de la farmacia* 39 (154–155) (1988): 73; Fernández-Carrión and Valverde, *Farmacia y Sociedad*, 61–64, 73.

19 AGNP Protocolos Siglo XVI 14 Ramiro Bote fols. 1629–1631 Juan de Bilbao dona a Pedro de Bilbao, su hijo, de una botica con todas sus botes y drogas 3 Aug. 1596. In fact Juan Bromley states that Juan de Bilbao inherited it from his father Francisco de Bilbao who had possessed the contract for the establishment of a pharmacy in the Hospital of Santa Ana (Juan Bromley, *Las viejas calles de Lima* (Municipalidad Metropolitana de Lima: Lima, 2005), 257–58.

The cost of establishing a *botica* was high. When a pharmacy was imported from Spain for the Hospital of Santa Ana in 1551, it was valued at 3,400 *pesos de oro*. However, details only exist for a shipment of 30 wooden chests containing medicines, equipment and books, valued at about 1,062 pesos (289,110 *maravedís*).[20] As with *boticas* in general, the medicines accounted for the most significant proportion of pharmacy assets, in this case about 65 percent.[21] The equipment, excluding medicinal books and manuals, was valued at only about 230 pesos (61,750 *maravedís*)(See Table 1). This was significantly less than the value of the equipment of a well-established private *botica*, such as that belonging to Bartolome Díaz Cabeza de Vaca. In 1608 this pharmacy contained stills (*alquitaras*), mortars, spatulas, sieves, ladles, a variety of boxes and phials, and an oven together valued at 2,217 pesos

TABLE 1 *Expenditure on the installation of the first pharmacy for the Hospital of Santa Ana, 1551.*

	Maravedís	Percentage
Materia medica	186,855	64.6
Equipment	61,750	21.4
Books	5,225	1.8
Import tax	27,204	9.4
Labour to build the furniture for the pharmacy and for the preparation and loading of goods for shipment	8,076	2.8
Total	289,110	100.0

Source: AGNP Protocolos Siglo XVI 63 Diego Gutiérrez fols. 167–88 El hospital de españoles y naturales compañía con Francisco de Bilbao 15 Oct. 1551.

20 AGNP Protocolos Siglo XVI 63 Diego Gutiérrez fols. 167–88 Hospital de los españoles y naturales compañía con Francisco de Bilbao de la botica 20 Oct. 1551; Miguel Rabí Chara, "La primera botica de los hospitales de la ciudad de Lima en el siglo XVI," *Asclepio* 52 (2000): 273–77.

21 AGNP Protocolos Siglo XVI 63 Diego Gutiérrez fols. 167–88 El hospital de los españoles y naturales con Francisco de Bilbao de la botica 9 Mar 1552. The detailed cost of individual items in the *botica* suggests it was worth 289,110 *maravedís* (1,062 pesos), of which 61,750 was spent on equipment. The figure of 289,110 *maravedís*, however, is difficult to reconcile with the stated overall cost of 3,400 *pesos de oro en plata ensayada.*

(See Table 2).[22] However, *boticarios* starting out on their pharmacy careers generally owned much more basic equipment. The weights, cooking pans, spatulas and mortars that the *boticario* Juan Sánchez took with him when he migrated to Lima in 1592 were valued at only 53 pesos.[23]

The total cost of the *botica* of the Hospital of Santa Ana of 3,400 *pesos de oro en plata ensayada*,[24] which was equivalent to 5,625 *pesos corrientes*, does not seem out of line with the value of private *boticas* in the city. The *botica*

TABLE 2 *Assets of the pharmacy belonging to Bartolomé Díaz Cabeza de Vaca, 1608.*

Pesos	Reals	Percentage	Item
596	6	4.4	Cash in a bag in the pharmacy
2,367	2	17.5	Simple and compound medicines
2,217		16.4	Equipment – stills (*alquitaras*), mortars, spatulas, sieves, ladles, a variety of boxes and phials, an oven and other essential tools
200		1.5	An old and infirm African slave
5,708	6	42.2	Client debts (of which 3,874 pesos were receipts for medicines supplied)
1,474		10.9	Owed by Francisco Martín Reyna part investor in the pharmacy
700		5.2	A calvary and linen
274	2.5	2.0	Furniture, including a desk (bureau) and chairs
13,538	0.5	100.0	

Source: AAL Testamentos leg. 5 exp. 1 fols. 175–178 Testamento de Bartolomé Díaz Cabeza de Vaca boticario y autos sobre su cumplimiento con el albacea Juan Manuel Carrasco, 1608–1626.

22 AAL Testamentos 5 leg. 1 fol. 165v. Memoria de las medicinas que tenía esta botica cuando la compraron... Bartolome Cabeza de Vaca 1608. The most expensive items were those made of metal – mortars, boiling pans and ladles –which were worth 250 pesos. Other items consisted of boxes, jars, pots and other receptacles containing pill, powders, syrups, cordials, oils, ointments and plasters. It is not clear whether the value of these receptacles included the medicines they contained.

23 AGI Contratación 1097 N5 fols. 199–202 Registro de Santa Catalina 1592.

24 AGNP Protocolos Siglo XVI 63 Diego Gutiérrez fols. 167–88 Hospital de los españoles y naturales compañía con Francisco de Bilbao de la botica 20 Oct. 1551; Rabí Chara, "Primera botica," 273–77. A *peso de oro* was equivalent to 450 maravedís and a *peso corriente* 272 *maravedís*.

commissioned by Diego de Tineo from Seville in 1555 cost 5,000 *pesos de buen oro*.[25] This pharmacy was designed to be exceptionally well-appointed with its installations fashioned by skilled artisans. A carpenter was instructed to create a ceiling for the pharmacy some twenty-two feet square on which were to be painted golden stars and a large sun with golden rays. Tineo indicated that the boxes to be used for the medicines were to be finely decorated with gilt and in place of labels (*medallas*) there were to be paintings of notable historical figures. One of these was to depict the legend known as "Virgil in a basket" (See Figure 2). In this legend, Virgil being enamoured with Nero's daughter, was enticed to lower himself from a window in a basket, but was suspended by her halfway thereby exposing him to public ridicule.[26] In revenge for his humiliation Virgil employed magic to extinguish the fires of Rome that could only be rekindled by holding a torch or candle to the woman's private parts. This popular legend was applied to many historical figures, including Hippocrates, and was attached to Virgil in the thirteenth century because of his association with magic and divination. Virgil was regarded as an authority by astrologers and alchemists and was even thought to possess the secret of the Philosopher's Stone.[27] The use of this image signifies the apothecary's interest in alchemy, which is further supported by his importation of six stills for use in distillation. The importance of alchemy to pharmacy practice of the time will be discussed further in Chapter 5.

Not all pharmacies were installed so extravagantly, but nevertheless, they commanded fairly high prices. In 1551 a *botica* sold to one Pero Lopes de Aguirre for 2,210 *pesos de buen oro*,[28] and in 1576, when the *botica* of Francisco de Alva was auctioned following his death, it netted a similar 2,400 pesos.[29] However, Alva's will was disputed by his children and five *boticarios* testified that it was actually worth more than 4,000 *pesos de buen oro*, because it was the best in the city and possessed very good equipment and medicines.[30] One of the

25 John Carter Brown Library (hereafter JCB) Mss codex Sp 136 Francisco Martínez y compañía obligación – Diego de Tineo y consortes 1555.

26 John Webster Spargo, *Virgil the Necromancer: Studies in Virgilian Legends* (Cambridge, Mass.: Harvard University Press, 1934), 136–55, 198–206.

27 Spargo, *Virgil the Necromancer*, 277–79.

28 AGNP Protocolos Siglo XVI 8 Simón Alzate fols. 841v–842 26 June 1551.

29 The *botica* had been purchased from Luis Núñez de Prado (AGNP Real Audiencia. Causas Civiles leg. 16 cuad. 81 fol. 7v., 16 Pleito…contra las bienes y herederos del bachiller Francisco de Alva, boticario morador en la Ciudad de los Reyes, 1576).

30 AGNP Real Audiencia. Causas Civiles leg. 16 cuad. 81 fols. 119–26, 687–715 Pleito… contra las bienes y herederos del bachiller Francisco de Alva, boticario morador en la Ciudad de los Reyes. 1576. The *boticarios* were: Guillermo Rodríguez aged 48, Francisco Velásquez aged 30, Blas de Medina aged 34, Diego de Tineo 70 and Doctor Franco, a physician.

FIGURE 2 *The poet Virgil in a basket. Lucas van Leyden, Woodcut, Leiden 1512.*
COURTESY OF THE METROPOLITAN MUSEUM OF ART, NEW YORK.

witnesses was Diego de Tineo, the owner of the aforementioned costly *botica*. Francisco de Alva was *familiar* to the Inquisition and was probably the most eminent *boticario* in the city at the time.[31] He was said to be worth 50,000 pesos *ensayados* owning several houses including one on the plaza. He also owned a shop where the *botica* was located and two "shop doors" (puertas de tiendas) where presumably he sold medicines.[32] The cost of establishing a *botica* can be compared to setting up a manufacturing enterprise, where the initial investment for a craft workshop might be between only 50 to 200 pesos, and for a manufacturing workshop 500 *pesos de a nueve*.[33] Clearly the high cost of establishing a *botica* limited the number of people who could purchase one outright.[34]

Due to the high cost of establishing a *botica*, some pharmacies were set up as joint enterprises. A common way of establishing businesses in sixteenth-century Lima was to form a *compañía* of two or more investors.[35] Apothecaries were able to attract investment from non-*boticarios*, because pharmacy was considered to be a profitable business.[36] In 1605 the *boticario* Bartolomé Díaz Cabeza de Vaca entered into a *compañía* with Francisco Martín Reyna for the purchase of a *botica* from Pedro and Joan de Tineo, which was located next to the cathedral.[37] Pedro and Joan de Tineo were nephews of the *boticario*

---

31 *Libros de cabildos de Lima*, 7: 549 Título familiar al bachiller Alba [sic] 21 Jan. 1574, *Libros de cabildos de Lima*, 8: 201 Familiar del santo oficio a Juan de Bilbao 24 Feb. 1576.

32 AGNP Real Audiencia. Causas Civiles leg. 16 cuad. 81 fol. 7v., 687 Pleito…contra las bienes y herederos del bachiller Francisco de Alva, boticario morador en la Ciudad de los Reyes. 1576. In 1599 the *boticario* Francisco Martín Reyna was renting a house under the *portales* of the plaza from the heirs of Marina Alva (*Libros de cabildos de Lima*, 13: 516–17 Francisco Martín Reyna 21 Jul. 1600).

33 Quiroz, *Artesanos y manufactureros*, 83. For similar shared investments in pharmacies in Mexico see: Hernández Sáenz, *Learning to Heal*, 161–62.

34 In the late eighteenth and early nineteenth centuries *boticas* in Mexico City were commonly valued at between one and two thousand pesos though might be considerably more in the case of the major hospitals For a hospital *botica* it was estimated that 6,000 pesos were required for installations and equipment only, and a further 14,000 pesos to stock it (Hernández Sáenz, *Learning to Heal*, 159–162; De Vos, "Art of Pharmacy," 72–73).

35 Quiroz, *Artesanos y manufactureros*, 84, and *Gremios, razas y libertad*, 71–82.

36 For example, in 1551 an *alcalde ordinario*, Gerónimo de Silva, sold a *botica* to Pero Lopes de Aguirre, with whom he had formed a *compañía*, neither of whom were referred to as *boticarios* (AGNP Protocolos Siglo XVI 8 Simón Alzate fols. 841v–842 26 June 1551). For Mexico City see: De Vos, "Art of Pharmacy," 72–74.

37 AAL Testamentos 5–1 fols. 12v–13 Testamento de Bartolomé Díaz Cabeza de Vaca 1608.

Diego de Tineo.[38] The agreed sum of 4,000 *pesos corrientes* was to be paid in two instalments, half to Pedro de Tineo at the end of April 1605 and the other to Joan de Tineo at Christmas the same year. The first instalment of 2,000 pesos was paid in cash on 10 May 1605, while the second was to be paid from the profits from the *botica*. In fact the latter was paid on 4 February 1606. Less than three years later on 15 April 1608, Díaz Cabeza de Vaca died. At that time after deducting expenditures and debts, the *botica* was worth 7,320 pesos, of which one-third belonged to his estate and the other two-thirds to Francisco Martín Reyna, who continued to own the *tienda* after his death.[39] It is unclear at this time whether Francisco Martín Reyna was a trained *boticario*. In the list of Limeños making a donation to the Crown in 1590 he is referred to as a *pulpero* and between 1595 and 1600 was supplying the Hospital of Santa Ana with honey, resins and burlap.[40] However, sometime later, in 1615 and 1618, he was importing *materia medica* from Spain and was referred to as a *boticario*.[41] It may be that his original interest in the pharmacy business derived from being a supplier of these products. Arrangements between artisans and those who supplied them with raw materials, or who traded products on their behalf, were commonplace in other enterprises in Lima.[42] In 1624 a *botica* was sold by one Pedro de Reyna to the *boticario* Bernardo Gil for 4,000 pesos.[43] The seller's surname suggests that this may have been the same *botica* that had been owned by Francisco Martín Reyna which he had inherited as a relative. If this was the case, then its value in 1624, which did not include medicines, had not increased since 1605. When Gil died in 1662 the *botica* was still valued at 4,000 pesos.

---

38 Diego de Tineo was still practising as a *boticario* in 1590 (AGI Lima 272 fols. 94–112v. Servicio hecho al rey nuestro señor en la ciudad de los reyes 1590).

39 AAL Testamentos 5–1 fols. 175–181v. Testamento de Bartolomé Díaz Cabeza de Vaca 1608.

40 AGI Lima 272 fols. 94–112v. Servicio hecho al rey nuestro señor en la ciudad de los reyes 1590; ABPL 9084 fols. 241–242v. Libro de cuentas de gastos del hospital de Santa Ana 1599–1600; ABPL 9085 fols. 59–64 Libro de la razón que toma Bartolomé de la Cueva escribano y veedor deste hospital de Santa Ana…desde el primero del mes de agosto 1595.

41 AGI Contratación 1162 N4 43–46 Registro del navío San Pedro 1615; AGI Contratación 1166 N1 403–406 Registro del navío El Espíritu Santo 1618.

42 Quiroz Chueca, "Compañías y gremios," and *Artesanos y manufactureros*, 85; Ruth Pike, *Aristocrats and Traders: Sevillian Society in the Sixteenth Century* (Ithaca, NY: Cornell University Press, 1972), 100.

43 AAL Testamentos 52–21 fols. 8v. Testamento de Bernardo Gil, boticario 1662.

The Premises

Clearly those seeking to acquire or establish *boticas* had to have access to considerable capital resources, but in addition they had to shoulder significant running costs. The value of pharmacies recorded in historical sources generally referred to its medicines, equipment and books and not the premises where it was located, since these were generally rented. A functioning pharmacy required at least three rooms: a front room where the *boticario* met clients and dispensed medicines; a back room known as a *trastienda* or *rebotica* where the medicines were prepared, the equipment kept, and accounts drawn up; and a store room or *aposento*.[44] In the case of Díaz Cabeza de Vaca, rent for the building in which his *botica* was located was 650 pesos a year; in addition he rented an *aposento* next door, for an unspecified amount, and where in fact he lived.[45] It was the tradition in Spain, even until recently, for *boticarios* to live on the premises, very often above the pharmacy which was on the ground floor.[46] This *botica* was situated in a prime location next to the cathedral and elsewhere rents were somewhat less.[47] In 1551 the physician, Alvaro de Torres was renting out a *tienda* and *trastienda* on the road running to Callao for the much smaller sum of 120 pesos a year.[48] Even hospitals might rent extra premises. In 1574 Rodrigo de Vargas, the *boticario* of the Hospital of Santa Ana, was renting two houses in the hospital's plaza for 60 pesos and 15 pesos a year.[49] Sometimes the cost of premises was paid for in kind, as was the case with Bernardo Gil who rented an *aposento* from the widow of the former *boticario* Melchor Malo de Molina for a notional 100 pesos a year for which he paid in kind through supplying her household with medicines.[50]

As for the location of a pharmacy, wind and humidity were thought to damage drugs,[51] so the advice to aspirant *boticarios* was that it should be "situated in a healthy place, dry (ajena de humedad) and free from dust, smoke and bad

44 Hernández Sáenz, *Learning to Heal*, 152; De Vos, "Art of Pharmacy," 88–89.

45 AAL Testamentos 5–1 fols. 14v., 181r. Testamento de Bartolomé Díaz Cabeza de Vaca 1608.

46 Herrera Dávila, "Boticas sevillanas de 1631," 72.

47 For an overview of rents in Lima in the seventeenth century see María Antonia Durán Montero, *Lima en el siglo XVII: Arquitectura, urbanismo y vida cotidiana* (Sevilla: Diputación Provincial, 1994), 166–74.

48 AGNP Protocolos Siglo XVI 8 Simón Alzate fols. 924–924v. 9 Jul. 1551.

49 ABPL 9080 fol. 102 Libro de cuentas, censos y haberes que corresponden al del hospital de Santa Ana 1575 a 1585.

50 AAL Testamentos 52–21 fol. 10–10v. Testamento de Bernardo Gil, boticario 1662.

51 Aguilera, *Exposición*, fol. 24. "…not windy or humid, or least very clear where there is a lot of sunshine."

odours."[52] The climate and elevation of Lima and its immediate surroundings did not vary significantly, so climatic factors were probably not an important consideration in the siting of pharmacies; of greater concern was the need to be close to the main market for medicines, which was among elites in the city centre.

Employing Pharmacy Workers

Licensed *boticarios*, including the *oficiales* referred to above, were probably only a small proportion of the numbers that assisted or even practised pharmacy. Many other pharmacists were Blacks and Indians who worked as salaried employees or slaves in privately-owned *boticas* and hospitals. As early as 1572 the *cabildo* of Lima was concerned that Blacks and Indians working in *boticas* were threatening the lives of patients by supplying medicines that did not comply with prescriptions, sometimes substituting items with ingredients that had been banned, such as opium, or selling mercury chloride or corrosive sublimate (*solimán*). It judged that the art of being an apothecary required scientific knowledge, skill and precision, which it was impossible for Blacks and Indians to possess; it was recognised that even Spaniards trained in the art occasionally made mistakes. The *cabildo* therefore ordered that Blacks and Indians should not work in *boticas* and that apothecaries found guilty of employing them should be fined 200 pesos and the workers exiled.[53]

Despite these concerns and attempts at regulation, to be discussed more fully in Chapter 5, Indians were assigned to work in pharmacies as forced labourers, while Blacks were widely employed there. In fact, *boticarios* observed that clients often sought advice and treatment from practitioners from the same ethnic background and thus they often employed Indians and Blacks to expand their customer base.[54]

Indian Forced Labourers

When the Spanish arrived in Peru, the native population of Lima was small and subsequently most Indians who resided there came from outside the city.

52 Suárez de Figueroa, *Plaza universal*, 302.

53 *Libros de cabildos de Lima*, 7: 268 Boticarios no tengan en boticas negros 28 Apr. 1572 and 270–72 Que no den medicinas negros en las boticas ni entren en ellas 2 May 1572. The art required "mucha ciencia e abilidad e fieldad."

54 Eguiguren, *Alma mater*, 256.

The 1613 census of Lima registered only about two thousand Indian residents;[55] these included urban migrants and those from outside the city who had been assigned there to undertake labour in public service. Arriving in Lima these forced labourers settled in El Cercado from whence, under an allocation system referred to as the *mita de plaza*, they were assigned by the *corregidor* to work in the maintenance of the city's infrastructure and elite houses or were allocated for agricultural labour. The Hospital of Santa Ana regularly received forced labourers to work in its pharmacy. For example, in 1596 the *corregidor*, Alonso de Mendoza Ponce de León, assigned Hernando Quispi from Yauyos in the hinterland of Lima to work for one year under Rodrigo de Vargas, the *boticario* of the Hospital of Santa Ana.[56] What tasks such workers may have performed is uncertain, though they were probably fairly unskilled since another *mitayo*, Juan Malca Atoc from Chinchacocha in the region of Junín, who was assigned there in 1593, was only twelve years old.[57] After 1572 *mitayos* were in theory to be remunerated for their labour at the rate of one real a day.[58] Such labour had the advantage of being cheap, but since it was constantly changing employers found it difficult to develop the skills of such workers.

Those who lived permanently in the city and others who were not required for public labour service might contract themselves to employers for fixed periods and wages. These contracts were supposed to be drawn up in the presence of a *corregidor*. Apart from fixed wages, workers on contract might receive clothing, shelter and medical care, conditions that might be attractive to those lacking any alternative means of subsistence. The minimum wage for contracted workers was 12 pesos a year, often plus lodging, food, clothing and medical care, though wages varied according to their skills.[59] In the Hospital of

55 Salinas y Córdova, *Memorial*, 245; Cook, *Indian Population*, 151.

56 AGNP Protocolos Siglo XVI 20 Rodrigo Alonso Castillejo fols. 271v.–272 1 Jul. 1596. There are a large number of contracts for the assignment (*asiento*) of indigenous labourers for different tasks in AGNP Protocolos Siglo XVI 18, 19, 20. For the employment of Rodrigo de Vargas in the Hospital of Santa Ana see: AGI Lima 122 fol. 36–37v. Información hecha... de la parte del hospital de los naturales 30 Oct. 1577; ABPL 9086 Visita al Hospital de Santa Ana fols.104–105 no date [1588].

57 AGNP Protocolos Siglo XVI 19 Rodrigo Alonso Castillejo fol. 575v. 16 Sep. 1593.

58 Lowry, "Forging an Indian Nation," 176–80; Paul Charney, *Indian Society in the Valley of Lima, Peru, 1532–1765* (Lanham, MD: University Press of America, 2001), 20–21. In 1563 the hospital de los espanõles (San Andrés) was employing a *yanacona* at 1 real a day (AGI Lima 131 Juan de Alvear en nombre del administrador y fundador del hospital despañoles 1563).

59 For Lowry, "Forging an Indian Nation," 195, 200–203. For an example see AGNP Protocolos Siglo XVI 51 Rodrigo Gómez Baeza fols. 285–285v. Blas de Medina concierto con Juan Real 8 Mar 1591.

Santa Ana ordinary workers, gardeners, or those who worked in the hospital's *chácara* were paid between 40 and 50 pesos a year.[60] In addition, some Indians were being employed in the *botica* on a casual basis for a few days, perhaps to undertake specific or skilled tasks, for which they were paid at the higher rate of 3 reals a day. Working in a hospital environment, Indians might acquire healing skills. One Indian worker in the Hospital of Santa Ana, Pedro Capicha, acquired sufficient skills as a barber and surgeon to be assigned to work on its hacienda in the valley of Jauja in order to treat the Indians there who were suffering from scrofula and other diseases, for which he was exempted from forced labour.[61]

Black Pharmacy Workers

Blacks vastly outnumbered Indians in Lima. In 1613 there were about 10,000 Blacks in the city,[62] and they were more commonly employed in all types of *boticas* than Indians. Most *boticarios* possessed several slaves, though many probably worked in their households rather than their pharmacies.[63] The will of one eminent apothecary, Francisco de Alva, who became *boticario* to the Inquisition,[64] owned six slaves, of whom one Matheo aged 30 worked in the *botica*.[65] This slave had clearly acquired considerable pharmacy skills, because following the *boticario's* death in 1576 when most of his possessions were auctioned, the "negro boticario" who remained with his second wife became

60 ABPL 9084 fols. 132–70 *passim*, fols. 241–242v. Libro de cuentas de gastos del hospital de Santa Ana año 1598. Charney (*Indian Society*, 21) gives wages of between 22 and 180 pesos a year for those contracted to work in agriculture.

61 ABPL 9095 Real Provisión 31 Aug. 1617. The text is rather ambiguous since it could refer to the general treatment of Indians or to illnesses which were specific to them. It reads: "lamparones y otras enfermedades de los indios naturales."

62 Salinas y Córdova, *Memorial*, 245.

63 There is evidence for the purchase of slaves by *boticarios* in the accounts of the slave trader, Manuel Bautista Pérez. Bernardo Gil bought a male slave from him in 1618 at the high cost of 655 pesos (AGNP SO CO Ca. 2 doc. 8 Venta e rendimento de duzemtas e vimte e sete peças de escravos 1618) and in 1625 Juan Matías de Vera, at that time *boticario* to the convent de los Descalços, purchased one for 580 pesos (AGNP SO CO Ca. 20 doc. 201 Venta de 76 pieças de esclavos 1625). In addition, the records of burials in the church of Sagrario attached to the cathedral, include payments, normally of 8 pesos, for the interment of adult slaves by Lima's most prominent *boticarios*, including Diego de Tineo, Francisco de Alva and Juan de Bilbao (AAL Libro Parroquial Sagrario Difuntos libros 1–3 1567–1609).

64 *Libros de cabildos de Lima*, 7: 549–550 Título familiar al bachiller Alba 21 Jan. 1574.

65 AGNP Real Audiencia. Causas Civiles leg. 16 cuad. 81 fol. 7v.-8 Pleito…contra las bienes y herederos del bachiller Francisco de Alva, boticario morador en la Ciudad de los Reyes. 1576.

central to a law suit between her and Francisco de Alva's children by his first marriage. The slave was deemed to be worth 1,000 *pesos ensayados* and able to bring in more than 250 to 300 pesos a year through being hired out.[66] It was common practice in Lima for widows to secure an income through hiring out their slaves. At that time, the "wages" earned by slaves averaged between about 8 and 12 pesos a month.[67] The skills acquired by Matheo were perhaps exceptional, but Blacks were routinely employed in *boticas*. Another skilled *boticario* slave belonging to Bernardo Gil was also judged to be worth more than 500 pesos.[68] Similarly, Francisco Martín Reyna purchased one Luis Biafra for 516 pesos for the *botica* he jointly owned with Bartolomé Díaz Cabeza de Vaca. By the time of the *boticario's* death in 1608, the slave had also died and been replaced by a "Matheo" from Mozambique. This slave, who also worked in the *botica*, had been purchased from father Pedro de Pareja, of the Compañía de Jesús, but was described as old and judged to be worth only 200 pesos.[69] In general, however, *boticarios* were advised not to allow employees to make medicines or to sell poisonous ones, especially if they were slaves in whom, it was judged, there was "always less trust and confidence."[70]

Not only private *boticas*, but also hospitals were staffed largely by African slaves. They mainly undertook routine tasks such as cooking, cleaning and washing, but might also serve as nurses and assist in surgery. In fact, in a *visita* of the Hospital of Santa Ana in 1587 it was recommended that "gente de color" be employed as nurses and assistants in preference to Indians due to their greater dedication to work.[71] Apart from undertaking routine tasks, there is

---

66 AGNP Real Audiencia. Causas Civiles leg. 16 cuad. 81 fols. 343, 346, 770 Pleito…contra las bienes y herederos del bachiller Francisco de Alva, boticario morador en la Ciudad de los Reyes. 1576.

67 Bowser, *African Slave*, 102–104.

68 AAL Testamentos 52–21 fols. 88–104 Testamento de Bernardo Gil, boticario 1662.

69 AAL Testamentos 5 leg 1 fol. 13v–14, 179 Testamento de Bartolomé Díaz Cabeza de Vaca 1608. At this time the average cost of a newly-arrived slave was between 570 and 600 pesos (Newson and Minchin, *From Capture to Sale*, 228–29). In 1625 the *boticario* Bernardo Gil similarly purchased an African slave for 570 pesos (AGNP SO CO Ca. 20 doc. 201 Venta de 69 peças de diferentes naciones que traxo Sebastian Duarte 1625). For the employment of Blacks in *boticas* in Cartagena see: Archivo General de la Nación, Colombia (hereafter AGNC) Médicos y Abogados 6 fols. 944v–945 Martín Sánchez de Velasco…informe sobre sus visitas a las boticas de Cartagena 1634.

70 ARJBM División 1, leg. 17 Prohemiales 8 Libro de examen de los simples medicinales Antonio de Robles Cornejo 1617.

71 Rabí Chara, *Hospital de Santa Ana*, 74. This is based on leg. 9086 in the ABPL, from which documents relating to this *visita* have since been lost.

some evidence that they were being trained in pharmacy. In the early seventeenth century, the post of *boticario* of the Hospital of San Andrés was held by an African slave, Juan Mandinga.[72] Somewhat later in 1639 one Antonillo Angola was working in the pharmacy of Santa Ana and despite being slave was earning 7 pesos a year.[73] Most renowned of all was Martín de Porres, a Mulatto and America's first Black saint, who combined the duties of nurse, barber-surgeon, and apothecary in the Dominican convent of El Rosario.[74]

Running a Pharmacy

Despite Lima being a particularly healthy city, life expectancy in the sixteenth century was low. Its residents were constantly preoccupied with their health which meant that they were willing to expend large sums of money on medical treatments.[75] A priest attached to the Cathedral who suffered from a long-term illness and had no financial resources was willing to take out a loan of 3,000 pesos from its *mayordomo* to pay for his treatment.[76] The high demand for medicines and willingness to pay high prices should have made pharmacy a profitable business, but a central feature of the way it was managed was that it ran largely on credit. This was problematic because often the patients died or migrated elsewhere before clearing their debts, so that *boticarios* were often owed several thousand pesos. On his death in 1636, a number of Pedro de Bilbao's clients were noted as living in Chile, Mexico, and Potosí.[77] Often relatives of the deceased or institutions had to be pursued through the law courts in order to recover the debts. This could be a protracted and costly process which meant that *boticarios* had to have significant financial backing in order to keep their businesses running on a day-to-day basis in the meantime.

Elite families generally made annual contracts with physicians and surgeons to treat their households. How much they charged varied, possibly according

72 AHIRA Maldonado A-III-306 fol. 115 Libros de egresos e ingresos del hospital de San Andrés 1612.

73 ABPL 9104 Libro de cuentas de gastos del hospital de Santa Ana 29 Aug. 1639.

74 Cussen, *Black Saint*, 55–56, 69–84.

75 For the general predisposition of Spaniards to spend large amounts of money on medicine see Lanning, *Royal Protomedicato*, 230. Colonial documents are replete with comments by officials on the state of their health.

76 AAL Causas Civiles leg. 20 exp. 5 Causa de acreedores a los bienes del padre Juan de Vargas y Mendoza que fue de la catedral de Lima 1618.

77 AGNP Protocolos Siglo XVII 1789 Sánchez Vadillo fols. 2068–2072 Inventario de bienes de Pedro de Bilbao 25 Aug. 1636.

to the number of patients they were required to attend. The New Christian surgeon, Tomé Cuaresma, charged an annual fee of thirty to forty pesos for a family, with payments occasionally paid in kind, for example in bags of wheat,[78] whereas in 1633 Licenciado Pedro Rodríguez Merchán contracted a surgeon, Luis de Molina Guzmán for the annual salary of 70 pesos to treat all persons in his household and *chácara*.[79] There is no evidence that similar contracts were made with apothecaries, but it is clear from legal cases that elite families generally acquired medicines from the same *boticario* and that bills were drawn up on an annual basis.[80] Rather than *boticarios* being contracted on a salary, the amount they were paid was dependent on the ailments they were treating and the medicines they dispensed. The slave trader, Manuel Bautista Pérez had agreements with at least two *boticarios,* namely Pedro de Bilbao and Alonso de Carrión, to treat his household and the African slaves on his *chácara*, but whether these were underpinned by legal contracts is unknown. For the year 1628 Manuel Bautista Pérez owed 1,052 pesos to Pedro de Bilbao, while in 1639 when Pérez was put to death by the Inquisition, he was in debt to Alonso de Carrión to the order of 1,710 pesos for medicines he had supplied at various times between 1635 and 1639.[81]

The procedure for charging for medicines and recovering debts was as follows. A physician would draw up a prescription (*receta*) for medicines which would then be supplied by a *boticario*. Generally the patient did not pay for the medicines immediately, but when the medicines were delivered he or she signed the prescription acknowledging that they had been received. The signed receipts or *vales*, which did not indicate the cost, were then kept by the *boticario* as evidence of the medicines he had supplied. Bartolomé Díaz Cabeza de Vaca kept his receipts for different patients and households on separate strings, which at the time of his death numbered 195. He often recorded the profession of his clients, from which it is apparent that he served a wide spectrum of individuals from the governor, members of the *cabildo*, priests,

78 AGNP SO CO Ca. 44 doc. 394 fols. 772–775 Conciertos de curar de Tomé Cuaresma 1623, 1625 and 1630.

79 AAL Causas Civiles leg. 43 exp. 2 Causa ejecutiva que sigue Luis de Molina Guzmán, cirujano, contra el Lic. Pedro Rodríguez Merchán 5 Oct. 1634.

80 AGNP SO CO Ca. 27 doc. 277 Pedro de Bilbao contra Don Juan Arévalo de Espinosa 1629). For the bills incurred by Manuel Bautista Pérez with Pedro de Bilbao and Alonso de Carrión see: AGNP SO CO Ca. 57 doc. 431 1629, 1635–1640.

81 For the bills incurred by Manuel Bautista Pérez with Pedro de Bilbao and Alonso de Carrión see: AGNP SO CO Ca. 57 doc. 431 1629, 1635–1640. When the value of the medicines was assessed by the physicians Doctors Gerónimo Andrés Rocha, Juan de la Vega and Manuel Pérez they were judged to be worth about half the amount claimed.

teachers, and merchants, to a variety of artisans, including silversmiths, a gilder, tailors, hat makers, shoemakers, a glovemaker, a confectioner, and a locksmith. Of the clients, 92 still owed him money at the time of his death. These debts averaged 43 pesos, but ten people owed over 100 pesos. While each of these debts might have been small, they totalled some 3,874 *pesos*, which together with other debts represented 42 percent of the assets of his *botica*.[82] (See Table 2).

In general large debts, that is mainly those over 100 pesos, were pursued through the courts. A significant number of legal cases relating to the recovery of sums owed to *boticarios* are to be found in the section Causas Civiles in the Archivo Arzobispal in Lima.[83] In presenting a claim a *boticario* would include a list of the medicines that he had supplied and the *vales* on which the list was based (See Figures 3 and 4). The legal procedure was that a physician or *boticario* would then be employed to assess the value of the medicines; in most cases they were valued at about half of what the *boticarios* claimed. *Boticarios* were infamous for overcharging for medicines, so that they rarely challenged the lower assessments.

Sometimes clients objected to the bills on the grounds that the *vales* had not been signed or were not authentic. They might also refuse to pay claiming collusion between the *boticario* and the witnesses who were called to testify. In response to a claim for 150 pesos brought by the *boticario* Bernardo Gil, the defendant a priest, Doctor Sebastián de Betanzos, claimed that two of the witnesses had "a close friendship and connexion with him [Gil] and are in his house and company and eat with him, and as such I think they are untrustworthy and challenge they should be witnesses or be believed."[84] Similarly, in a redhibition case brought against Manuel Bautista Pérez by one Doña Francisca de Guzmán y Quintana for the sale of a slave who was suffering from severe stomach pains and unable to work, she claimed that all the witnesses, which included two surgeons, were his "servants, compatriots, [and]

---

82 AAL Testamentos Leg 5 exp. 1 fols. 27v–35v. Testamento de Bartolomé Díaz Cabeza de Vaca 1608.

83 For seventeen cases between 1613 and 1651 see: AAL Causas Civiles leg. 14 exp. 35, leg. 17 exp. 18, leg. 20 exp. 5, leg. 32 exp 14, leg. 33 exp. 19, leg. 39 exp 5, 16, leg. 41 exp. 15, leg. 42A exp. 2, 3, 10, leg. 48A exp. 11, 20, leg. 50 exp. 6, 16, 24, 38, leg. 60 exp.6. For overcharging see also: Biblioteca del Palacio Real, Madrid II/546 fols. 119–119v. Cuaderno de algunos papeles…Marqués de Montesclaros 1 Apr. 1612.

84 AAL Causas Civiles leg. 48A exp. 20 Autos seguidos por Bernardo Gil, boticario, contra el Dr. Sebastián de Betanzos, clérigo de menores órdenes, por 150 pesos de medicinas adquiridas en su botica. 20 Oct. 1639.

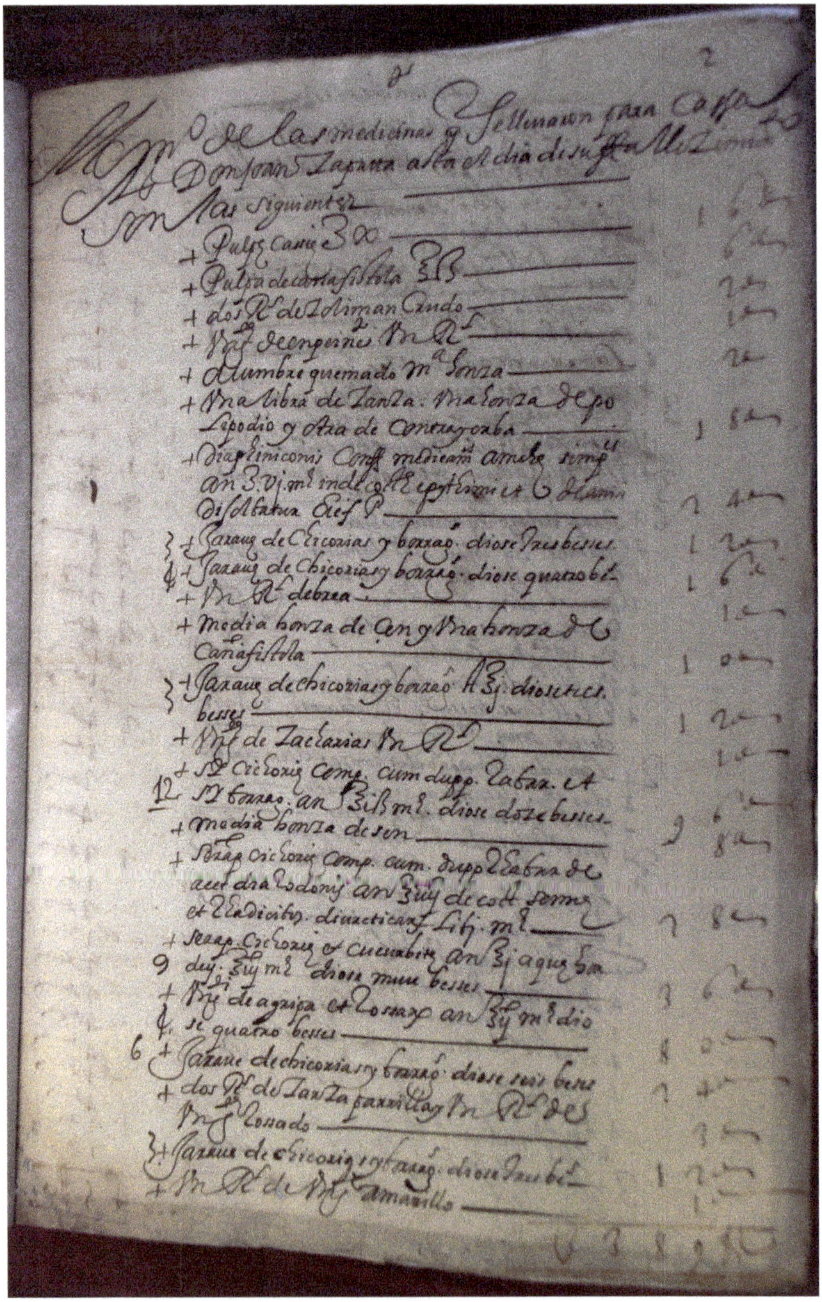

FIGURE 3 *List of medicines supplied by Bernardo Gil to Don Juan Zapata 1640.*
Source: AAL Causas Civiles legajo 50 exp. 16 Autos seguidos por Bernardo Gil, boticario, contra los bienes y albacea del Lic. Don Juan Zapata, clérigo presbítero. 1640.
COURTESY ARCHIVO ARZOBISPAL, LIMA. PHOTO: AUTHOR.

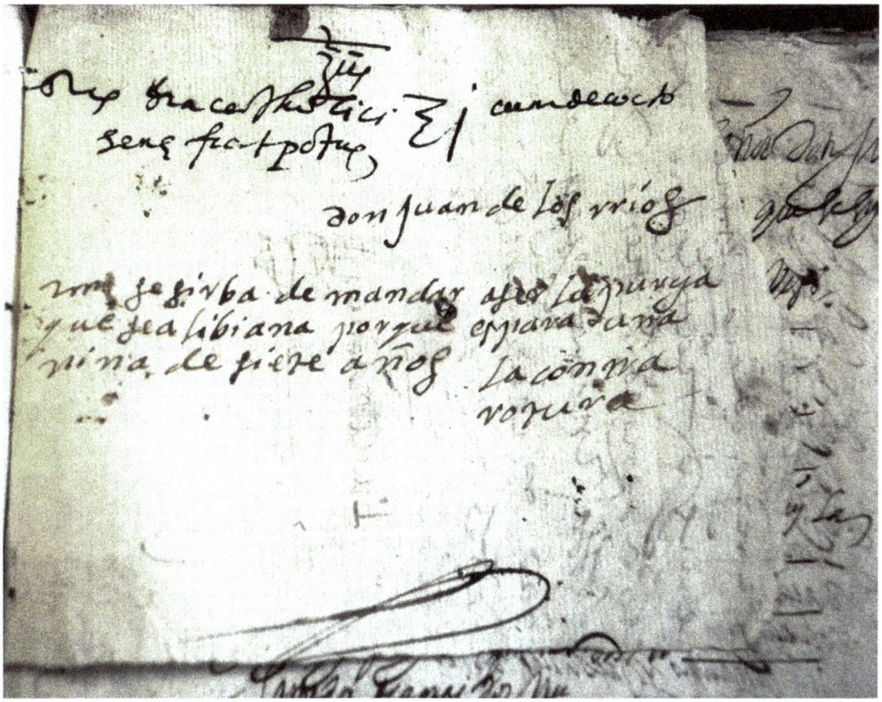

FIGURE 4 *Receipt for medicines supplied to Don Juan de los Ríos by Bernardo Gil, 1633.*
… diacatholici ʒi cumdecocto sene fiat potus
[one ounce of diacatolicon with a decoction of senna to make a potion].
don juan de los rríos
… se sirva de mandar aser [hacer] la purga que sea libiana [liviana] porque es para una niña de siete años. La contrarotura.
[…it serves to order to make a purgative which should be light because it is for a girl of seven years old. The plaster.]
Source: AAL Causas Civiles Leg. 42A, exp. 2 Causa seguida por Bernardo Gil, dueño de botica, contra el bachiller Don Juan de los Ríos, clérigo presbítero. 1633.
COURTESY ARCHIVO ARZOBISPAL, LIMA. PHOTO: AUTHOR.

close relatives."[85] Clients occasionally asserted that the *boticario* had supplied the medicines free of charge. Hence, Doña Beatriz de Garay stated that even though she had signed the *vales* presented by the *boticario* Mateo Pastor, this was only to eliminate fraud and that he had not charged for medicines, indeed she claimed that she could even request medicines by word of mouth. She said this was because he was a "paysano" of her mother and he treated her as family because "in this kingdom those who are from the same place in Spain regard

85 AGNP Real Audiencia Causas Civiles leg. 70 cuad. 263 Autos seguidos por Doña Francisca de Guzmán y Quintana contra Manuel Bautista Pérez sobre la redhibitoria de un esclavo 1626.

each other as *parientes* (relatives)."[86] In other cases, the prebend and teacher of the Cathedral claimed he was unable to pay because of the high cost of living in Lima and because due to a long illness he had run up bills with three doctors, a surgeon and a *boticario*.[87] On the other hand, *boticarios* or their executors defended their right to include unsigned *vales* or those signed by a relative because the patients were too ill to sign or had sent their slaves to the *botica* to collect the medicines on their behalf, explanations that were often accepted.[88] Clearly in many cases debts could not be recovered, though there is evidence of relatively small amounts being pursued over several decades.

The largest debts were incurred by monasteries, convents, hospitals, and the Inquisition. It was practice for the *cabildo* to make contracts with specific *boticarios* to supply medicines to all the city's monasteries and from the outset it had to tackle the issue of overcharging.[89] Up to the beginning of the seventeenth century this contract, while substantial, was never worth more than 5,000 *pesos corrientes* and it was handled by one *boticario*. However, in 1604 the contract was divided between two *boticarios* and by 1606 the cost had risen to 13,015.5 pesos, a figure that the *cabildo* considered unacceptable. In order to reduce the cost, the contract was put out to tender and in 1610 it was split between two other *boticarios*: Gerónimo Pujadas, who was to supply the Dominican, Franciscan, and Jesuit monasteries, and Pedro de Bilbao the Augustinian. Pujadas's pharmacy and contract were later taken over by Francisco de Sandoval about whom there were constant complaints. It was claimed he committed many frauds, altering and adding to prescriptions, such that one claim that he submitted for supplying medicines between 1619 and 1620 was reduced significantly

86 AGNP SO CO 132–1199 Don Diego Laurencio Valenzuela administrador del patronato de Mateo Pastor de Velasco contra doña Beatriz de Garay 14 Jun. 1663.

87 AAL Causas Civiles leg. 20 exp. 5 Causa de acreedores a los bienes del padre Juan de Vargas y Mendoza que fue de la catedral de Lima 1618.

88 See for example, AAL Causas Civiles leg. 41 exp. 15 Autos ejecutivos seguidos por Gabriel de España, boticario, contra el Lic. Diego Cabrera albacea y tenedor de bienes del Lic. Francisco Juárez Salgado 1632; AAL Causas Civiles leg. 42A exp. 3 Lima. Causa seguida por Bernardo Gil, contra los bienes que quedaron por fin y muerte del padre Luis Nieto Palomino 1632/1633; AAL Causas Civiles leg. 31 exp. 14 Causa ejecutiva seguida por Antolín Reynoso, boticario, contra Alonso Martínez Pastrana, contador mayor, albacea y tenedor de bienes de don Juan Velásquez 1627. In this case they were signed by his sister or niece, witnesses differ, but the argument was that in his will he indicated in detail the debts he owed. See also AGNP SO CO 125–1091 Gaspar de Calderón sobre medicinas que dio a Juan de Turiçes difunto 18 Sep. 1656.

89 AGI Lima 112 Tristán Sánchez, contador, 1588.

by the assessor to 1,840 pesos. Aware that fraud had also been perpetrated for four years previously, receipts for this period, which amounted to some 13,025 pesos, were also reviewed and assessed by the *boticarios, licenciado* Antonio de Robles and Antolín Reynoso, who valued them at 3,511 pesos, an amount to which Sandoval agreed.[90] Subsequently, it was suggested that a limit should be placed on the amount that a *boticario* could claim; both 3,500 and 2,200 pesos were suggested. It was noted that there were other more suitable *boticarios*, who had demonstrated "more satisfaction and knowledge" who could undertake the commission.

Francisco de Sandoval not only supplied several monasteries, but also the female Convent of La Concepción. In 1625 he initiated a legal case against the Convent on behalf of the widow of Gerónimo Pujadas from whom he had assumed charge of the *botica*. This was over the payment of 10,087 pesos relating to medicines that had been supplied between 1619 and 1622.[91] However, the claim was disputed by the Convent on the grounds that not all the *vales* for items listed had been presented and others were unsigned. It also asserted that some of the items had not been claimed for within three years, as was apparently required by law, and were therefore invalid. In addition the Convent argued that the assessment by the *boticario*, Juan Ximénez Villayzan, had included many invalid *recetas* and a number of arithmetic errors. The assessor himself acknowledged that the task of reviewing some 5,912 *recetas* had been daunting and had been hard work over two months. The Convent then raised the question of whether the widow of Gerónimo Pujadas had a legal claim. In attempting to resolve the issue, a reassessment was ordered on the basis of 3,987 agreed receipts and in the end, in 1629, a value of 5,788 pesos 2 reales was imposed by the physician and later *protomédico*, Juan de la Vega. The latter amount appears to have been paid, but there were ongoing disputes between Sandoval and the Convent over medicines that he had supplied on his own account.[92] As late as 1650, the legitimate son of Gerónimo Pujadas

90 AGI Lima 97 Consulta a la Real Audiencia sobre las medicinas donadas a los conventos 12 May 1621; Carta de la Real Audiencia al rey sobre limosna de medicinas a los conventos 6 May 1622. Most likely Antonio de Robles was the physician Antonio de Robles Cornejo, who was the author of several manuscripts on the use of medicinal simples to be discussed in Chapter 5.

91 AAL Monasterio de la Concepción leg. 2 exp.18 Causa seguida por Francisco de Sandoval, boticario de Lima, contra el Monasterio de la Concepción 1626–1629.

92 AAL Monasterio de la Concepción leg. 3 exp. 19 Causa seguida por Francisco de Sandoval, boticario de Lima, contra el Monasterio de la Concepción por las medicinas que sacaron de su botica 1627.

was still trying to claim for the *recetas* on which agreement had not been reached.[93]

Even if claims were valid, often clients lacked the resources to meet them. The Convent of La Concepción claimed that it was poor and did not receive government support, unlike male monasteries. Although in theory it received substantial dowries for the entry of women into the Convent, very often they were paid from *censos*. These were annuities paid from the income of a house or other property, usually of five percent, which were pledged for a number of years or in perpetuity.[94] The problem from the perspective of the Convent was that often householders could not afford repairs to their properties so that many fell into disrepair such that the income they generated declined.[95] Apart from the *censos*, the only other source of income was from an *obraje* producing serges and rope and this was deemed insufficient to pay for medicines and for the service of medical practitioners. The Convent admitted that it owed "a large amount of pesos" which it was unable to pay.[96]

Francisco de Sandoval may have been running a particularly fraudulent business, but supplying medicines to convents required substantial financial resources and incurred some risk for anyone accepting the contract. When Pedro de Bilbao died in 1636, he was not only owed 8,211 pesos for prescriptions supplied to 156 individuals, plus 76 pesos supplied to the Inquisition's prison, but also for medicines supplied over nine years to the Convent of La Merced and for four years to the Convent of San Agustín.[97] In addition he was owed 28,000 pesos worth for medicines he had supplied to the Convent of Nuestra Señora de la Encarnación, which was subject to a law suit.[98] This was not all,

---

93 AAL Monasterio de la Concepción leg. 9 exp. 60 Causa seguida por Ignacio de Pujadas, hijo legítimo de Gerónimo de Pujadas, contra el Monasterio de la Concepción por 11,000 pesos de medicinas que le dio de su de su botica y le han quedado debiendo 1651.

94 Arnold J. Bauer, "The Church in the Economy of Spanish America: Censos and Depósitos in the Eighteenth and Nineteenth Centuries," *Hispanic American Historical Review* 63(4) (1983): 715–17; Brian R. Hamnett, "Church Wealth in Peru: Estates and Loans in the Archdiocese of Lima in the Seventeenth Century," *Jahrbuch für Geschichte Lateinamerikas* 10 (1973): 115–16.

95 AGI Lima 215 N 6 R 1 fol. 3 Informaciones del monasterio de la Concepción de Lima 18 Mar 1603.

96 AGI Lima 215 N 6 R 1 fol. 11v. Informaciones del monasterio de la Concepción de Lima 18 Mar 1603.

97 This excluded the cost of other medicines that had been supplied previously.

98 AGNP Protocolos Siglo XVII 1789 Sánchez Vadillo fols. 2068–2072 Inventario de bienes de Pedro de Bilbao 25 Aug. 1636.

because another *boticario*, Alonso de Carrión who was not listed amongst his debtors, also certified that he owed Pedro de Bilbao 650 pesos.[99]

One may wonder why *boticarios* allowed clients to run up such large debts, even if they were exaggerated. It suggests that their businesses were sufficiently profitable to bear the debts and that the amounts that they were owed did not affect the day-to-day operation of their pharmacies. There are no instances of *boticarios* being made bankrupt or evidence of them complaining of being poor. Such debts were common in many businesses at this time.

In the early modern period, business success depended on reputation and trust.[100] In Europe the largest debts were incurred by elites, such as government officials and lawyers. Apothecaries felt obliged to treat them even if they anticipated that they would not be paid, since they feared that refusal would damage their reputations.[101] Debts, though incurring a cost, might actually bring some benefits. They could be used to develop or maintain relations of patronage-clientage that could ensure continued custom. Large contracts with major public and charitable institutions could only be fulfilled by *boticarios* with considerable financial backing and a good reputation. Although as shown above a good reputation was not always justified, the award of such contracts served as an indicator of their high economic status. Supplying medicines on credit to charitable institutions might also reflect on the religious piety and social standing of a *boticario*. For this reason, Paula De Vos argues that *boticarios* in Mexico City allowed charitable institutions to run up debts and rarely called them in.[102] The evidence for Lima indicates that convents were similarly allowed to build up debts over considerable periods of time, but in contrast to Mexico some *boticarios* in Lima did pursue some convents through the courts, as was the case of Pedro de Bilbao against the Convent of La Encarnación. Furthermore, in common with other *boticarios* he did not demure from overcharging charitable institutions. It suggests that there were limits to the support that *boticarios* gave to charitable causes in the form of reduced prices for medicines. After all, as will be shown in Chapter 7, they could demonstrate their

99 AGNP Protocolos Siglo XVII 1789 Sánchez Vadillo fols. 1375–1378v Alonso de Carrión to Juan de Sanmillán 28 Jul. 1636.

100 Daviken Studnicki-Gizbert, *A Nation upon the Ocean Sea: Portugal's Atlantic Diaspora and the Crisis of the Spanish Empire, 1492–1640* (Oxford: Oxford University Press, 2007), 84–86.

101 Brockliss and Jones, *Medical World*, 324; Robert Ralley, "Medical Economies in Fifteenth Century England," in *Medicine and the Market in England and its Colonies, c.1450–1850*, eds. Mark Jenner and Patrick Wallis (New York: Palgrave Macmillian, 2007), 29–30.

102 De Vos, "Art of Pharmacy," 185–86.

charity and piety in other ways, such as through membership of *cofradías*, private donations, and charitable works.[103]

Conclusion

The manner in which *boticas* were established and apothecaries and ran their businesses did not differ significantly from those in Spain. However, the personnel they employed and their interaction with charitable institutions seems to have differed. In Lima as in Spain, it was the case that *boticarios* had to have substantial financial resources to own a *botica*. Although detailed financial accounts kept by *boticarios* are lacking, the fact that pharmacy could attract non-*boticario* investors suggests that it was regarded as a profitable business. As will be shown in Chapter 7, the most prominent apothecaries amassed significant personal fortunes and could pay large dowries for their daughters. While *boticarios* in Lima were certainly interested in making profits, at the same time they were not entirely capitalistic for they did not invest their gains in expanding their businesses, but rather expended them on charitable activities that confirmed their social standing and assured them of an advantageous position in the afterlife.

Because substantial financial resources were needed to establish a *botica*, few were established in the city. Yet there were many others who were practising pharmacy as salaried employees, forced labourers, or slaves in both private pharmacies and hospitals. While the *cabildo* issued ordinances against *boticarios* employing non-licensed workers, especially Blacks and those of mixed race, they were not vigorously enforced. While this reflected in part the predominance of African slaves in the city's population, the seeming preference for the employment of slave labour may also have derived from their status of permanent servitude that enabled *boticarios* to invest in developing the pharmacy knowledge and skills of these workers in a way that was not possible with Indian forced labourers who were generally assigned for short periods and were constantly changing.

The ethnic diversity of those employed in *boticas* provided opportunities for the exchange of medicinal knowledge. However, the hierarchical structure of employment in *boticas* and the slave status of many employees would have discouraged a two-way exchange. Employees would have been taught how to prepare medicines by a licensed *boticario*, who invariably had been trained in humoral medicine and on whom they depended for their job, livelihood, and

103 For Mexico City see De Vos, "Art of Pharmacy," 171–81.

treatment. Nevertheless, there is evidence that Blacks and Indians were altering prescriptions, as attested by *cabildo* attempts to control their employment in pharmacies. Whether such changes to prescriptions were the result of error, insufficient guidance, or deliberate changes based on experience is unknown. However, the general absence of native botanical materials in pharmacy inventories and prescriptions, to be discussed in Chapter 6, suggests that the exchange of knowledge between ethnic groups with different healing traditions with the context of a functioning pharmacy was limited.

Pharmacies whether established in hospitals or run as private businesses did not operate as independent enterprises. They were subject to regulation by the *protomedicato* and municipal authorities, which among other things arranged for the inspection of pharmacies. This was primarily to ensure that out-of-date and dangerous drugs were not being prescribed and to prevent overcharging. However, they also paid attention to the types of medicines they contained, so that as will be shown, the inspections served to promote orthodox humoral practice. The regulation of pharmacies will be explored in Chapter 5 where competing approaches to medical practice are considered. Since considerable insight into medical practices in Peru can be gleaned from an analysis of the cargoes of *materia medica* and books imported from Spain and traded locally, the study turns first to an analysis of the transatlantic and local trading networks through which these commodities moved.

CHAPTER 4

Trading Medicines and *Materia Medica*

> The Indies have been rewarded better with respect to plants than other goods: for those that have come to Spain are few and do not grow well: those that have passed from Spain are many and they yield well.[1]

∴

Materia medica represented the most significant ongoing cost for apothecaries. The medicinal products prescribed by physicians in the New World emanated from geographically diverse sources and were handled by an equally broad range of traders that included large scale merchants, local producers and sellers, and even *boticarios* themselves. Consistent with Spanish humoral beliefs and with state policies controlling medical practice in the early modern period, many medicinal products employed in Peru were imported from Spain many having been acquired from merchants based in Lisbon, Antwerp or Venice.[2] However, as will be shown, these imports were unable to meet local demand and were often of poor quality and expensive. Hospitals and *boticarios* alike therefore purchased simple medicinal products in local markets, while hospitals and convents also established their own herb gardens and sometimes their own pharmacies for the preparation of compound medicines. This chapter therefore examines the extent of the dependency of medical practice in Lima on the transatlantic trade, showing how inadequacies in provision should have fostered an interest in local *materia medica* and encouraged innovation.

---

1 José de Acosta, *Historia natural y moral de las Indias* in *Obras del Padre José de Acosta*. Biblioteca de autores españoles 73 (Madrid: Ediciones Atlas, 1954), lib. 4 cap. 31: 125. ["Mejor han sido pagadas las Indias en lo que toca a plantas que en otra mercaderías: porque las que han venido a España son pocas y danse mal: las que han pasado de España son muchas y danse bien."].

2 Francisco Guerra, "Drugs from the Indies," 44–48; Mercedes Fernández-Carrión and José Luis Valverde, "Research Notes on Spanish American Drug Trade," *Pharmacy in History* 30–31(1988–1989): 29–30; De Vos, "Art of Pharmacy," 90–101; Stefan Halikowski-Smith, "'Profits Sprout Like Tropical Plants': A Fresh Look at What Went Wrong with the Eurasian Spice Trade c. 1550–1800," *Journal of Global History* 3(2008): 390–409.

 | DOI 10.1163/9789004351271_005

Organisation of the Transatlantic Trade

In order understand the availability of *materia medica* from Spain it is necessary to have knowledge of the way that the transatlantic trade was organised. To ensure that Spain and its citizens enjoyed the economic benefits that derived from the acquisition of its American colonies, the Crown instituted a monopolistic trading system. This involved the establishment of the *Casa de Contratación* (Board of Trade) in Seville in 1503, which among other things registered all out going cargoes and levied appropriate taxes. For their protection, from 1526 ships were not permitted to sail to the Americas alone but were required to travel in convoys; from the 1560s a fleet system was established with each fleet operating on a two-year cycle and focused on a limited number of ports in both Spain and the Americas. One fleet – the *Flota* – was destined for New Spain and Central America, and the other – the *Galeones* – focussed on the South American mainland and the ports of Cartagena and Nombre de Dios (later Portobello). All goods, including medicines, destined for Peru passed through the latter ports.[3] The fleets were protected by *armadas* – armed vessels – which were paid for by a tax known as the *avería* which was based on the value of the cargo being shipped. The fleet system reached its apogee in the 1580s when the *Galeones* fleet comprised over 90 vessels, after which sailings became less regular and the number of ships comprising each fleet declined.

Consistent with Spain's policy of retaining its monopoly of trade with the Indies, all ships and their cargoes sailing to the New World had to be registered in Seville where bonds had to be provided. All merchandise, which included medicines and books, was subject to the same procedures of registration and inspection, while cargoes of books were additionally scrutinised by officials of the Inquisition to ensure that they did not include any heretical or banned works. The ships themselves were inspected three times: first, to ensure that ships were seaworthy and possessed the necessary equipment, including arms and artillery for their defence; and on the second and third occasions, to ensure they possessed sufficient crew and provisions for the journey.[4]

Each vessel carried a bewildering variety of goods ranging from domestic and foreign cloth and clothing, hardware, furniture, bedspreads, books,

3 For the operation of the fleet system see: Clarence H. Haring, *Trade and Navigation between Spain and the Indies* (Cambridge, Mass.: Harvard University Press, 1918), 201–30; John Lynch, *Spain under the Habsburgs* (Oxford: Blackwell, 1965) 1: 147–66.

4 Enriqueta Vila Vilar, *Hispanoamérica y el comercio de esclavos: los asientos Portugueses* (Seville: Escuela de Estudios Hispanoamericanos, 1977): 143–44; Newson and Minchin, *From Capture to Sale*, 23–25.

medicines, wine and oil among other things, all contained in array of boxes, chests, sacks, bundles, flasks, and bottles. Items were registered by both generalist and specialised traders, as well as by passengers travelling with the vessel. An itemised list of all goods contained in each box or bale needed to be presented to officials of the *Casa de Contratación* for tax evaluation purposes. Medicines destined for the Americas were for the most part subject to the same taxes as other commodities, and these taxes were considerable. First there was a customs duty known as the *almojarifazgo*. Initially this *ad valorem* tax was set 7.5 percent, with 2.5 percent levied in Seville and 5 percent in the Americas, but in 1566 both were doubled and a further 2.5 percent was collected on arrival in American ports, thereby giving a total tax of 17.5 percent.[5] Duties exacted in the Americas were based on prices there rather than in Spain, so that goods destined for Peru first paid *almojarifazgo* in Panama, and then on arrival in Callao a further 5 percent based on the increase in the value of the commodity between the isthmus and Peru.[6] In addition, all cargoes were subject to the *avería* which was a tax imposed to pay the cost of convoying and defending the fleets. This tax, which was collected in Seville, varied between 1 and 12 percent according to the perceived foreign threat at the time and therefore the need for defence.[7] Books were subject to the *avería*, but they were exempt from the *almojarifazgo*. It was observed that merchants were making a good profit from trading books since there was a growing demand from schools and universities and it was suggested that in order to raise revenue the Crown might impose the *almojarifazgo* on all books, except those destined for churches, convents, and monasteries. Nevertheless, the exemption remained in force.[8]

From 1553 all goods destined for the support of hospitals and convents, which included medicines, while being subject to the *avería*, were exempt

5 Rafael Antúnez y Acevedo, *Memorias históricas sobre la legislación y gobierno del comercio de los españoles con sus colonias en las Indias occidentales* (Madrid: Imprenta de Sancha, 1797), 211–18; Haring, *Trade and Navigation*, 84. This level of duty continued to 1660.

6 See AGNP Cajas Reales: Lima H-3 leg. 4 lib. 24-a fols. 1–30 Aranceles para cálculo del almojarifazgo (1617). *Almojarifazgo* was also collected in Lima on intercolonial trade, but at the lower rate of one percent (Borah, *Early Trade and Navigation*, 112–13).

7 Haring, *Trade and Navigation*, 67–82.

8 AGI Lima 143 fol. 35 Memorial y apuntamiento sumario de las cosas muy importantes 1612; *Recopilación* lib.8 tit. 15 ley 27 4 Nov 1548; Antúnez y Acevedo, *Memorias históricas*, 218; Irving A. Leonard, *Books of the Brave: Being an Account of Books and of Men in the Spanish Conquest and Settlement of the Sixteenth-century New World* (Berkeley and Los Angeles: University of California, 1992), 133; Carlos A. González Sánchez, *New World Literacy: Written Culture across the Atlantic, 1500–1700*, trans. Tristan Platt (Lewisburg: Bucknell University Press, 2011), 49–50.

from the *almojarifazgo*.[9] However, this exemption does not appear to have been applied consistently. The fleet of Juan Flores Rabanal dispatched in 1618 contained four cargoes of medicines, one for each of the three hospitals of Santa Ana, San Andrés and La Caridad in Lima, and the other for the royal hospital in Potosí.[10] While the *avería* was levied on all four cargoes, at the rate of about 1.1 percent, only the consignment for the hospital in Potosí appears to have been exempt from *almojarifazgo*. This seems to have been because the *boticario* in Seville who was remitting the medicines had previously paid this duty. As for the other three consignments, taxes were levied on rounded values of cargoes resulting rates varying between 5.8 and 8.0 percent.[11] In some cases officials did not exempt the cargoes of medicines for hospitals, but reduced their value and levied the regular duty on the lesser amount.[12] In contrast to medicines destined for hospitals, which in theory should have been exempt from the *almojarifazgo*, the shipping records of the *Casa de Contratación* in Seville suggest that at the end of the sixteenth century private merchants were paying *almojarifazgo* at 5 percent.[13] In order to avoid paying taxes, therefore, it was said that *boticarios* resorted to hospital pharmacies to acquire medicines at cheaper prices, despite the fact that hospital ordinances specifically forbade them to sell medicines on pain of a fine of twenty pesos and for the second offence forty pesos and loss of office. It was the intention that the *materia medica* destined for hospitals should be used to treat the poor and sick.[14] Nevertheless,

9 *Recopilación* lib.8 tit. 15 ley 28 15 Dec. 1553. When the first *botica* for the hospital of Santa Ana was shipped to Lima in 1551, 5,829 *maravedís* were paid in *averías*, which at this time was calculated on the basis of tonnage, calculated to be 7 and 9.5 twelfths tons. A further 4,500 *maravedís* customs duty ("aduana de indias") were levied (AGNP Protocolos Siglo XVI 63 Diego Gutiérrez fols. 185–186 Hospital de los españoles y naturales compañía con Francisco de Bilbao de la botica 20 Oct. 1551).

10 AGI Contratación 1166 N4 277–80, 297–298 Registro del navío San Francisco 1618; AGI Contratación 1166 lib.10 fols. 353–56 Registro del navío San Pedro 1618; AGI Contratación 1166 lib. 1 fols. 459–65 Registro del navío El Espíritu Santo 1618.

11 It is not clear why different levels of taxes were imposed, especially since they were all assigned to hospitals and assessed for tax by the same officials.

12 AGI Contratación 1126 lib. 1 fols. 323–24 Registro del navío Nuestra Señora de las Angustias 1597–1598. In this consignment to the hospital of Portobello, the assessed value of the medicines was reduced by 25,000 maravedís from 103, 989 *maravedís* and 5 percent *almojarifazgo* paid on 78,989.

13 See for example, AGI Contratación 1127 lib. 2 fols. 91–94, 263–64, 433–36 Registro del navío San Cristóbal 1597–1598.

14 AGNP Protocolos Siglo XVI 63 Diego Gutiérrez fol. 186 Hospital de los españoles y naturales compañía con Francisco de Bilbao de la botica 20 Oct. 1551; Ordenanzas y constituciones del hospital real de Santa Ana, Lima 28 Feb. 1609 (Ordinance 45).

in the early seventeenth century officials of the royal exchequer in Lima grew concerned that the Crown was losing tax revenue because hospitals were selling medicines to private *boticarios*, while apothecaries were making large profits and paying their *oficiales* high salaries.[15] Although, as will be shown below, there is evidence that on occasions physicians acquired medicines from hospitals, these were generally not imported products but rather botanical materials that were produced by the hospitals themselves in their gardens and on their haciendas.

As with many financial transactions at the time, there was considerable scope for fraud in the assessment of taxes. With the exception of boxes containing books, the cargoes themselves were not inspected, rather those registering them attested on oath to the nature, quantity, and value of their contents. It was common practice for merchants to undervalue cargoes, it was said by as much as a fifth or a quarter, in order to reduce the taxes payable. As a consequence, taxes were generally based on a total figure that was marginally higher than the total value of items given on the registration list. On other occasions a corrupt official might be bribed to lower the valuation. Another source of fraud was that passengers' personal effects were exempt from *almojarifazgo*, so that immigrants often took large quantities of merchandise for sale, even though they were specifically forbidden to do so.[16] The Crown realised that the lack of detailed inspections of cargoes left the way open for fraud, but the cursory procedure that was adopted minimised delays in dispatching ships and facilitated the acquisition of profits for merchants on whom it depended to generate revenue for its exchequer.[17]

Although individual boxes of merchandise were not inspected, detailed lists of their contents had to be presented to officials. These lists contain information on the amount and value of each individual medicinal product contained in the shipment. Such detailed lists survive for cargoes sent with the fleets between 1583 and 1624.[18] After that date detailed cargo lists for each box were not required, rather just a general description of their contents using generic names, such as "medicinas," with taxes assessed by weight rather than value.[19]

15 AGI 143 Memorial y apuntamiento sumario de algunas cosas muy importantes al reparo y aumento de las rentas de Su Magestad 1612.

16 Antúnez y Acevedo, *Memorias históricas*, 211, 236; Lutgardo García Fuentes, *Los peruleros y el comercio de Sevilla con las Indias, 1580–1630* (Sevilla: Universidad de Sevilla, 1997), 56.

17 Haring, *Trade and Navigation*, 90–91; Lynch, *Spain under the Habsburgs*, 2: 181.

18 AGI Contratación 1080 to 1174. These legajos also include the *armadas* destined for New Spain.

19 Joseph de Veitia Linage, *Norte de la contratación de las Indias Occidentales* (Sevilla: Imprenta de Francisco de Blas, 1672), lib. 1 cap. 18, para 6, pp. 118–19; Haring, *Trade and Navigation*, 91.

Since details of individual consignments are not generally available for shipments after 1624, the analysis here focuses primarily on the period prior to this date, though it includes some information for a few boxes dispatched up to 1640.[20]

In addition to providing exceptional insight into the nature and value of the *materia medica* being exported from Spain, these registrations reveal information about the merchants, factors, and other individuals involved in the trade. The preamble to the registration of each cargo gives the name of the person registering it, often their residence, and occasionally their occupation. It goes on to provide the name of the ship, shipmaster, and fleet with which it is to travel and its destination. It then specifies the person to whom the goods are to be delivered. Normally more than one person is named in the event that the specified recipient should be absent. This was a common occurrence since merchants and factors were often on the move and slow communications between Spain and the Indies meant that knowledge of people who had moved or died was often imperfect. Sometimes the final recipient of the goods was a named *boticario* in Lima or Cartagena, but in other cases it was a factor who would ensure their onward transmission or alternatively a merchant who was to arrange their sale. Sometimes traders entrusted boxes of medicines to passengers travelling on ship and even to the shipmaster or another member of the crew, with instructions about their delivery. The precise ownership of the goods and who would bear the risk in the event of loss were also specified (Figure 5). Thus, the cargo registrations make it possible to identify who was trading medicines and therefore reconstruct trading networks not only between Spain and the Americas, but to some extent within the colonies.

Apothecaries, Pepperers and Spicers

The medicine trade in Spanish America is best understood in the context of changes occurring in the structure of the trade in Europe. In the thirteenth century pharmacy began to split from medicine, encouraged by the specialisation of the two sciences and by the emergence of professionals who were more involved in trading drugs and preparing medicines than practicing medicine *per se*. Those trading medicines included apothecaries, pepperers and spicers. These traders initially handled similar products, namely drugs, spices and other commodities such as sugar and dried fruit, but later began to specialise. In England the pepperers tended to be wholesale merchants and shippers, such

20 Of the 249 shipments, only 13 fall in the period 1624 to 1634.

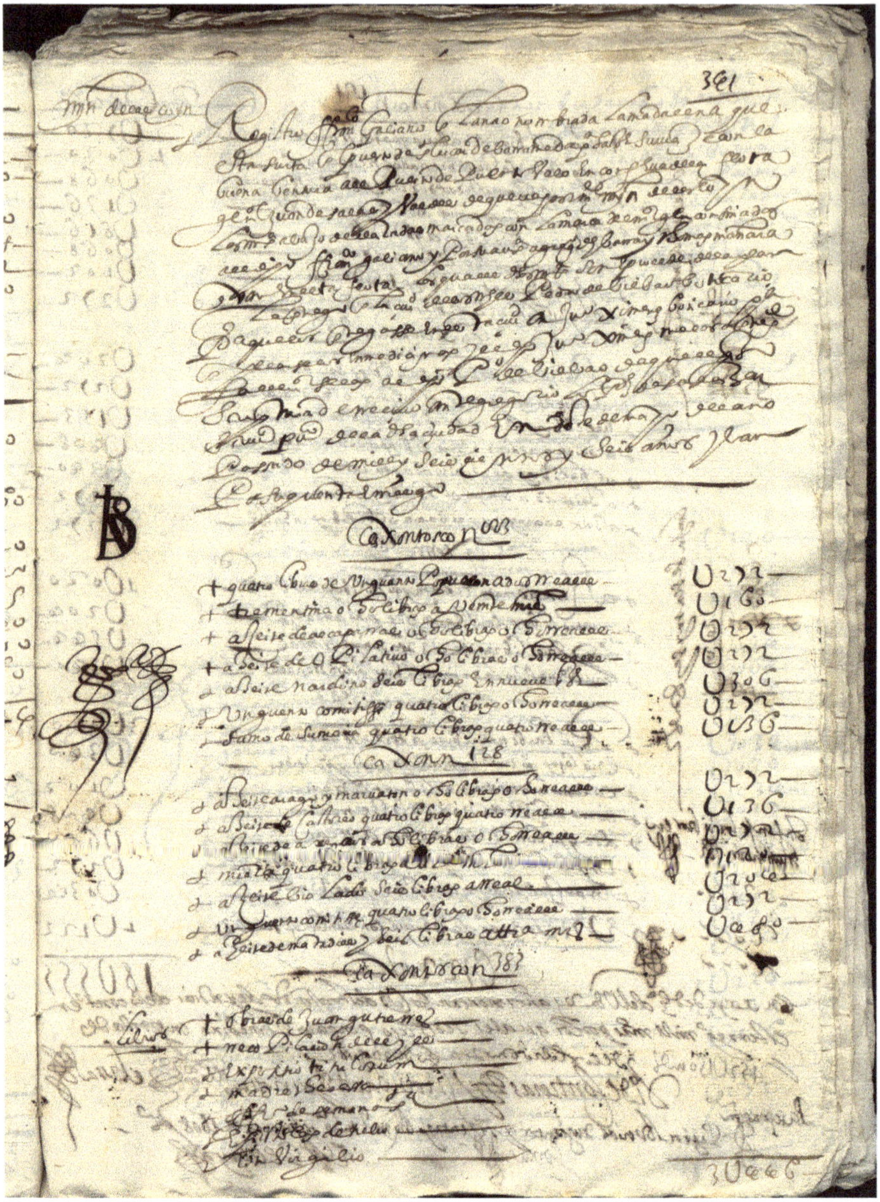

FIGURE 5 *Register of cargoes shipped by Francisco Galiano to Pedro de Bilbao, 1607.* Source: AGI Contratación 1151B N1 Image 683 Registro del navío La Magdalena 1607.

COURTESY ARCHIVO GENERAL DE INDIAS, SEVILLE.

Francisco Galiano registers in the ship called La Magdalena, which is anchored in the port of Sanlúcar de Barrameda to make the journey with good fortune to Puerto Belo in company of the fleet of general Juan de Salas y Valdés, of which the shipmaster is Martín de Lazcoain, the goods declared below marked with the mark in the margin consigned to Francisco Galiano and in his absence to Gregorio de Ibarra and Thomas Manara who travel with this fleet, which are those goods which are and proceed from the silver that the apothecary Pedro de Bilbao gave him in the city of the kings [Lima] to deliver to the apothecary Juan Ximénez in this city [Seville] for medicines and the said Juan Ximénez has given them to me to take to the said Pedro de Bilbao from whom he presents a notarised receipt before Gregorio López de Salazar public notary of the city on the twelfth of May last year of 1606 and they are carried on his account and risk.

Large Rough Box Number 123

Four pounds of *populeón* [complex] ointment at two reals	272 [maravedís]
Turpentine eight pounds at 20 maravedís [a pound]	160
Oil of capers eight pounds eight reals	272
Opilativo oil [to remove obstruction] eight pounds eight reals	272
Oil of spikenard six pounds nine reals	306
Comitisse ointment [unidentified] four pounds eight reals	272
Fumitory syrup four pounds four reals	136

Large Box Number 128

Aragon and marciaton [complex] oil eight pounds 8 reals	272
Oil of castoreum four pounds four reals	136
Oil of wormwood eight pounds eight reals	272
Juniper oil four pounds at thirty maravedís [a pound]	120
Violet oil six pounds at one real [a pound]	204
Comitisse ointment [unidentified] four pounds eight reals	272
Oil of *mata* [unidentified shrub] sixteen pounds at thirty maravedís	480

Large Rough Box Number 383
Books
Works of Juan Gutiérrez
Recopilación de leyes
Expósitio titulorum
Mother Theresa
Service for Holy Week [*Oficio de semana santa*]
Letters of [Marco] Tulio
A Virgil

that they came to be known as grocers, that is selling goods *en gros* (in gross amounts). Spicers on the other hand worked in the retail trade and like apothecaries prepared medicines. In medieval England the distinction between apothecaries and spice merchants was practically non-existent; both belonged to the grocer's company or guild until 1617 when the apothecaries broke away and formed a separate company.[21] In Italy the apothecaries formed separate guilds from about 1300,[22] but in France, as in England, the apothecaries, spicers, druggists, herbalists, and wax chandlers formed a single guild.

In Spain, early legislation regulating the activities of spicers (*especieros*) and *boticarios* often referred to them together.[23] Hence inspections of *boticas* and spice shops were conducted on a similar basis and at the same time every two years, primarily with the objective of ensuring that poor quality medicines

21 T.D. Whittet, "Pepperers, Spicers and Grocers – Forerunners of the Apothecaries," *Proceedings of the Royal Society of Medicine* 61 (1968): 25–26; Paul Freedman, *Out of the East: Spices and the Medieval Imagination* (New Haven: Yale 2008), 61–62, 116–22.

22 *Kremers and Urdang's History of Pharmacy* (Madison WI: American Institute for the History of Pharmacy, 1986), 56–58, 67–90; David Gentilcore, *Healers and Healing in Early Modern Italy* (Manchester: Manchester University Press), 60–61.

23 In the fourteenth and fifteenth centuries pharmacists in the Barcelona region were at different times known as spicers and apothecaries and sometimes the same person was referred to by both titles (Javier Sorní and José María Suñe, "Barcelona. Baja edad media. Especieros o boticarios?" *Boletín de la sociedad española de historia de la farmacia* 34 (Number 136) (1983): 139–46). Like physicians, surgeons and *boticarios*, in the fifteenth century Spain *especieros* were also supposed to demonstrate *limpieza de sangre* before being allowed to conduct their business (Valverde and López Andújar, "Disposiciones legales," 257), but possibly because of the growing distinction between the two professions and the increased control over the training of *boticarios* and preparation of medicines, the requirement for spicers to demonstrate *limpieza de sangre* does not appear in later legislation (*Recopilación de las leyes, pragmáticas reales* (Muñoz), cap. VI ordinance 1: 71–72).

and spices were not being sold.[24] *Boticarios* and *especieros* were, however, distinguished from *drogueros*. From 1588, *boticarios* were ordered not to sell *drogas*, except those that might contain opium or *confecciones de alquermes* or *jacintos*.[25] At that time *especieros* were making medicines such as syrups and electuaries,[26] but from the end of the fifteenth century draft ordinances for the medical profession began to distinguish between the responsibilities of *boticarios* and *especieros*, suggesting that medicines such as canafistula, rose and violet oil, *carne de membrillo* (quince cheese), rose sugar and waters made through distillation were to be the preserve of *boticarios*.[27] Nevertheless, the specific medicines that could be prepared by apothecaries and spicers do not appear to have been included in general legislation but rather left to city ordinances and the judgement of inspectors. In the early seventeenth century, *especieros* in Antequera in Spain were being fined for selling senna (*sen*), polypody (*polipodio*) and poppy (*adormideras*),[28] but generally they expressed more concern about the quality of goods they dispensed than their type. While spicers did prepare medicines, they became primarily traders of spices, wax, honey, grain, fat, tallow, pitch, among other products.[29] Spicers included confectioners (*confiteros*) who focused on products made of sugar

24 *Novísima recopilación de las leyes de España*, lib. 8 tit. 13, ley 2: 107 3 Mar. 1477 and 4: 108–109 7 Nov. 1617; *Recopilación de las leyes, pragmáticas reales* (Muñoz), cap. XV ordinance 1: 187. In 1527 the city ordinances of Seville referred to *boticarios* and *especieros* together. Those found guilty of selling poor quality drugs would have them confiscated on the first occasion, pay double on the second, and suffer one hundred lashes on the third (*Recopilación de los ordenamientos de Sevilla* (Sevilla: Juan Varela de Salamanca, 1527), p. 237v. Online at: http://fondotesis.us.es/fondos/libros/332/13/ordenancas-de-sevilla/ [Accessed 2 Jul. 2016].

25 *Recopilación de las leyes, pragmáticas reales* (Muñoz), cap. XIV ordinance 16: 184 1588. *Confecciones de alquermes* or *jacintos* were complex medicines. The former was based on the Kermes insect (Luis de Oviedo, *Methodo de la colección, y reposición de las medicinas simples, de su corrección, y preparación* (Madrid: Luis Sánchez, 1622), 153) and the latter on a variety of precious and semi-precious stones (Oviedo, *Methodo*, 163).

26 *Recopilación de las leyes, pragmáticas reales* (Muñoz), cap. XIV ordinance 17: 184 1588. See also, González Arce, "Proyectos de ordenanzas," 224.

27 Archivo de Simancas, Cámara de Castilla, Diversos de Castilla leg. 1 doc. 56 fols. 1–11 Ordenanzas que sus altezas deben mandar ordenar acerca de los físicos, boticarios y especieros [1498] Online at: http://pares.mcu.es/ParesBusquedas/servlets/Control_servlet?accion=3&txt_id_desc_ud=2224592&fromagenda=N [Accessed 2 Nov. 2015].

28 Mercedes Fernández-Carrión and José Antonio Pérez Romero, "Visitas de inspección sanitaria en especierías, confiterías, cererías y aguardenterías de Antequera durante el siglo XVII," *Baética* 8 (1985): 444.

29 See the city ordinances of Seville: Recopilación de los ordenamientos de Sevilla, 1527, p. 237v.

and flour, but who along with the wax chandlers (*cereros*) also traded wax.[30] A similar pattern of specialisation was evident in Lima in the sixteenth century where *confiteros* sold spices, sugar, dried fruit and honey, and *cereros* wax and candlewicks.[31] The guild of *cereros* was one of the first guilds to be established in Lima in 1551. At that time, it included *confiteros*, of which there were 18 in the city in the 1630s.[32] However, towards the end of the colonial period the *cereros* allied with the tallow-chandlers (*veleros*), who not only made candles, but also traded oils, wax and ointments.[33] As will be seen later in this chapter, the trade in products required for the preparation of medicines in Lima shows some degree of specialisation, though *boticarios* also traded some raw medicinal products, such as wax.

The Transatlantic Trade in *Materia Medica*

The analysis of the transatlantic shipment of *materia medica* presented here is based on an examination of the cargoes of 350 ships sailing with the galleon fleets to Tierra Firme that were destined for Panama (Nombre de Dios, Portobello, and Panama), as well as to Cartagena de Indias, and Peru (Lima, Potosí and Cuzco) between 1583 and 1640.[34] These vessels contained 249 consignments exclusively of medicines, of which 61 were specified as being destined for Peru, the majority to Lima (53).[35] However, it is likely that a good proportion of the goods arriving with the Tierra Firme galleons, especially those destined for Nombre de Dios, or after 1597 Portobello, would have found their way to Lima.[36] Only cargoes composed wholly of medicines and medical equipment

---

30 Fernández-Carrión and Pérez Romero, "Visitas de inspección," 438–39.

31 Quiroz Chueca, *Artesanos y manufactureros*, 85.

32 Salinas y Córdova, *Memorial*, 257; Bromley, "Ciudad de Lima," 285. Earlier in the sixteenth century there were nine workshops of *cereros* (wax workers), *confiteros* (confectioners) and *candeleros* (candle makers) together, and seven of *pasteleros* (pastry-cooks) and *molineros* (millers) (Quiroz Chueca, *Artesanos y manufactureros*, 95).

33 Quiroz Chueca, "Gremios en Lima," 506–507, 511–12.

34 AGI Contratación leg. 1094–1172, 1180 *passim*.

35 The value of eight of these cargoes is not specified, so they are not included in calculations here, which are based on the value of 241 cargoes.

36 The terminus of the Tierra Firme fleet was moved from Nombre de Dios to Portobello in 1597 due to the shallow nature of the bay, difficult navigation, and the unhealthy climate (Haring, *Trade and Navigation*, 184–85; Enriqueta Vila Vilar, "Las ferias de Portobelo: apariencia y realidad del comercio con Indias," *Anuario de estudios americanos* 39 (1982): 279–80; Christopher Ward, *Imperial Panama: Commerce and Conflict in Isthmian America, 1550–1800* (Albuquerque: University of New Mexico Press, 1994), 103–37.

have been analysed here. Most ships contained only one or two consignments exclusively of medicines, while many had none. These consignments were destined primarily for hospitals, *boticarios* or large scale merchants. However, small quantities of medicines were often included in mixed cargoes, since many merchants were generalist traders.[37] These have not been included in the analysis, neither have the small quantities of medicines that passengers took with them for their personal use. Finally, the analysis excludes certain categories of *materia medica* that were generally handled by other traders. They include oil, spices, notably saffron, pepper, cumin, clove, and cinnamon, which could be used for culinary purposes as well as medicines, though small amounts of these products were sometimes included in consignments of medicines as well. The organisation of the trade thus reflects the growing, but still blurred, distinction between spice merchants and *boticarios* in Spain at the time.

The largest cargoes of medicines were destined for hospitals or prominent *boticarios*. Responsibility for the supply of medicines for hospitals lay with their *mayordomos*, who were themselves normally traders rather than persons with medical expertise.[38] Rodrigo Arias de Buiça, who was *mayordomo* of the Hospital of Santa Ana in the mid-1590s,[39] was also a prominent merchant. Born in Salamanca, he settled in Lima and made at least two journeys to Spain trading in other commodities as well as medicines.[40] According to Ordinance 30 of the Hospital of Santa Ana, the *mayordomo* was permitted to send 500 pesos a year to Spain for the purchase of medicines. For this purpose a list of medicines had to be drawn up and signed by the hospital's physician, surgeon and *boticario*, who were then to check off the medicines when they arrived ensuring that they were of good quality.[41] There were criticisms that merchants and *boticarios* in Spain often sent medicines that were past their best.[42] In fact

37 Lockhart, *Spanish Peru*, 79–80.

38 Ibid. 91.

39 ABPL 9085 Libro de la razón que toma Bartolomé de la Cueva, escribano veedor del hospital de Santa Ana, 1595–1596. While *mayordomo* his slaves were treated in the hospital (fols. 10–11).

40 Garcia Fuentes, *Los peruleros*, 111, 114–15. He also supplied the hospital of San Andrés with medicines (AGI Contratación 1145A Registro del navío San Vicente y San Juan 1605; AGI Contratación 1145B Registro del navío San Pedro 1605). For his petition to travel to Peru see AGI Contratación 5233 N23 Expediente de información y licencia de pasajero a indias de Rodrigo Arias de Buiça 14 Jul. 1590. Arias de Buiça died in Lima 1609. For his estate see: AGI Contratación 518 N1 R2 Autos sobre los bienes de Rodrigo Arias de Buiça 1617).

41 *Constituciones y ordenanzas del hospital real de Santa Ana*, Lima.

42 ARJBM División 1, legajo 17 Prologo. Libro de examen de los simples medicinales Antonio de Robles Cornejo 1617.

not all consignments reached the permitted value of 500 pesos. For example, the cargoes of medicines registered for the Hospitals of San Andrés and Santa Ana by the Peruvian merchants Alonso González de la Canal y Juan Bautista Crespo in 1618 were valued at only 91 pesos and 31 pesos respectively.[43]

The procedure followed by hospitals in acquiring medicines was that the *mayordomo* would entrust a quantity of pesos, normally in the form of silver bars, to a merchant or other individual, generally a resident of Peru, who would travel to Spain and arrange for the purchase of medicines.[44] The purchased medicines would then be shipped back to Lima with the next fleet.[45] The same process of remitting silver with a factor was employed by private *boticarios* in Lima. Sometimes, the contract specified the particular person from whom the medicines were to be purchased.[46] The cost of transport and the taxes imposed meant that the value of medicines that arrived was often considerably below the value of the silver that had been remitted. Hence, although in 1633 the Hospital of Santa Ana sent 585 pesos of silver to Spain for the purchase of medicines, those purchased were valued at only 250 pesos, the rest being absorbed in unspecified "gastos."[47] Such a shortfall was not unusual. In 1639 the *mayordomo* of the same hospital similarly entrusted 600 pesos to Francisco de Arano to acquire a list of specified medicines through Pedro de Roças in Spain, but the medicines he obtained were valued at only 438 pesos, the remainder being consumed by taxes and expenses.[48] A partial explanation for the reduced value of the medicines is that they were undervalued in order to reduce tax liabilities, but the costs of shipping were also high. Commonly transport costs and taxes on silver and other goods shipped to Spain were about 10 percent of

---

43 AGI Contratación 1166 N4: 277–78, 297–98 Registro del navío San Francisco 1618. There was also a consignment for Juan de Bilbao as well as for a *boticario* in Panama. Alonso González de la Canal was one of the most prominent merchants in Lima who in 1630 was Prior of the Consulado de Mercaderes in Lima (Bromley, "Ciudad de Lima," 280, 284). He was also responsible for funding the reconstruction of the church of Nuestra Señora de Montserrat and its opening ceremony (Suárez, *Desafíos transatlánticos*, 190). He died on 3 June 1635 having been given a purgative by a doctor (Suardo, *Diario de Lima* 2: 82).

44 For example, ABPL 9083 Libro mayor de rentas del hospital de Santa Ana fol. 75v. 1615; ABPL 9104 no folio Escritura de recibo de Francisco de Arana 27 May 1637.

45 See also, the registration of cargoes for the hospitals of San Andrés and Santa Ana respectively by the Peruvian traders Alonso González de la Canal and Juan Bautista Crespo (AGI Contratación 1166 N4: 277–78 and 297–98 San Francisco 1618).

46 For example, AGI Contratación 1151B N6: 429–30 Registro del navío Santa Cruz 1607 and 1163 N1 R2 81–3 Registro del navío La Visitación de Nuestra Señora 1615.

47 AGI Contratación 1180 N2 R7 fols. 217–18 Registro del navío Nuestra Señora de la Concepción 1634.

48 ABPL 9104 no folio Escritura de recibo de Francisco de Arana 27 May 1637; AGI Contratación 1184 N2 R2 fols. 51–52 Registro del navío Nuestra Señora de la Candelaria 1640.

their value.[49] The cost of transport of only six boxes of medicines from Portobello to Lima in 1650 was an exorbitant 125 pesos.[50] In addition, payments to factors could reach 30 percent of the value of the goods they handled.[51] It is perhaps not surprising that Gaspar de Perales, who among his other trading activities supplied the Hospital of San Andrés with medicines, was described in 1595 as one of the richest merchants in Peru.[52]

Temporally the volume of the medicine trade peaked in the 1590s and early 1600s after which it declined (Figure 6). This reflected trends in the galleon trade in general which shrank in the early seventeenth century as Spain's inability to supply its colonies with merchandise led first to the fleets sailing every two years rather than annually, and later from the 1620s to becoming even more irregular.[53] The increase in the volume of medicines being shipped to Peru probably reflected the general penetration of *peruleros* into the galleon

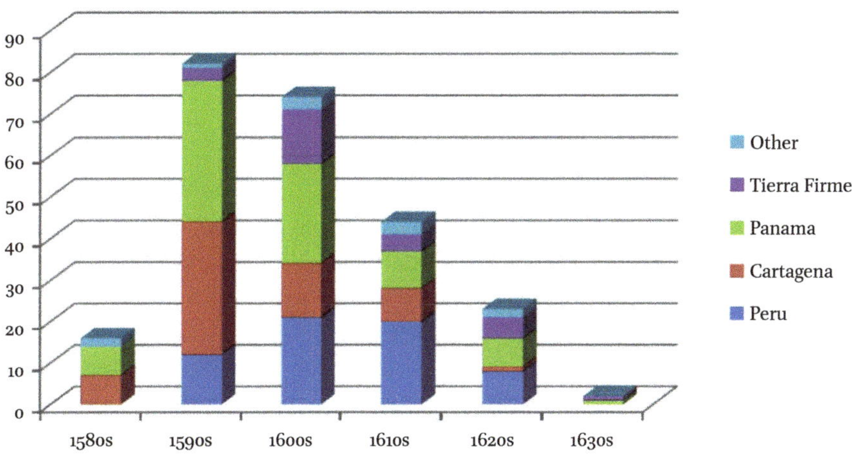

FIGURE 6 *Number of consignments of medicines to different destinations, 1580s to 1630s.* SOURCES: AGI CONTRATACIÓN LEG. 1094–1172, 1180.

49 García Fuentes, *Los peruleros*, 41–43.

50 ABPL 9100 Libro de cuentas del hospital de Santa Ana 1649–1651 Juan de Valverde 5 Mar. 1650.

51 Margarita Suárez, *Comercio y fraude el en Perú colonial: las estrategias mercantiles de un banquero* (Lima: Banco Central de Reserva del Perú, Fondo Editorial: Instituto de Estudios Peruanos, 1995), 58.

52 AGI Indiferente General 2102 N100 (Image 2) Información de Gaspar de Perales 12 Sep. 1595; AGI Contratación 1115 N4 676–77 Registro del navío la Magdalena 1596 and 1116 N6: 348–49 Registro del navío San Gregorio 1596.

53 Haring, *Trade and Navigation*, 208; Huguette and Pierre Chaunu. *Séville et l'Atlantique, 1450–1650* (Paris: A. Colin, 1956), 6_2: 586–587; Lynch, *Spain under the Habsburgs*, 2: 200–206; Suárez, *Comercio y fraude*, 28–40.

trade.[54] *Peruleros* were traders who had been born or resided in Peru who acted as factors for local residents, travelling to Spain taking with them silver bars entrusted to them for the purchase of specific goods.[55] Initially, merchants in Spain often sent their sons to the New World, where they became partners and factors.[56] While many of these early traders may have aspired to return to Spain in fact they often stayed and a second generation of merchants emerged who were based in Peru. Their involvement in the galleon trade reached its peak in the 1590s.[57]

Of the 249 shipments of medicines from Spain with the Tierra Firme fleets between 1583 and 1640, about one quarter were destined specifically for Peru. Of the 61 Peruvian cargoes, 34 were registered by *peruleros*, 8 by merchants, and 10 by *boticarios* based in Seville.[58] Merchants were always on the move so that although the ships' registers recorded the domicile of the trader it was not always the case that this was their place of birth. Equally an individual might be described on different occasions as being a resident of different places. Hence the merchant, Pedro de Avendaño, was at times described as a *vecino* of Seville having lived there for some time, when in fact he had been born in Peru. The registers therefore constitute an imperfect guide to the birth place of merchants and should be used in conjunction with other documentary and published sources.[59]

Keeping this limitation in mind, an analysis of 61 consignments of medicines to Peru suggests that those remitted by *peruleros* averaged 150 pesos and were substantially smaller than those dispatched by merchants and *boticarios* based in Seville whose shipments averaged nearly 400 pesos and 250 pesos respectively (See Table 3). Of the thirteen *boticarios* in Seville who were involved in the galleon trade to the Tierra Firme between 1583 and 1622, only four were involved in shipping medicines direct to Peru (See Table 4), though they, and the other nine, may also have supplied medicines in small quantities to merchants who included them as part of their general cargoes.[60] Those *boticarios*

54 García Fuentes, *Los peruleros.*

55 García Fuentes, *Los peruleros*; Guillermo Lohmann Villena, *Plata del Perú, riqueza de Europa: los mercaderes peruanos y el comercio con la metropolí en el siglo XVII* (Lima, Fondo Editorial del Congreso del Perú, 2004); Suárez, *Comercio y fraude*, 33–40.

56 Lockhart, *Spanish Peru*, 88–91.

57 García Fuentes, *Los peruleros*, 56.

58 The residences of the remaining nine are unknown.

59 See especially García Fuentes, *Los peruleros.*

60 In addition to the thirteen listed in Table 4, Fernández-Carrión and Valverde (*Farmacia y sociedad*, 71) name another four *boticarios* who were trading with the Indies between 1580 and 1584.

TABLE 3 *Consignments of medicines to Peru by trader type, 1591–1622.*

Registered by	Number of consignments from Seville	Total value in pesos	Average value in pesos
Sevillian merchants	18	5,676	315
Apothecaries	*10*	*2,499*	*250*
Non-apothecaries	*8*	*3,177*	*397*
Peruleros	34	5,100	150
Unknown	9	665	74
Total	61	11,441	188

Sources: AGI Contratación legs. 1094–1172.

in Seville who were involved in the Atlantic trade represented only a quarter of the total number in Seville, where according to official inspections of *boticas* conducted in 1594 and 1631 there were only 42 and 49 respectively.[61]

The two most important *apothecaries* in Seville who registered medicines for Peru were Juan Sánchez and Juan Martínez de Ortuçar. In 1618 Juan Sánchez dispatched one large cargo worth 244,664 *maravedís* (900 pesos) for the royal hospital of Potosí and three years later sent a much smaller consignment worth 30 pesos to the Hospital of San Andrés in Lima.[62] He also supplied a number of *boticarios* in Lima, namely Juan de Bilbao, Francisco Martín Reyna, and Gerónimo Pujadas, while he also shipped medicines to Panama for sale through intermediaries. Juan Sánchez possessed a *botica* in the Cathedral district of Seville[63] and his appointment as an inspector of the city's *boticas* in 1614 suggests he was a *boticario* of some standing.[64] Juan Martínez de Ortuçar seems to have traded mainly with *boticarios.* He was a major supplier not only to Juan de Bilbao in Lima, but also to *boticarios* in Cartagena and Nombre de Dios.[65] Cristóbal de Castro and Francisco Marroquín also sent small quantities to Lima for sale.[66]

61 AMS Sección 13 Siglo XVI Tomo 1 documento 114. Sevilla 20 June 1594; AMS Varios Antiguos 370 Visitas de boticas 1631; Herrera Dávila, "Boticas sevillanas," 63–81.

62 AGI Contratación 1166 N1: 459–65 Registro del navío El Espíritu Santo 1615; AGI Contratación 1171 N13: 181–82 Registro del navío Nuestra Señora de la Candelaria 1621.

63 Herrera Dávila, "Boticas sevillanas," 79.

64 AMS Varios Antiguos 414 Visita de las boticas 12 Apr. 1614.

65 They were the *boticarios* Andrés Gonzales and Francisco Suarez in Cartagena and Agustín de Bonilla in Nombre de Dios.

66 AGI Contratación 1127 N2: 91, 263 Registro del navío San Cristóbal 1598.

TABLE 4 *Apothecaries in Seville trading with the Tierra Firme and Peru, 1583–1634.*

Registered by apothecary in Seville	Dates of dispatch of cargoes	Number of fleets with medicines	Total number of consignments of medicines
Cristóbal de Castro	1583–1600	4	7
Pedro del Castillo	1586–1593/4	2	0
Juan Martínez de Ortuçar	1586–1601	5	9
Francisco de Ribera	1586–1601	7	16
Francisco Rocha Araña	1593–1615	6	6
Hernando de Medina	1593–1594	1	1
Francisco Marroquín	1597–1598	1	1
Cristóbal Gómez	1601	1	1
Francisco Galán Ramírez	1604–1619	6	8
Francisco de Ortega	1609–1622	4	7
Gerónimo Méndez	1612	1	1
Francisco Meléndez	1615–1618	2	0
Juan Sánchez	1615–1622	5	10
Miguel Fernández	1634	1	1
Francisco Ramírez Galán	1634	1	1
Total		47	69

Sources: AGI Contratación legs. 1080–1172, 1180, 1184.

Those *boticarios* and merchants based in Seville who wished to sell medicines on their own account rather than fulfil orders sent by American clients, generally sent them to Panama or Cartagena, since their closer geographical location meant that sales could be effected and the profit remitted to Spain more promptly. Hence, Francisco de Ribera, who sent large quantities of medicines to Tierra Firme on several occasions, generally registered them for Cartagena, specifying that they should be entrusted to the merchant, Luis de Espliega, and the *boticario*, Andrés Gonzales, who were to sell them for the best prices for cash and not on credit. He further instructed that they were to remit the revenue to Spain in the form of "gold, assayed silver or reals" and not in any other merchandise.[67] It is worth noting that Doctor Melchor de Amusco, a

67 For example, AGI Contratación 1094 N4: 175 Registro del navío San Pedro 1591, AGI Contratación 1095 N1: 465 Registro del navío Santa Catalina Mártir 1592. The gold, silver or pesos were to be remitted in ships that were carrying silver to Spain, presumably because

Number of boxes consigned to Lima	Total value of medicines dispatched in pesos	Value of medicines dispatched to Lima in pesos
1	255	45
0	296	0
2	2,534	852
0	5,548	0
0	386	0
0	33	0
1	121	121
0	255	0
0	798	0
0	276	0
0	268	0
0	89	0
6	2,146	1,482
0	45	0
0	36	0
10	13,086	2,500

priest as well as physician who became *protomédico* in Lima in 1616, similarly acted as an agent for the *boticario* Cristóbal de Castro and the merchant Juan Bautista Dorta when he was based in Panama.[68]

they would be guarded by the *armadas*. Andrés Gonzales was a prominent *boticario* in Cartagena, who in 1620 was accused of settling in Cartagena without license and trading illegally (AGI Escribanía de Cámara 589B Denuncia contra Andrés Gonzales, boticario, Portugués 1620).

68 A citizen of Toledo, Melchor de Amusco graduated with a doctorate in medicine from the University of Seville in 1583. In the same year, he also petitioned for licence to travel to Tierra Firme with his wife, children and servants, his medical credentials being certified by the king's chamber physician, Andrés Zamudio de Alfaro (AGI IG 2093 N177 Expediente de concesión de licencia para pasar a Tierra Firme, a favor del doctor Melchor de Amusco 1583; AGI IG 1952 lib.2 fol. 166v. Real Cédula...para que dejen pasar a Tierra Firme al doctor Melchor de Amusco 28 Dec. 1583).

TABLE 5 *Apothecaries in lima receiving consignments of medicines from Spain, 1583–1640.*

Apothecaries in Lima receiving consignments	Value of medicines in pesos	Dates of dispatch of medicines	Number of consignments	Percentage of total medicines dispatched by value
Juan de Bilbao	1,532	1591–1601	4	38.1
Pedro de Bilbao	849	1607–22	7	21.1
Gerónimo Pujadas	457	1618–19	2	11.4
Blas de Medina	286	1601	1	7.1
Luis de Alfaro	278	1597–98	1	6.9
Francisco Martín Reyna	202	1615–18	2	5.0
Francisco Crespo	180	1601	1	4.5
Bartolomé Díaz Cabeza de Vaca	164	1597–98	1	4.1
Juan Sánchez	73	1592	1	1.8
Graviel de España	unspecified	1640	1	
	4,021	1597–98	21	

Sources: AGI Contratación legs. 1080–1172, 1180, 1184.

Of the medicines sent to Peru, about ten percent went to hospitals and half to *boticarios*. Only ten *boticarios* are named as recipients of medicines, of whom Juan de Bilbao and his son, Pedro de Bilbao, together accounted for about 60 percent of the total value of medicines received by *boticarios* in Lima (Table 5). Only a few *boticarios* in Lima seem to have had the financial resources to become involved directly in the Atlantic trade; most probably acquired their medicines from these few *boticarios*, merchants or other local sources. Given their prominence as apothecaries in Lima, it is worth describing in detail how Juan de Bilbao and Pedro de Bilbao arranged one shipment. In April 1599, they entrusted six bars of silver valued at 1,771 pesos ensayados to the *perulero*, Pedro de Avendaño. He was instructed to acquire a list of specified medicines from Juan Martínez de Ortuçar, a *boticario* in Seville, who lived in the parish of Santa Catalina, or in his absence from Juan Gutiérrez, a *boticario* who resided in the barrio of Alfalfa, or else from Pedro or Rodrigo del Castillo, who were also *boticarios*.[69] Avendaño was to be paid 30 pesos for the service. That year, Avendaño sailed to Spain with the *flota* of General Don Luis Fajardo. In February 1601 Juan Martínez de Ortuçar registered two cargoes of medicines on the Nuestra Señora de la Concepción and San Nicólas, which sailed in the fleet commanded by Don Francisco del Corral y Toledo. Martínez de Ortuçar specified that on arrival in America the medicines were to be entrusted to Pedro de Avendaño, who was probably travelling with one of the vessels, and who was to arrange for their transfer to Lima. The medicines were valued at 852 pesos on which taxes of 5 percent *almojarifazgo* and 2 percent *avería*, amounting to 67 pesos together, were paid. The value of the medicines was thus significantly lower than the amount of silver indicated in the contract between Bilbao and Avendaño, yet in 1602 Bilbao declared himself satisfied that the contract had been fulfilled. It is perhaps worth noting that these medicines represented less than five percent of the total value of all the cargoes registered for that fleet by Pedro de Avendaño in Seville, calculated at 19,140 pesos *ensayados*.[70]

The total value of 241 consignments of medicines carried to Tierra Firme in fleets between 1583 and 1634 was about 46,379 pesos, of which 11,441 pesos worth were destined specifically for Peru (Table 6).[71] These figures probably underestimate the true value of the medicines slightly, since as indicated above they tended to be undervalued in order to minimise the amount of tax paid. However, it suggests that imports of medicines only averaged about two to three hundred pesos a year and as such accounted for only a small proportion of the medicines actually consumed in Lima. In 1570 the Hospital of Santa Ana was treating 80 to 90 Indians and claimed it was spending 3,000 pesos a year on medicines.[72] The annual accounts and more general reports for the hospital suggest it spent a rather more modest sum of about 500 to 800 pesos a year,[73] such that the figure of 3,000 pesos was probably exaggerated to obtain some

69 AGNP Protocolos Siglo XVI 15 Ramiro Bote fols. 2239–2239v. Juan de Bilbao da poder a Juan Martínez de Ortuçar, boticario de la ciudad de Sevilla 16.4.1599; AGNP Protocolos Siglo XVI 15 Ramiro Bote fols. 3006v-3008 Pedro de Avendaño, residente en Lima entrega recibo a Juan y Pedro de Bilbao 27.4.1599.

70 García Fuentes, *Los peruleros*, 111.

71 The value of eight of the 249 consignments is not given in the sources.

72 AGI Lima 122 fols. 19v., 37v. Del hospital de los naturales de la ciudad de los reyes sobre que se haga alguna merced con que los yndios se puedan curar atento a su pobreza 16 Jan. 1570.

73 ABPL 9080 fols. 254–57 Libro de cuentas del Hospital de Santa Ana 1575 a 1585; ABPL 9083 fols. 73–74, 122–23 Libro mayor de rentas del hospital de Santa Ana desde 1593 a 1629; ABPL 9084 fols. 167–68, 241–242v. Libro de cuentas de gastos del hospital de Santa Ana 1598; ABPL 9085 fols. 59–64 Libro de la razón…deste hospital de Santa Ana de los gastos que

TABLE 6 *Destination and value of cargoes of medicines by trader type, 1583–1634.*

Destination	Value of *materia medica* in pesos registered by		Total value of consignments in pesos	Percent registered by apothecaries	Total number of consignments
	Merchant/ passenger	Apothecary in Seville			
Lima	6311	1600	7911	20.2	53
Potosí	160	899	1059	84.9	3
Cuzco	2471	0	2471	0	5
Cartagena	9130	7451	16581	44.9	61
Nombre de Dios	4052	1076	5128	21.0	33
Portobello	4477	1096	5572	19.7	32
Panama	2353	379	2732	13.9	17
Tierra Firme	3184	513	3697	13.9	26
Other	1192	37	1229	3.0	11
Total	33330	13049	46379	28.1	241

Sources: AGI Contratación legs. 1094–1172, 1180.

charitable funding from the Crown. Meanwhile in 1563 the Hospital of San Andrés for Spaniards was spending two pesos a day for about half that number of patients;[74] by 1600 the expenditure had risen to over 1,000 pesos a year.[75] As for the prison, the annual expenditure on medicines was judged to be 50 pesos a year.[76] Apart from public institutions, the large debts that were run

hace Rodrigo Arias de Buyca [Buiça] mayordomo…desde el primero del mes de agosto 1595.

74 AGI Lima 131 Juan de Alvear en nombre del administrador y fundador del hospital despañoles 1563. At that time the hospital had 44 Spanish sick patients and 7 who were mentally ill. In 1619 it was said that the hospital was treating 1,800 people a year (AGI Lima 301 Relación de las ciudades, villas y lugares, parrochias y doctrinas que ay en este Arçobispado de Lima 20 Apr. 1619).

75 ABPL 8427 Libro de cuentas que llevo el Licenciado Don Sebastián Clemente, mayordomo del hospital real de San Andrés, que corre desde el día 15 de abril del año 1600. In 1619 it was said that the hospital was treating 2,000 people a year (AGI Lima 301 Relación de las ciudades, villas y lugares, parrochias y doctrinas que ay en este Arçobispado de Lima 20 Apr. 1619).

76 AGI Lima 134 Cofradía de los pobres de las cárceles de la ciudad de los reyes 1598.

up by some private clients, which have been discussed in Chapter 3, indicate the high consumption of medicines in the city. Even though more medicines would have been imported than is evident from the analysis above, it is reasonable to conclude that imports of *materia medica* from Spain failed to meet the demand in Lima. As such products had to be acquired from alternative and more local sources.

The Intercolonial Trade in *Materia Medica*

The discussion hitherto has focused on the importation of medicinal products from Spain, however, some were imported from other American regions. Perhaps the most notable source was Mexico from which the most important imports were resins, balsams, and oils, which were generally unavailable on the coast of Peru. They included liquidambar, *copal, tacamahaca*[77] and *sangre de drago*. Other products imported were mechoacan, *guaiacum* (lignum-vitae), cacao, tobacco, and Mexican pepper.[78] The fact that all these products appear in the schedule of taxes to be paid on goods coming from New Spain, and that the resins in particular are common in pharmacy inventories and prescriptions, suggests they were traded on a significant scale.

In the 1590s, between thirty five and forty vessels were plying the South Sea.[79] Although in the early seventeenth century the Crown ramped up restrictions on trade between Mexico and Peru, largely to control the outflow of silver through Mexico to China and the influx of silk to Peru, the number of vessels involved in intercolonial expanded and contraband trade flourished.[80] Apart from ships trading directly with Mexico, other smaller vessels plied the Pacific coast, with some delivering sarsaparilla from Ecuador.[81] It is impossible to ascertain the overall volume of medicines imported to Lima from American regions, but it was probably significant in the case of the resins and oils.

77 De Vos, "Art of Pharmacy," 101–10. Tacamahaca was an American gum-resin extracted from a variety of trees whose name derived from the Aztec term, tecomahiyac (Francisco Hernández, *Quatro libros* (Mexico: Casa de la viuda de Diego Lopez Daualos, 1615) parte 1 lib. 2 cap. 16 fol. 16v.; Monardes, *Historia medicinal*, fols. 2–3.

78 See AGNP Cajas Reales: Lima H-3 Leg. 4 Libro 24-a fols. 34–39 Tasación de las mercaderías que vienen a esta ciudad de los reyes de las provincias de México (1617).

79 Lawrence A. Clayton, "Trade and Navigation in the Seventeenth-Century Viceroyalty of Peru, Journal of Latin American Studies 7 (1): 6–7.

80 John Lynch, *Spain Under the Habsburgs. Volume Two: Spain and America 1598–1700* (Oxford: Basil Blackwell, 1981), 225–26.

81 Marie Hellmer, "LeCallao (1617–1815)," *Jahrbuch für Geschicte von Staat, Wirtschaft und Gesellschaft Lateinamerikas* 2 (1965): 181.

However, evidence from medicinal products found in pharmacies and contained in prescriptions, to be discussed in Chapter 6, suggests that the *range* of products imported was limited. Thus inadequacies in the Spanish trading system do not seem to have stimulated a flourishing interregional exchange of products, knowledge and practices, though the evidence is limited. Such exchanges that existed probably occurred though the personal contacts of *boticarios* and by Jesuits travelling through the network of Colleges they had established in the Viceroyalty.[82]

Acquiring *Materia Medica* Locally

Apart from imports from Spain, the sources from which *materia medica* were acquired were diverse. A few substitutes were sought in native plant assemblages and local mineral resources, while European plants were grown in local gardens, especially those belonging to the hospitals and religious orders. Ordinances for the Hospital of Santa Ana in 1609 clearly show how the authorities tried to ensure that it was supplied with fresh medicines at moderate prices. However, it seems that in the first instance hospitals relied on imports from Spain and only when the expected cargoes did not arrive or when there were shortages did they acquire *materia medica* locally.[83] The *mayordomo* of the Hospital of Santa Ana was exhorted to buy medicines in bulk when they were available rather than in small quantities when the need arose and prices were higher.[84]

The account books of the Hospitals of Santa Ana and San Andrés provide some insight into the suppliers of *materia medica*, since they often indicate the persons from whom purchases were made, though unfortunately not on a consistent basis.[85] About one third (150) of the 460 entries found in hospital

82 Martín, *Intellectual Conquest*, 97–118.

83 In 1641 the surgeon of the Hospital of Santa Ana reported that he had been forced to buy some medicines from Graviel de España because the expected shipment from Spain had not arrived (ABPL 9127 fol. 95v. Pedro Garcia Capicha 8 Jul. 1641).

84 *Constituciones y ordenanzas del hospital real de Santa Ana*, Lima.

85 For the Hospital of Santa Ana see: ABPL 9080 Libro de cuentas, censos y haberes que corresponden al hospital de Santa Ana, 1575–1584; ABPL 9083 Libro Mayor de rentas del Hospital de Santa Ana, 1575–1633; ABPL 9084 Libro de cuentas de gastos del hospital de Santa Ana, 1597–1599; ABPL 9084 Libro de cuentas de gastos del hospital de Santa Ana, 1597–1599; ABPL 9085 Libro de la razón que toma Bartolomé de la Cueva, escribano veedor del hospital de Santa Ana, 1595–1596; ABPL 9127 Protocolo de cuentas del hospital de Santa Ana, 1640–1736. For the Hospital of San Andrés accounts indicating the source

TABLE 7 *Sources of* materia medica *supplied to the hospital of Santa Ana 1575–1600 and 1640–1641.*

	Number of entries	Total value in pesos	Percentage of value	Percentage of total entries
Apothecary (*boticario*)	70	9,476.5	42.6	46.7
Merchant (*mercader*)	23	5,617.5	25.2	15.3
Physician (*médico*)	7	1,906	8.6	4.7
Shop (*tienda*)	21	1,680	7.5	14.0
Indian (*indio*)	3	976	4.4	2.0
Shopkeeper (*pulpero*)	6	588	2.6	4.0
Farm (*chácara*)	7	580	2.6	4.7
Keeper of cellar or tavern (*bodeguero*)	1	384	1.7	0.7
Kitchen garden (*huerta*)	4	333	1.5	2.7
Store (*casa*)	1	248	1.1	0.7
Confectioner (*confitero*)	2	240	1.1	1.3
Surgeon (*cirjuano*)	2	151	0.7	1.3
Auction (almoneda)	1	36	0.2	0.7
Barber (*barbero*)	1	20	0.1	0.7
Native market (*tianguez*)	1	18	0.1	0.7
	150	22,254	100.00	100

Sources: ABPL 9080 Libro de cuentas, censos y haberes que corresponden al hospital de Santa Ana, 1575–1584; ABPL 9083 Libro mayor de rentas del hospital de Santa Ana, 1575–1633; ABPL 9084 Libro de cuentas de gastos del hospital de Santa Ana, 1597–1599; ABPL 9085 Libro de la razón que toma Bartolomé de la Cueva, escribano veedor del hospital de Santa Ana, 1595–1596; ABPL 9086 Protocolo de expedientes del hospital de Santa Ana, 1599–1607; ABPL 9127 Protocolo de cuentas del hospital de Santa Ana, 1640–1736.

accounts kept irregularly between 1575 and 1600 and from 1640 and 1641 indicate the name and occupation of the supplier or the location of the source of supply (See Table 7). As might be expected, the more detailed entries tend to relate to bulk purchases of goods. Over eighty other persons are named in the accounts but unfortunately no further details are given.

of medicines are only available for 1600 (ABPL 8427 Libro de las cuentas que llevó el licenciado don Sebastián Clemente mayordomo del hospital de San Andrés, 1600).

What is striking is the variety of types of traders and producers from whom the hospitals acquired *materia medica* and the degree to which the suppliers specialised in particular products (Figure 7). Thirteen *boticarios* based in Lima at different times sold medicines to the Hospital of Santa Ana. Francisco de Alva, Juan de Bilbao, Francisco Martín Reyna, Pedro de Bilbao, Graviel de España and Juan Matías de Vera all supplied medicines to the hospital over several years, following each other chronologically rather than supplying them simultaneously.[86] Apart from the items purchased from Juan de Bilbao that were recorded only as "medicinas," the products most commonly supplied by *boticarios* were senna, rhubarb, scammony and a number of complex electuaries, most commonly *diafinicón* and *diacatolicón*.[87] However, it seems that *boticarios* were sometimes unwilling or reluctant to supply hospitals because they feared that because of their limited resources they would not be paid.[88]

Large quantities of unspecified medicines were also supplied by several individuals referred to only as "mercaderes," including Francisco Rodríguez Soria and Alonso de Paz. Apart from unspecified medicines, the only items that were consistently traded by merchants were spices – saffron, clove and cinnamon – and some mineral products, namely *cardenillo* (verdigris or acetate of copper), *alumbre* (alum or potassium aluminium sulphate) and *solimán* (mercury chloride). Merchants also acted as agents for other suppliers. So for example, although physicians were explicitly forbidden to prepare medicines,[89] in 1600 one Doctor Figueroa, supplied the Hospital of San Andrés with 680 pesos worth of unspecified medicines, through the merchant, Juan López de Altopica.[90] In most cases, physicians seem to have been supplying simples, mainly purgatives such as rhubarb, scammony and canafistula rather than complex medicines.[91]

Hospitals acquired similar products from "tiendas." *Tiendas*, such as the "tienda de Anaya", were probably retail outlets for goods imported by

86 Others were Francisco Crespo, Francisco Cano, Mateo Pastor, Blas de Medina, Gerónimo Pujadas, Juan Bécquer Cantero, Juan Matías de Vera, and Alonso de Carrión.

87 For the composition of diafinicón and diacatolicón see Oviedo, *Methodo*, 88, 122. The former was based on dates and the latter on senna, canafistula and tamarind.

88 AGI Lima 214 N19 R1 fol. 14v. Información…del hospital de San Andrés 1602.

89 *Recopilación de las leyes, pragmáticas reales* (Muñoz), cap. XIII ordinance 3: 159–60 1537.

90 ABPL 8427 Libro de las cuentas que llevó el licenciado don Sebastián Clemente mayordomo del hospital de San Andrés, 1600.

91 ABPL 9080 Libro de cuentas, censos y haberes que corresponden al hospital de Santa Ana, 1575–1584; ABPL 9084 Libro de cuentas de gastos del hospital de Santa Ana, 1597–1599. The physicians supplying the Hospital of Santa Ana were Licenciados Hidalgo and Pedro de Vergara. Lic. Hidalgo also supplied some unspecified medicines.

peruleros.[92] They supplied similar items to merchants, but normally in smaller quantities. One supplier was the "casa de Valladolid", which specialised in the supply of wax for making ointments. This store probably belonged to the Valladolid family of wax chandlers whose shop was located on the Calle de Valladolid.[93] In addition to these specialised suppliers of wax, it was also traded by *boticarios*. The testament of the *boticario* Juan Matías de Vera indicated that he possessed barrels, blocks, and boxes of white wax from Castile which were worth 1,400 pesos, which was almost equivalent to the value of all the *materia medica* he owned, which was estimated at 1,500 pesos. Although the *boticario* may have been using some wax for ointments himself, the large quantity suggests that he may have been trading it as a simple product,[94] as well as

FIGURE 7 *Pharmacy of the convent of Los Descalzos, Lima.*
PHOTO AUTHOR

92 See AGI Contratación 1145B N4: 303–11 Registro del navío Santa Maria del Juncal 1605 for imports by one Juan Núñez de Anaya. See also, García Fuentes, *Los peruleros*, 84.
93 Bromley, *Viejas Calles de Lima*, 316.
94 AAL Testamentos 25–1 fol. 6 Testamento de Juan Matías de Vera, boticario, 1641.

making it into wax-based medicines such as ointments and plasters for sale, including to the Hospital of Santa Ana.[95]

Other specialist suppliers were confectioners, who sold honey, and a *bodeguero* who supplied oil. Only five sellers referred to as *pulperos* appear in the hospital accounts and they were mainly trading gums and resins – *almaciga* and *incensio* – and some canafistula. In addition the hospital purchased some items in the marketplace, occasionally from Indians or Africans. For example, it acquired large quantities of canafistula from one Jaime Canpi.[96] It is likely many other purchases were made from unnamed persons, but in quantities too small to warrant entry into the pharmacy accounts. In the 1650s at least the hospital was buying large quantities of sugar direct from the Jesuits,[97] most likely from their haciendas at Surco located just southwest of Lima or from their estates to the north in the Huara Valley.[98] Finally, most of the named persons whose occupation is not specified appear only once or twice in the accounts. The exceptions are four persons who were trading almost exclusively in gums, resins, honey, and saffron,[99] suggesting that there was some specialisation amongst traders. It is worth noting that of the 150 named entries, only four were women, including one African and one Indian. While this probably underestimates the number of local people supplying the hospital with *materia medica*, it may not apply to Andean traders, especially native women, who tended to specialise in the supply of food, drink or clothing rather than unprocessed products.[100]

Despite the diverse range of sources from which hospitals acquired *materia medica*, because of difficulties of timely supply, the poor quality of medicines, and probably cost, most hospitals and monasteries established their own herb gardens. When the Hospital of Santa Ana was inspected in the 1588, it possessed no gardener, only an Indian who it was said knew nothing about gardening, which resulted in it having to acquire herbs from outside at considerable

95 ABPL 9127 fol. 63 Memoria de lo gastado en la botica [hospital de Santa Ana] 1640–1641.

96 ABPL 9080 Libro de cuentas del hospital de Santa Ana 1575 a 1585 for accounts for 1575.

97 ABPL 9100 Libro de cuentas del hospital de Santa Ana, 1625–1690.

98 Nicholas P. Cushner, *Lords of the Land: Sugar, Wine, and Jesuit Estates of Coastal Peru, 1600–1767* (New York: SUNY Press, 1980), 29–38, 63–68.

99 These were: Juan de Guevara (12 entries), Antonio Herrera (7), Bartolomé Rodríguez (6) and Pedro Marqués (5).

100 Jane E. Mangan, *Trading Roles: Gender, Ethnicity and the Urban Economy in Colonial Potosí* (Durham, NC: Duke University Press, 2005), 76–105, 139; Ilana Aragón Noriega, "El pan, el vino, y otros negocios: aspectos laborales de la mujer," in *Lima en el siglo XIV*, ed. Laura Gutiérrez Arbulú (Lima: Pontificia Universidad Católica del Perú, Instituto Riva-Agüero, 2005), 265.

expense. The *mayordomo* was therefore ordered to ensure that "lettuces, cabbages, borage, parsley, and other herbs" were always available for the treatment of the sick.[101] A decade later the hospital was employing two native gardeners and others in its *chácara*,[102] In 1599 the *mayordomo*, then the *perulero* Rodrigo Arias de Buiça, was able to report that he had overseen the cultivation of

> …lettuces, cabbages, radishes, aubergines, aval and other vegetables as well as medicinal plants such as borage, mallow, manzanilla, and others for the pharmacy and treatment of the poor.[103]

Even so, an inspection of the hospital in 1607 found that many of the drugs in the pharmacy were in a poor state.[104] Worth mentioning here, but discussed more fully in Chapter 6, is the substantial garden and pharmacy of the Jesuit College of San Pablo, which became a wholesale centre for other pharmacies in Lima and received orders from all over the Viceroyalty.[105]

The local trade in medicines was not one-way. Hospitals were forbidden to sell medicines,[106] but they did so in small quantities along with larger quantities of certain products, such as roses and fat, which were produced in their gardens and *chácaras*.[107] In 1575, the Hospital of Santa Ana was supplying huge quantities of roses from its *chácara* to three of Lima's eminent *boticarios*, Juan de Bilbao, [Juan de] Alva and [Diego de] Tineo, as well as to three doctors and the Hospital of San Andrés.[108] Juan de Bilbao purchased about half of the 834 pounds of roses that were sold by the hospital that year.[109] Not insignificant

101 ABPL 9806 fol. 271v. Visita al hospital de Santa Ana 26 May 1588.

102 ABPL 9084 fols. 132–68 Libro de cuentas de gastos del hospital de Santa Ana, 1597–1599.

103 ABPL 9086 fols. 321–23 *Visita* al hospital de Santa Ana 22 Oct. 1599 in Protocolo de expedientes del hospital de Santa Ana, 1599–1607. Part of this volume (fols. 271–74) which relates to an important *visita* conducted on 26 May 1588 is no longer in the archive.

104 ABPL 9806 fols. 320–23 Visita al Hospital de Santa Ana, cargos que resultan contra el bachiller Juan Manuel Carrasco mayordomo del hospital de los naturales 17 Jan. 1607.

105 Martín, *Intellectual Conquest*, 102.

106 Ordinance 45 Constituciones y ordenanzas y del hospital real de Santa Ana, Lima 28 Feb. 1609.

107 See for example, AAL Hospitales leg. 2 exp. 3 Solicitud presentada por Francisco López de Cepeda, mayordomo y administrador del hospital de los naturales de Santa Ana [1622]; ABPL 9127 Memoria de lo gastado en la botica [hospital de Santa Ana] 1640–1641.

108 ABPL 9080 Libro de cuentas del hospital de Santa Ana 1575 a 1585 fols. 60–71 Cargo de lo que ha de haver este hospital de lo que procede de la chácara 1575.

109 In 1627 the hospital of San Andrés similarly purchased 382 pounds roses: see BNL 1577 Cuentas que da Hernando de Herrera mayordomo que fue deste hospital de San Andrés Agosto 1627.

quantities were also sold to African women who traded them in the city. Roses were valued for their astringent properties and were used to prepare different types of medicines, such as rose water, rose oil, rose honey, rose conserve, and rose syrup.[110] Notwithstanding the widespread use of roses in pharmacy, the quantity purchased by Bilbao seems enormous and it maybe that he sold them on to other *boticarios* or used them to manufacture products for sale in his pharmacy. In any case, it suggests the large scale of his business.

Hospitals not only sold simple products, but on occasions also provided medicines for private individuals. This is evident from a claim made in 1622 by the *mayordomo* of the Hospital of Santa Ana, Francisco López de Çepeda, on the estate of the physician, Doctor Diego Ramírez, for 652 pesos worth of medicines it had supplied.[111] These medicines included syrups, electuaries and ointments, but also a purge for a female African slave.

Much less in known about the way in which owners of private phramacies obtained *materia medica* locally. Probably most was acquired from small retailers or in the market. In Spain *boticarios* often possessed their own private gardens where they experimented with new plants from overseas. Well known are the botanical gardens of Nicólas Monardes and Simón de Tovar in Seville.[112] There is no evidence for apothecary gardens in Lima, though it is known that private gardens existed. However, contemporary observations suggest that these gardens were dominated by plants imported from Europe. The anonymous *Descripción del Virreinato* written at the beginning of the seventeenth century recorded that most gardens were watered by canals and were filled with flower pots with carnations (*claveles*), stocks or wallflowers (*alhelies*), bay (*albahaca*) and common cress (*mastuerzos*).[113] Of course, such observations need to be viewed with caution for they were generally made by Europeans who may have been unfamiliar with native plants.

110 Enrique Laval, *Botica de los Jesuitas de Santiago* (Santiago: Asociación Chilena de Asistencia Social, 1953), 172.

111 AAL Hospitales leg. 2 exp. 3 Solicitud presentada por Francisco López de Cepeda, mayordomo y administrador del hospital de los naturales de Santa Ana, sobre los 652 pesos quedo debiendo Doctor Diego Ramírez, difunto, al dicho hospital 1622. See also the sale of *polvos de coloquíntidas* to Doctor Don Luis Çegarra (ABPL 9127 fol. 45 Memoria de lo gastado en la botica [hospital de Santa Ana] 1640–1641).

112 Barrera-Osorio, *Experiencing Nature*, 212–26. Findlen, *Possessing Nature*, 256–61 makes the same general observation.

113 Lewin, *Descripción*, 38. It is generally recognised that the author of this was a Jewish "spy" Pedro de León Portocarrero.

Conclusion

The transatlantic trade in medicines was dominated by merchants and prominent *boticarios* in Seville, who controlled their prices. Apart from the high cost of the products themselves, the price of imported medicines in Peru was inflated by high state-imposed taxes and transport costs, and the profit that *boticarios* expected to make. Not only did the high cost of imported medicines restrict access to them, but weaknesses in Spain's monopolistic trading system meant that over time the supply declined and became unpredictable. Increasingly imported medicines were criticised as being not only expensive but also of poor quality. Hence, the trading system failed to support the type of medical practice that regulation and the Church promoted, reducing colonial dependency on Spain and encouraging the emergence of other sources of supply. This did not result, however, in an active search for alternative products that might have encouraged experimentation and the adoption of native *materia media*, which might ultimately have challenged humoral practices. It has been shown that local sources of supply, for example, hospital gardens, focused on the cultivation of Old World plants, and as Chapter 6 will demonstrate, the few native products that were adopted were substitutes for those that were central to the practice of humoral medicine.

In order to explain why there was limited experimentation, the following chapter examines the changing nature of medical practice in Spain and Europe more broadly, including debates about the efficacy of local versus imported medicinal products. In particular, it explores the effectiveness of the Inquisition and system of regulation in supporting orthodox humoral practices.

CHAPTER 5

Selecting *Materia Medica*

> The New World laid bare the limitations of classical texts, but those texts continued to provide necessary models for the intellectual and actual appropriation of the New World.[1]

The *materia medica* that were imported from Spain and traded locally clearly reflected the natural philosophy that underpinned medical practice at the time. In sixteenth and seventeenth centuries healing practices in Europe were dominated by humoral beliefs based on the study of classical texts, but new experimental methods were emerging that challenged this approach and focused on new *materia medica*, notably chemicals. This was therefore a period when different approaches to medical practice existed. This chapter will show how orthodox medical practice was promoted by the regulatory framework that the Spanish were developing for medical practice and by the Inquisition, though these had their limitations and may even have had potentially counterintuitive effects. Moreover, the two approaches were not always opposed, but could be combined or work in parallel. Hence, a number of *protomédicos* such as Antonio Sánchez Renedo and Antonio de Robles Cornejo, while generally promoting humoral practices, also possessed an interest in native medicinal plants, an approach that was inherently empirical, and even experimental. Nevertheless, it will be shown how orthodox medical practices generally prevailed, though it will be argued that their prevalence cannot be attributed solely to effective regulation and control. It will demonstrate that where empirical methods were employed, the aim was primarily to find solutions to old problems rather than challenge the natural philosophy that underlay medical practice.

Humoralism

Medical practice in sixteenth-century Spain was dominated by the views of Galen and Hippocrates, which during the Renaissance attracted renewed

1 Nieto Olarte, "Scientific Practices," 147.

 | DOI 10.1163/9789004351271_006

interest as humanist scholars turned back to Ancient Greece for their inspiration and began to examine the original Greek texts. Galen and Hippocrates saw illness as a function of an imbalance in the humours or bodily fluids – blood, phlegm, black bile and yellow bile – which, like all substances, were regarded as hot or cold, moist or dry.[2] Hence blood was hot and wet, phlegm cold and wet, yellow bile hot and dry, and black bile cold and dry. Humoral medicine thus sought to correct any humoral imbalance, which it did through diet, purging, vomiting, and bleeding. Imbalances might derive from inappropriate behaviour, such exhaustion or excessive consumption, or from particular temperature, air or water conditions that could upset a body's hydraulics. Treatments were personalised according to an individual's complexion or temperament, as well as their gender and age.[3] However, they were applied on the general principle that an imbalance in the humours could be restored through prescribing medicines or foods that possessed properties contrary to those being displayed by the patient. A person with a phlegmatic complexion might therefore be treated with substances deemed to be "hot." Humoral beliefs were entrenched in popular thought to the extent that in the absence of doctors lay people often applied treatments according to these principles. Soldiers on military campaigns, for example, might make and administer purgatives, and even let blood.[4]

Astrology, or the movement of the planets and stars, was believed to influence the four elements – earth, fire, water and air – which in turn affected the humours, and consequentially human complexions and temperaments. Such beliefs were held by Christians, Jews, and Muslims alike.[5] The cosmos was

---

2 Foster, *Hippocrates' Latin American Legacy*, 2–4; George Foster, "Relationship between Spanish and Spanish-American Folk Medicine," *Journal of American Folklore*, 66 (1953): 201–02; Siraisi, *Medieval and Early Renaissance Medicine*, 102–09; Roy Porter, *The Greatest Benefit to Mankind: The Medical History of Humanity from Antiquity to the Present* (London: HarperCollins, 1997), 55–62, 73–77, 168–86; Lindemann, *Medicine and Society*, 12–15, 68–69.

3 John Henry, "Doctors and Healers: Popular Culture and the Medical Profession," in *Science, Culture and Popular Belief in Renaissance Europe*, ed. Stephen Pumfrey, Paolo L. Rossi and Maurice Slawinski (Manchester: Manchester University Press, 1991), 199–200; Brockliss and Jones, *Medical World*, 111; Lindemann, *Medicine and Society*, 12, 15; Luis García Ballester, *La búsqueda de la salud: sanadores y enfermos en la España medieval* (Barcelona: Ediciones Peninsula, 2001), 129–33, 145, 176–80. For a recent succinct introduction to humoral medicine in Spain and Spanish America see Earle, *Body of the Conquistador*, 19–41.

4 Vargas Machuca, *Indian Militia*, 64–65.

5 Siraisi, *Medieval and Early Renaissance Medicine*, 97–114; Jorge Cañizares-Esguerra, "New World, New Stars: Patriotic Astrology and the Invention of Indian and Creole Bodies in Colonial Spanish America, 1600–1650," *The American Historical Review* 104(1) (1999): 36–37, 50; García Ballester, *Búsqueda de la salud*, 176–80; Monica Azzolini, "Reading Health in the Stars: Politics and Medical Astrology in Renaissance Milan, *Horoscopes and Public Spheres: Essays*

thought to influence not only human bodies, but all natural things. Hence, the virtues of plants found in particular places were considered to be the most appropriate for the treatment of people and diseases to be found there; some even argued that they had been placed there by God to meet local needs.[6] Not only were the movements of the planets and stars seen to affect the humours, but also indicate the specific times or critical days that plants should be collected and therapies applied.[7]

Views on the suitability of local versus imported medicines varied, however. Arguments in favour of employing local plants often stressed the excessive cost and often poor quality of imported medicines. Sometimes these arguments were imbued with patriotic sentiments that took pride in local remedies. Others, such as Paracelsus, who objected to the universalism of Galenic theory and promoted the employment of empirical methods, also advocated the use of local plants. Conversely, merchants who saw any preference for local *materia medica* as a potential threat to their business expressed more sceptical views, as did Galenic physicians who generally focussed on fitting newly-discovered *materia medica* into a humoral framework. The arguments became more complex as a debate emerged as to whether imported plants grown in a new environment were as effective as when they were raised in their place of origin.[8] Much of this debate was conducted at an intellectual level in Spain. The issue does not appear to have been widely discussed in Lima. In Peru, there was an overriding belief in the therapeutic value of plants of Old World origin, but there is no evidence of a discussion from the perspective of cosmological influences

in the History of Astrology, eds. Günter Oestmann, H. Darrel Rutkin, and Kocku von Stukrad (Berlin: Walter de Gruyter, 2005), 187–89.

6 Andrew Wear, "The Early Modern Debate about Foreign Drugs: Localism versus Universalism in Medicine," *The Lancet* 354 (1999): 150; Richard Drayton, *Nature's Government: Science, Imperial Britain and the 'Improvement' of the World* (New Haven: Yale University Press, 2000), 12–13.

7 Siraisi, *Medieval and Early Renaissance Medicine*, 135–36; Anthony Grafton and Nancy Siraisi, "Between the Election and My Hopes: Girolamo Cardano and Medical Astrology," in *Secrets of Nature: Astrology and Alchemy in Early Modern Europe*, eds. William R. Newman and Anthony Grafton (Cambridge, MA: MIT Press, 2001), 78–86; Azzolini, "Reading Health in the Stars," 197–201.

8 Cañizares-Esguerra, "New World, New Stars," 49–53; Cooper, *Inventing the Indigenous*, 21–50; Wear, "Early Modern Debate," 150–51 and *Knowledge and Practice in English Medicine, 1550–1680* (Cambridge University Press, Cambridge, 2000), 72–77, 86–92. For the case of sarsaparilla and see Anne E Winterbottom, "Of the China Root: a Case Study of the Early Modern Circulation of *Materia Medica*," *Social History of Medicine* 28 (1)(2015): 34–36.

of the relative merits of growing them in Europe as opposed to raising them locally. The main concern was with the quality of medicines with imports being deemed inferior due to their long journey from Spain. The predominant belief in the efficacy of plants of Old World origin was evident in the approach that naturalists and medical practitioners took to native *materia medica* where their emphasis was on finding products with the same healing qualities that could act as effective substitutes for those that were generally used in Europe.

Scholarly Scientific Explorations

Whatever views were held at the time, they demonstrated a general interest in acquiring *materia medica* from the New World. This quest was driven in part by commercial interests, though actually very few American plants were adopted in Europe in the early modern period. At the same time a genuine intellectual curiosity prevailed about the lands and peoples of the New World, knowledge of which was deemed important for "good government." Thus, Philip II himself sponsored the well-researched scientific expedition to Mexico by the physician, Francisco Hernández. He also initiated an empire-wide enquiry into the nature of its peoples, their environments, and history, a task that was overseen by the Spanish cosmographer Juan López de Velasco and produced responses known collectively as the *relaciones geográficas*.[9] In this survey local officials were instructed to inquire about the medicinal and poisonous plants to be found their jurisdiction and on their use by the local people to treat the sick.

The Crown did not sponsor a scientific expedition to Peru that was comparable to that of Hernández, but Antonio Sánchez Renedo, who arrived in Lima as royal *protomédico* in 1569, took a particular interest in local herbs and plants.[10] His research in this area was known to López de Velasco in Spain, who reported to the Crown that Sánchez Renedo was writing a "Historia natural de la yervas y plantas y cosas notables de aquellas partes" and recommended

---

9 Goodman, *Power and Penury*, 68–71, 234–38; Barrera-Osorio, "Knowledge and Empiricism," 231–32. For the text and context of Francisco Hernández' expedition see Simon Varey, ed., *The Mexican Treasury: The Writings of Dr. Francisco Hernández*, trans. Cynthia L Chamberlain and Simon Varey (Stanford: Stanford University Press, 2000) and Varey, Chabrán and Weiner, eds., *Searching for the Secrets of Nature*. For the *relaciones geográficas* of Peru see: *Relaciones geográficas de Indias. Peru*, ed. Marcos Jiménez de la Espada. Biblioteca de autores españoles 183–85. 3 vols. (Madrid: Ediciones Atlas, 1965).

10 Lanning, *Royal Protomedicato*, 29–30; Matallana, *Historia de la enseñanza médica*, 76.

that he should be rewarded to encourage his labours. Unfortunately no copies of this text have been found.[11] Shortly after in 1587, the Salamancan physician Antonio de Robles Cornejo, who had met Hernández in Madrid, migrated to Peru. He was generally supported by wealthy relatives, but in 1594 he was provided with a large salary by the city of La Paz that enabled him to live there for a year.[12] By the beginning of the seventeenth century Robles Cornejo had travelled extensively in Peru and was a good friend of Andrés Gonzales the rich and experienced *boticario* in Cartagena referred to in the last chapter as one of that city's main traders in medicines.[13] His experience formed the basis of a manuscript he wrote entitled, "Simples medicinales indianos." Its publication was approved by the Crown in 1617, when he was given license to take four printers to Peru to publish and sell the book.[14] However, the Crown declined to provide him with a salary or with financial assistance to publish the work, as had been the case with Francisco Hernández. In compensation he made him *protomédico* of Tierra Firme, with which he remained dissatisfied. During the return journey to Peru, his manuscript and the printing equipment, which he had paid for at his own expense, were seemingly lost in a shipwreck off Panama. Reporting on the disaster, he claimed that he possessed no copy and had lost 25 years work.[15] However, a manuscript of 709 pages entitled "Examen de los simples medicinales," which was similarly approved for publication

---

11 AGI IG 1086 Lib. 6 fol. 200v. (Image 405) Juan López de Velasco 28 May 1578; Matallana, *Historia de la enseñanza médica*, 27, 76. On Juan López de Velasco see: María M. Portuondo, *Secret Science: Spanish Cosmography and the New World* (Chicago: University of Chicago Press, 2009).

12 AGI IG 2096 N 94 Expediente de concesión de licencia para pasar a Perú a favor de Antonio de Robles Cornejo, 1586; ARJBM División 1, leg. 17 fol. 197 Libro de examen de los simples medicinales Antonio de Robles Cornejo 1617; Miguel Colmeiro, "Noticias acerca de un manuscrito perteneciente al licenciado Antonio Robles Cornejo," *Revista de los progresos de las ciencias exactas, físicas y naturales* 9 (7) (Madrid, 1859): 440–44.

13 ARJBM División 1, leg. 17 fol. 263 Libro de examen de los simples medicinales Antonio de Robles Cornejo 1617; Quintín Chirlone and Carlos Mallaína, *Historia de la farmacia* (Madrid: Imp. José M. Ducazcal, 1865), 292. For the shipment of medicines to Andrés Gonzales see: Contratación 1080–1163 *passim* 1583–1612. Between 1582 and 1612 he handled 18 of 54 cargoes destined for Cartagena, which were worth about one-third of the value of all medicines imported through Cartagena.

14 AGI IG 428 Lib. 35 fols. 15v.–16 Licencia al Licenciado Antonio de Robles Cornejo 4 Nov. 1617; AGI Contratación 5363 N10 fols. 17, 52–52v. Informaciones de pasajeros a Indias 23 Apr. 1618.

15 AGI Panama 17 R3 N34 Antonio de Robles Cornejo 24 Jun. 1619; Mar Rey Bueno, "Concordias medicinales de entrambos mundos." *Revista de Indias* 66 (no. 237) (2006): 349.

in Madrid in 1617, survives in archive of the Real Jardín Botánico in Madrid.[16] (Figure 8) Clearly written by Robles Cornejo it is not clear what relationship this manuscript may have had to the one that was lost, though it would appear that several decades later a manuscript by him was in the hands of the Sevillian physician Gaspar Caldera de Heredia.[17]

Even though some intellectuals showed an interest in native plants, they often viewed them through the lens of humoralism and attempted to fit them within a humoral framework. In his manuscript Robles Cornejo considered the characteristics of each plant in turn, discussing their origins, names, where they were to be found, and what had been written about them by other authors. In outlining his approach to describing flowers and fruits he recorded that he took advice from living and past masters and consulted diligently books by the most esteemed ancient and modern authors.[18] In the text he refers mainly to traditional authors including Dioscorides, Galen, Mesüe, Avicenna, Laguna, Serapion, Platearius and Pliny, though not without some criticism, but also to the contemporary naturalists Nicólas Monardes, Cristóbal de Acosta, Garcia de Orta, and Carolus Clusius. He also refers to these authors in the marginalia. Both the illustrations and marginalia are entered in a different coloured ink from the text and they appear to have been inserted by a different hand; in fact blank spaces for illustrations are left above the description of some plants suggesting that the manuscript was probably incomplete. Although Robles Cornejo mainly cites traditional texts, he sometimes refers to "los modernos" who were experimenting with plants, though often to criticise them. His criticisms were largely about the procedures used in the preparation and application of particular medicines; he did not reject the employment of empirical methods *per se*. In fact it is clear from the text that he himself was conducting experiments

16 ARJBM División 1, leg. 17 Libro de examen de los simples medicinales Antonio de Robles Cornejo 1617. For notes on the biography of Robles Cornejo see: Chirlone and Mallaína, *Historia de la farmacia*,, 292–300; Colmeiro, "Noticias," 440–44; Mar Rey Bueno, "Los proyectos sobre materia médica peruana de Antonio de Robles Cornejo y Matías de Porres (1617–1621)," in *Rotas da Natureza: Cientistas, vaigens, expedições, instituições*, eds. Ana Leonor Pereira and João Pita (Coimbra: Universidade de Coimbra, 2006), 27–31. For a summary account of the contents of the manuscript see: Quintín Chirlone and Carlos Mallaína, "Biografía Antonio Robles Cornejo," *El restaurador farmacéutico*, ano 11 no. 8 20 Mar. 1855. On line at: http://bibliotecavirtual.ranf.com/es/catalogo_imagenes/grupo.cmd?path=2010122 [Accessed 11 May 2016].

17 Rey Bueno, "Los proyectos," 29–30.

18 ARJBM División 1, leg. 17 fol. 260 Libro de examen de los simples medicinales Antonio de Robles Cornejo 1617.

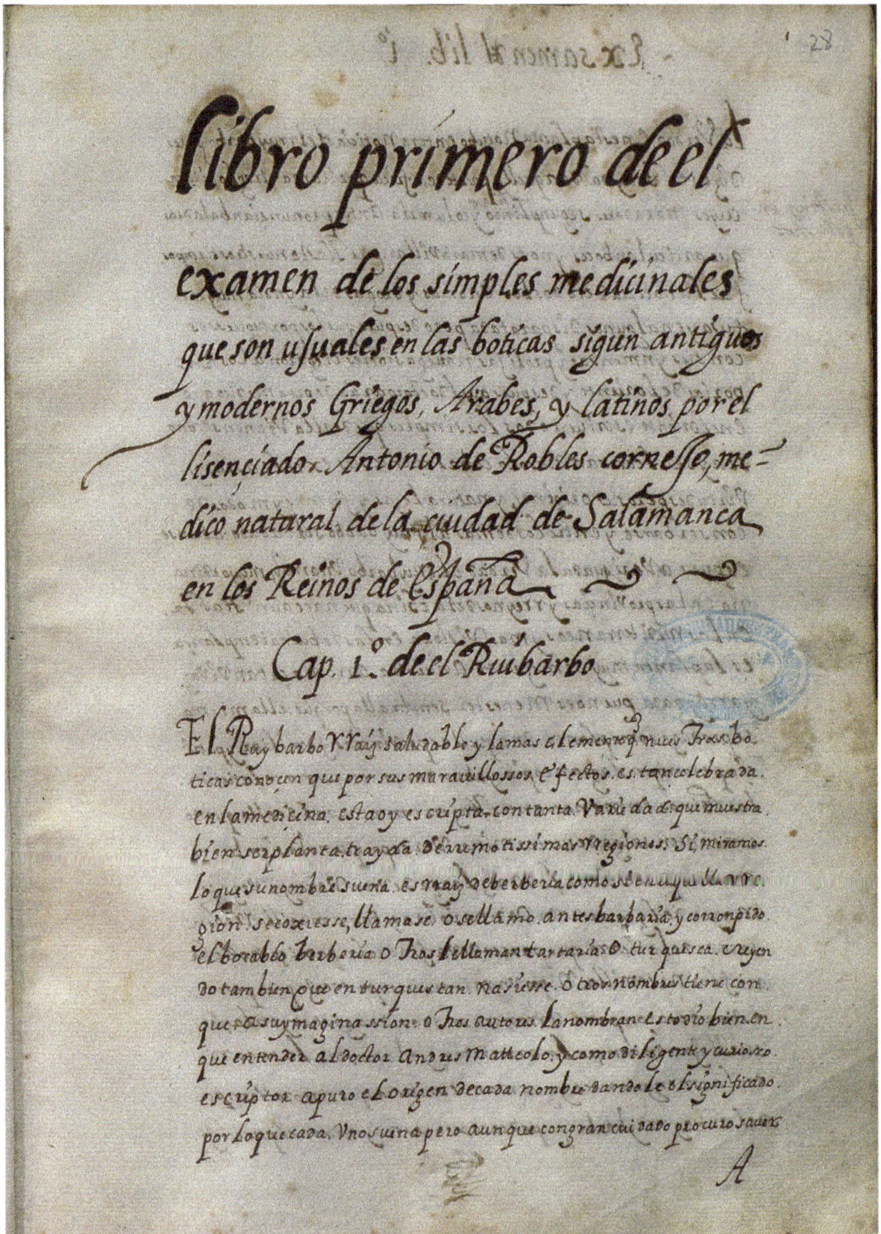

Libro primero de el examen de los simples medicinales que son usuales en las boticas sigun antiguos y modernos Griegos, Arabes, y latinos por el lisenciado Antonio de Robles cornesso, medico natural de la ciudad de Salamanca en los Reinos de España

Cap. 1.º de el Ruibarbo

FIGURE 8 *Front page of Antonio de Robles Cornejo's "Examen de los simples medicinales," 1617.* COURTESY SPAIN. MINISTERIO DE CULTURA. ARCHIVO DEL REAL JARDÍN BOTÁNICO, MADRID. DIVISIÓN 1, LEG. 17 FOL. 1.

with a number of native plants.[19] Robles Cornejo does not name specific "modernos" or modern authors, but in his exposition on minerals he indicates an awareness of the writings of Paracelsus.[20]

Scholars were interested in nature from the perspective of demonstrating the magnificence of God and his works. Paradoxically, however, this Christian spur to reveal the secrets of nature led to observations in the New World that sometimes challenged prevailing scientific and cosmological beliefs derived from biblical texts.[21] Among the scholar/naturalists who travelled extensively in Peru were the Jesuits José de Acosta and Bernabé Cobo, who were at different times at the College of San Pablo.[22] José de Acosta's *Historia natural y moral de las Indias* was published in 1590 and Bernabé Cobo's *Historia del Nuevo Mundo* followed in 1653, although it was based largely on his observations in Peru and the southern Andes in the first three decades of the seventeenth century.[23] Cobo based his account on direct observations and his descriptions of plants and animals are exceptional in their exactitude. Acosta, however, wrote much of his *Historia* in Spain and paid relatively little attention to medicinal plants. His *Historia* was essentially a natural philosophy that focussed on the principles and causes of natural phenomena.[24] Cobo underlined differences

19 See for example his comparison of use of *juncia* (Gencian root) by traditional and modern *boticarios* in ARJBM División 1, leg. 17 fol. 94 Libro de examen de los simples medicinales Antonio de Robles Cornejo 1617.

20 Ibid., fols. 589, 591.

21 Barrera-Osorio, *Experiencing Nature*, 103. For Acosta see Barrera-Osorio, *Experiencing Nature*, 114–20; Saul Jarcho, "Origin of the American Indian as Suggested by Fray Joseph de Acosta (1589)," *Isis* 50 (4) (1959): 430–38; Thayne R. Ford, "Stranger in a Foreign Land: José de Acosta's Scientific Realisations in Sixteenth-century Peru," *Sixteenth Century Journal* 29 (1) (1998): 19–33; Sabine McCormack, *On the Wings of Time: Rome, the Incas, Spain and Peru* (Princeton: Princeton University Press, 2009), 153–65; Andrés I. Prieto, *Missionary Scientists: Jesuit Science in Spanish South America 1570–1810* (Nashville: Vanderbilt University Press, 2011), 143–68. For Cobo see: Luis Millones-Figueroa, "La historia natural del padre Bernabé Cobo: algunas claves para su lectura," *Colonial Latin American Review* 12 (1) (2003): 85–97 and Prieto, *Missionary Scientists*, 99–115. For a comparison of their approaches and methods see Prieto, *Missionary Scientists*, 8, 102–05, 169–94.

22 Acosta went to Peru in 1571 and he returned to Spain via Mexico in 1587, where he died in 1600. Cobo arrived in Peru in 1599 and stayed in the Andean region until 1630, when he went to Mexico. He subsequently returned to Peru in 1650 where he died in 1657.

23 Acosta, *Historia natural y moral*, 1–247; Cobo, *Obras*.

24 Acosta, *Historia natural y moral*, 3–4; Barrera-Osorio, *Experiencing Nature*, 115. See also, Sabine Anagnostou, "Jesuits in Spanish America: Contributions to the Exploration of the American Materia Medica," *Pharmacy in History* 47 (1) (2005): 5–10.

between the New World and the Old seeing them as evidence of the God's omnipotence, whereas Acosta adopted an Aristotelian approach which encouraged him to focus on their similarities and on the general principles underlying natural phenomena. Often scholars struggled to reconcile their observations with prevailing cosmological thought, sometimes leading them into convoluted reasoning. When this failed, they were often forced to resort to divine intervention or miracles for an explanation.

Other scholars in Peru who had a particular interest in the influence of the stars and planets on healing practices included the Creole Augustinian Antonio de la Calancha. Calancha argued that being located in the southern hemisphere and coming under a different constellation from the north, Peru was particularly blessed.[25] However, another non-medical resident of Peru, Juan de Figueroa, who among other things served on Lima's *cabildo* and was *familiar* to the Inquisition, disagreed. He had a passionate interest in astrology and argued against the specificity of the southern stars. In his *Opusculo de astrología en medicina* (1660) he contended that the influence of the stars was universal and he adopted a more traditional approach paying particular attention to the identification of the critical days for treatments, including the letting of blood.[26] Although his book was published in Lima and medical astrology was taught in the university,[27] Claudia Brosseder argues that by the time his book was published the discourse on astrology had shifted towards astronomy and as such his work had little impact on medical practice or treatises.[28]

Knowledge of critical days was also seen as essential to the Mercedarian historian, Martín de Murúa, who worked for a period as a missionary in Peru and in 1611 published his *Historia General del Perú*. While he was impressed by native healers' knowledge of the virtues of medicinal plants, he was concerned that they did not know "the necessary methods to apply them, and further the

25 Antonio de la Calancha, *Corónica moralizada del orden de San Agustín en el Perú* (Barcelona: Pedro Lacavalleria, 1639); Cañizares-Esguerra, "New World, New Stars," 50–51, 59; Claudia Brosseder, "Astrology in Seventeenth-century Peru," *Studies in History and Philosophy of Biological and Biomedical Sciences* 41(2) (2010): 147.

26 Juan de Figueroa, *Opusculo de astrología en medicina* (Lima: No Publisher, 1660), 101–92; Pedro M. Guibovich Pérez, *Censura, libros e inquisición en el Perú colonial, 1570–1754* (Seville: Universidad de Sevilla, Secretariado de publicaciones, 2004), 259–61; Brosseder "Astrology," 147–50.

27 Eguiguren, *Alma mater*, 257; Matallana, *Historia de la enseñanza médica*, 76.

28 Brosseder, "Astrology," 151, 155.

most advantageous times."[29] The extent to which physicians used astrology in their individual treatments probably varied, but the influence of the stars and cosmos was embedded in early modern thought from the elite to the popular level. It was, in Jorge Cañizares-Esguerra's words something that "everybody took for granted."[30]

Initially the development of science was driven by the desire of scholars to find new answers to old problems rather than to challenge the existing natural philosophy.[31] Hence, native plants were placed within a Galenic framework and were classified as being hot, cold, dry, or humid to different degrees. This involved testing them for their smell and taste, and in assessing their therapeutic function and efficacy.[32] The process of assessment was a major undertaking given the number of native plants that needed to be observed and categorised and, as the following chapter will demonstrate, it often produced conflicting outcomes. Indeed the difficulties of agreeing on how the American plants fitted into the existing humoral framework was a significant factor delaying their adoption in Europe.[33] What is significant, however, is that the procedures that were being used clearly embodied inherently contradictory approaches. While the aim was to support a humoral approach to medical practice, at the same time the empirical methods of observation and experiment that were being employed were key features of the scientific approach to medicine that was being advocated in Europe by Paracelsus among others.

---

29 Martín de Murúa, *Historia general del* Perú (Madrid: Instituto Gonzalo Fernández de Oviedo, 1964 [1616]), 103 "por no aplicarse por el modo necesario, ni aun en los tiempos convenientes."

30 García Ballester, *Búsqueda de la salud*, 177; Goodman, *Power and Penury*, 28. Cañizares-Esguerra, "New World, New Stars," 37. On p. 45 he notes that Franciscan friar Bernardino de Sahagún, the physician, Francisco Hernández, and Philip II himself were firm believers in the influence of the constellations on peoples and their environments. See also: Henry, "Doctors and Healers," 201–03.

31 Edward Grant, "When Did Modern Science Begin?" *The American Scholar* 66 (1997): 111.

32 Guenter B. Risse, "Transcending Cultural Barriers: the European Reception of Medicinal Plants from the Americas," in *Botanical Drugs of the Americas in the Old and New Worlds: Invitational Symposium at the Washington-Congress, 1983*, ed. by Wolfgang-Hagen Hein (Stuttgart: Wissenschaftliche Verlagsgesellschaft, 1984), 32, 34–35; Paula Ronderos, *El dilema de los rótulos: Lectura del inventario de una botica santafereña de comienzos del siglo XVII* (Bogotá: Pontificia Universidad Javeriana, 2007), 131. For native plants classified according to Galenic principles, such as wet or dry, hot or cold see: RGI 2: 7–9 Relación de la ciudad de Cuzco, Vasco de Contreras y Valverde 1 Jan. 1650, and ARJBM División 1, legajo 17 Libro de examen de los simples medicinales 1617 *passim*.

33 Estes, "European Reception," 11–13.

Paracelsianism

Paracelsus argued for an experimental approach to the diagnosis and treatment of illness and advocated the employment of chemical methods that were being used by alchemists. Alchemists were concerned with the search for the legendary Philosopher's Stone, which it was thought could turn base metals into gold and prolong life. In the search for this substance they focussed on the transformation of minerals containing mercury and sulphur using methods of distillation, calcination and sublimation. By the late thirteenth and early fourteenth centuries these processes had produced the discoveries of alcohol and mineral acids, notably sulphuric acid (oil of vitriol) and nitric acid (*aqua fortis*).[34] These methods were also applied to other *materia medica*, particularly herbs. Distillation and mineral acid dissolution aimed to produce purer forms of waters or oils – "fifth essences" or quintessences – that were seen to have healing properties.[35] However, these medicines, like simple herbal remedies, were applied within a Galenic humoral framework. Even so, the experimental methods by which they were produced began to challenge the existing natural philosophy based on scholasticism. They questioned the humoral interpretation of illness and saw body malfunctioning to be the result of a chemical reaction caused by an external agent.[36] Chemistry was seen to be the key to understanding nature and central to medical treatments. As such Paracelsus promoted the *internal* application of chemically prepared medicines. Traditionally such medicines had been used quite extensively, but because of their

34 Allen G. Debus, *The Chemical Philosophy: Paracelsian Science and Medicine in the Sixteenth and Seventeenth Centuries* vol. 1 (New York: Science History Publication, 1977), 20–25; Robert P. Multhauf, "The Significance of Distillation in Renaissance Medical Chemistry," *Bulletin of the History of Medicine* 30 (1956): 330–33 and *Chemistry, and the Scientific Revolution* (Cambridge, Mass.: Harvard University Press, 2005), 12–25. For the relationship between alchemy and chemistry see: William R. Newman, "From Alchemy to 'Chymistry'," in *The Cambridge History of Science*, eds. Katharine Park and Lorraine Daston (Cambridge: Cambridge University Press, 2006), 3: 497–517.

35 Robert P. Multhauf, "Significance of Distillation," 333–36 and "John of Rupescissa and the Origin of Medical Chemistry," *Isis* 45(2)(1954): 359–61; Mar Rey Bueno and Maria Esther Alegre Pérez, "Renovación en la terapéutica real: los destiladores de su magestad, maestros simplicistas y médicos herbolarios de Felipe II," *Asclepio* 53 (1)(2001): 33.

36 Debus, *Chemical Philosophy*, 51–61; Allen G. Debus, *The English Paracelsians* (New York: Franklin Watts, Inc., 1968), 26–29 and "Chemists, Physicians and the Changing Perspectives on the Scientific Revolution," *Isis* 89 (1) (1998): 71–74; Bruce T. Moran, "The *Herbarius* of Paracelsus," *Pharmacy in History* 35(3) (1993): 99–127 and *Distilling Knowledge. Alchemy, Chemistry, and the Scientific Revolution* (Cambridge, Mass.: Harvard University Press, 2005), 71–79.

toxic properties had only been employed *externally*.[37] While distillation was still essential to their experiments, the Paracelsians, if not Paracelsus himself, became more interested in precipitates and residues.[38] Inorganic salts, often based on mercury, sulphur and antimony, and spirits (vitriol, laudanum and tartar), which had traditionally been used by alchemists to prepare quintessences, thus began to feature more prominently in medicines.[39]

José María López Piñero has argued that the adoption of chemical methods in Spain did not necessarily involve the acceptance of a new natural philosophy.[40] Indeed, Phillip II himself was interested in the process of distillation and established three distilleries, the first at Aranjuez in 1564. However, the medicines they produced were simple waters and oils that were based on botanical materials in which he had a special interest.[41] In fact Paracelsus's work was not well known in Spain, where the chemical methods that were employed developed more directly from the alchemic tradition embodied in the works of Ramón Llull, Arnau de Villanova, and John of Rupescissa, which had an influence beyond Spain in the fifteenth and sixteenth centuries.[42] In fact knowledge of distillation methods dated back to classical times having

37 Multhauf, "Significance of Distillation," 340–41; George Urdang, "How Chemicals Entered the Official Pharmacopoeias," *Archives internationales d'histoire des sciences* 7 (1954): 304. Paracelsus added salt to mercury and sulphur as the three essential material and symbolic components of matter.

38 Robert P. Multhauf, "Medical Chemistry and 'the Paracelsians'," *Bulletin of the History of Medicine* 28(1954): 125–26; Multhauf, "Significance of Distillation," 344–46; Debus, "Changing Perspectives on the Scientific Revolution," 73–74; Debus, *Chemical Philosophy* 1: 112–17. While vitriol was used by both Paracelsus and the Paracelsians, the former used it as a medicine itself, while his followers were more interested in it as a reactant to make other products.

39 Multhauf, "Medical Chemistry and the 'Paracelsians'," 110–20; De Vos, "From Herbs to Alchemy," 154–60.

40 José M. López Piñero and Francesc Bujosa Homar, "Tradición y renovación en la medicina española del siglo XVI," Asclepio vol. 30–31 (1978–1979): 285–306.

41 Francisco Javier Puerto Sarmiento, "Alquimia aúrea: alquimia y destilación en la corte de Felipe II (1527–1598)," *Dynamis* 17 (1997):118–26; Mar Rey Bueno and María Esther Alegre Pérez, "Los distiladores de Su Magestad: distilación, espagiria y paracelsismo en la corte de Felipe II," *Dynamis* 21 (2001): 331–47 and "Renovación," 27–55. In fact Philip II' s interest in chemical processes was inconsistent with orthodox Catholic beliefs which linked alchemy to witchcraft. This may help explain why the Inquisition was inconsistent in its approach to these practices (Puerto Sarmiento, "La farmacia renacentista española," 126).

42 Miguel López Pérez, "La influencia de la alquimia medieval hispana en la Europa moderna," *Asclepio* 54 (2002): 211–29; López Piñero, "Paracelsus and his Work," 113–41; Debus, "Paracelsus and the Delayed Scientific Revolution," 148–50.

been described by Dioscorides and Mesuë, whose texts were among the most commonly consulted by apothecaries in the early modern period.[43]

It has been argued that the development of Paracelsian medicine in Spain was delayed by the Counter Reformation. Following the Reformation, many countries in Europe began to censor the publication of heretical books and control the introduction of foreign literature. Some controls were introduced in Castile in 1502, but the main impetus came in 1558 when the Inquisition in Spain was ordered to compile an Index of prohibited books.[44] It included sections of Paracelsus's work, which being based on chemical principles were linked to witchcraft. This censorship coupled with the ban a year later on Spaniards studying abroad has traditionally been seen as having a detrimental impact on the development of medicine, and indeed science in general, delaying the Scientific Revolution in Spain until the late seventeenth century, behind that in England, France, and Italy.[45] However, recent research has questioned the efficacy of these controls, showing that few Spaniards sought training abroad and the law was often circumvented or ignored.[46] It has also demonstrated that chemical methods were being employed quite widely having developed from the practices used by alchemists, though generally within a Galenic framework, while works by Paracelsus and his followers circulated in Spain despite the prohibitions.[47] Furthermore, it has been noted that medical knowledge need not depend on the diffusion of printed texts, but might be disseminated in more informal ways.

43 Richard Palmer, "Pharmacy in the Republic of Venice in the Sixteenth Century," in *The Medical Renaissance of the Sixteenth Century*, eds. A. Wear, R.K. French and I.M. Lonie (Cambridge: Cambridge University Press, 1985), 114–15.

44 Kamen, *Spanish Inquisition*, 103–34.

45 López Piñero, *Ciencia y técnica*, 141–44; López Piñero, "Paracelsus and his Work," 119–21; Debus, "Paracelsus and the Delayed Scientific Revolution," 148, 160–61; Jonathan I. Israel, "Counter Reformation, Economic Decline, and the Delayed Impact of the Medical Revolution in Catholic Europe," in *Health Care and Relief in Counter-Reformation Europe*, eds. Ole Peter Grell, Andrew Cunningham, with Jon Arrizabalaga (London: Routledge, 1999), 40–55.

46 Goodman, "Philip II's Patronage," 50–53 and *Power and Penury*, 220–21; Kamen, *Spanish Inquisition*, 104–08. López Piñero, *Ciencia y técnica*, 121–22 notes that many works by Paracelsus and his followers circulated in Spain despite prohibitions at the end of the sixteenth century.

47 Rey Bueno, "Paracelsistas," 41–55. For recent evaluations of the impact of Paracelsus in Spain see: Rodríguez Guerrero, "Censura y Paracelsismo"; López Pérez, "Spanish Paracelsus Revisited."

Maintaining Medical Orthodoxy

During the sixteenth and early seventeenth centuries traditional humoral explanations for illness and their treatments were therefore being challenged in Europe, but what about Spanish America and specifically Peru? In the New World several procedures operated to maintain orthodox medical practice. Spain regulated medicine more than any other nation in the early modern period and it sought to transfer the regulatory framework that it was developing in Castile to the New World. This endeavored to regulate the training and licensing of medical practitioners, in part through the founding chairs in universities for teaching medicine and by establishing medical boards, the *tribunales del protomedicato*, under the authority of the Crown (See Chapter 2).[48] Even though the Crown did not regulate the trade in medicines, some official guidance was given on the products that a well-stocked pharmacy should possess that was based on the needs of humoral medical practice.[49] Furthermore, the Crown introduced a system of inspections to ensure that poor quality and dangerous drugs were not being sold.[50] In some regions of Europe the regulation of pharmacies was supported by physicians who sought to maintain their authority within the medical profession by limiting the activities of apothecaries to the preparation of only those medicines that were composed of ingredients specified in official pharmacopeia.[51]

Crown attempts to regulate medical practice in the New World were supported by the Inquisition which discouraged progressive developments through the censorship of books deemed to be heretical and through controls on the introduction of foreign literature. While these activities promoted orthodox medical practice, in fact they might have been counter-productive in creating a conservative intellectual environment in Spain that might encourage

48 Risse, "Medicine in New Spain," 15; Lanning, *Royal Protomedicato passim*; De Vos, "Art of Pharmacy," 6.

49 Charles Davis and María Luz López Terrada, "Protomedicato y farmacia en Castilla a finales del siglo XVI: edición crítica del *Catálogo de las cosas que los boticarios han de tener en sus boticas*, de Zamudio de Alfaro, protomédico general (1592–1599), *Asclepio* 62 (2) (2010): 579–626. In fact new medicines could be added, if they were judged to be effective within the context of Galenic medicine (Clouse, *Medicine*, 75–90).

50 *Novísima recopilación*, lib. 8 tit. 13, ley 2: 107 3 Mar. 1477 and ley 3: 107 2 Aug. 1593; *Recopilación de las leyes, pragmáticas* (*Muñoz*), cap. 15 art. 2: 188; Clouse, *Medicine, Government and Public Health*, 114–17.

51 Harold J. Cook, "Physicians and Natural History," in *Cultures of Natural History*, eds. Nicholas Jardine, James A. Secord and Emma C. Spary (Cambridge: Cambridge University Press, 2008), 96–97.

some practitioners of a more progressive persuasion to develop their careers in the New World where surveillance was more limited. The remainder of this chapter therefore explores the successes and failures of official attempts to regulate medical practice, in particular the activities of *boticarios*, first directly through regulation and second indirectly through the actions of the Inquisition. The particular influences that these procedures may have had on medical practice will be viewed and assessed in the following chapter through an examination of the medicines that were actually prepared and employed in treatments in Lima.

The Regulation of Pharmacies

In Spain the regulation of apothecaries and their pharmacies originally lay with the municipal authorities. However, with the establishment of the *protomedicato* in 1477 control over the training, examining, and licensing of medical practitioners, that included *boticarios*, passed to specially appointed *protomédicos*, thereby reducing the role of municipalities in the regulation and inspection of *boticas*.[52] According to the Pragmatic of 1593, inspections were to be conducted annually by *corregidores* accompanied by two *regidores* and an approved physician; those conducted in Seville also included *boticarios*.[53] These *visitas* scrutinised the titles of *boticarios* and inspected the pharmacies' medicines, weights, and measures. The inspections often resulted in fines and the destruction of poor quality and potentially dangerous drugs; in extreme cases a *botica* could be closed down.[54] A city-wide inspection of *boticas* in

52 José Luis Valverde, "Regulaciones de farmacia contenidas en las ordenanzas municipales de las ciudades españolas," *Sonderdruck aus Veröffentlichungen der Internationalen Gesellschaft für Geschichte der Pharmazie* [*Die Vörträge der Hauptversammlung in Luxemburg*]: 36: 135–57. Stuttgart, 1970.

53 *Recopilación de las leyes, pragmáticas* (Muñoz), cap. 15 art. 2: 188; *Novísima recopilación*, lib. 8 tit. 13 ley 3: 107 2 Aug. 1593. For sixteenth and early seventeenth century inspections in Seville see: AMS Varios Antiguos 414 Visitas de boticas 12 Apr. 1614, 1625, 1631; AMS Sección 13 siglo XVII vol. 2 doc. 6 fols. 50–51v. 1631; Fernández-Carrión and Valverde, *Farmacia y sociedad*, 21–28; M. Fernández-Carrión and D. Martín-Castilla, "Francisco de Ortega y la profesión farmacéutica sevillana en el primer tercio del siglo XVII," *Boletín de la Sociedad Española de Historia de la Farmacia* 37 (148) (1986): 293–99; Herrera Dávila, "Boticas sevillanas," 73–78.

54 Lanning, *Royal Protomedicato*, 243–44. Accounts of Sevillan *visitas* do not describe the contents of *boticas*, but refer to items as being absent or in a bad state.

Seville in 1631 resulted in over half of its 49 *boticarios* being fined.[55] It was an imperfect system that was open to fraud, bribery, and favouritism.[56] During an earlier inspection of the Seville's *boticas* in 1594 by the Madrileño *boticario*, Juan Pérez de Prado, several *boticarios* had complained that he had levied illegal payments for the inspection and imposed arbitrary and excessive fines, noting that he had dined in the houses of several *boticarios*, which was against the law.[57]

Not only were the inspectors required to ensure the quality of the medicines, but also that pharmacies contained particular essential *materia medica*. Although several pharmacopoeias were published in Spain in the early sixteenth century,[58] it was not until 1593 that the Crown ordered that a "general pharmacopeia" should be drawn up by Spain's *protomédicos*, three physicians, and three *boticarios* that would serve as guidance for apothecaries and pharmacy inspectors.[59] It was previously thought that a pharmacopeia such as this was not produced until the eighteenth century.[60] However, a list entitled "Catálogo de las cosas que los boticarios han de tener en sus boticas," that was drawn up in the late 1590s by Andrés Zamudio de Alfaro who was *protomédico general* in Spain, has recently been uncovered by Charles Davis and María Luz López Terrada.[61] The *materia medica* included in his extensive list of over 400 items is contained in Appendix B. The extent to which inspectors, or indeed *boticarios* in Spain, used this list is not known. Nevertheless, there were other similar guides, such the *Directorio de boticarios* produced by the *boticario* Francisco de Ortega who inspected *boticas* in his home city of Seville in 1625.[62] Probably more commonly used, including in Peru, were pharmacy manuals written by

---

55 Herrera Dávila, "Boticas sevillanas," 76, 78–81. Actually 53 percent.

56 Lanning, *Royal Protomedicato*, 248–49.

57 AMS Sección 13 Siglo XVI vol. 1 doc. 114. Sevilla 20 Jun. 1594. See also Herrera Dávila, "Boticas sevillanas," 76–77 for a case brought against Francisco de Ortega.

58 Davis and López Terrada, "Protomedicato y farmacia," 589. See Chapter 6 for a discussion of pharmacopeias.

59 *Novísima recopilación de las leyes de España*, lib. 8 tit. 13, ley 3: 107 2 Aug. 1593.

60 Lanning, *Royal Protomedicato*, 233; Clouse, *Medicine*, 140.

61 This list was found among the papers of a *boticario*, Pedro de Brines, in the Archivo Histórico de Protocolos de Madrid 1744 fol. 793–94 and has been studied by Davis and López Terrada, "Protomedicato y farmacia."

62 Francisco de Ortega, *Directorio de boticarios, en que se ponen ocho advertencies…* (Sevilla: Matías Clavijo), 625. For the activities of Francisco de Ortega in Seville see: M. Fernández-Carrion and D. Martín-Castilla, "Francisco de Ortega," 289–301.

the Arabic scholar Mesuë (Yūhannā ibn Māsawayh),[63] or those based on his writings such as the Franciscan Bernardino de Laredo's, *Modus faciendi* (1527) and Luis de Oviedo's *Methodo de la colección, y reposición de las medicinas simples* (1581).[64] In any case, evidence from the contents of six *boticas* in Seville in the sixteenth and early seventeenth centuries shows a very clear correspondence between the types of products recommended by Zamudio de Alfaro and those kept in pharmacies in both Spain and Lima.[65]

The regulation of *boticas* in Peru followed the same procedures as in Spain and similarly included the inspection of hospital pharmacies.[66] The Pragmatics issued in 1477, 1491 and 1498 governing the operation of *protomedicatos* in Spain formed the basis for the regulation of medicine in America,[67] but their establishment in the New World was slow and incomplete. Initially the *cabildo* of Lima appointed *protomédicos* with the authority to licence doctors and inspect *boticas*[68] and when the Crown appointed Antonio Sánchez Renedo as *protomédico general* and president of the *Tribunal* in Lima in 1568,[69] it was reluctant to cede responsibility to him. It complained that Sánchez Renedo

63 For a study and revaluation of the importance of Mesuë see: Paula S. De Vos, "The 'Prince of Medicine': Yūhannā ibn Māsawayh and the Foundations of the Western Pharmaceutical Tradition," *Isis* 104 (4) (2013): 667–712.

64 Bernardino de Laredo, *Modus faciendi: cum ordine medicandi* (Seville: Jacobo Cromberger, 1527); Oviedo, *Methodo*. See also Chapter 6.

65 Fernández-Carrión and Valverde, *Farmacia y sociedad,* 80–96. See Appendix B.

66 Lanning, *Royal Protomedicato*, 52, 245–51.

67 In the mid-eighteenth century, Miguel Eugenio Muñoz, an *oidor* of the *audiencia* of Valencia and deputy of the *protomedicato* was appointed by the Crown to bring together the laws and decrees that had been issued in Spain relating to the *protomedicato*, which included these early pragmatics. Miguel Eugenio Muñoz, *Recopilación de las leyes, pragmáticas, reales decretos, y acuerdos del Proto-medicato* (Valencia: Imp. De la Viuda de Antonio de Bordazar, 1751). Online at: https://books.google.co.uk/books/about/Recopilacion_de_las_leyes_pragmaticas_re.html?id=JRe_Ab6MzzoC&redir_esc=y [Accessed 1 Mar. 2017]. The legislation is also found in *Novísima recopilación de las leyes de España* (Madrid, 1805) on line at: http://fama2.us.es/fde//ocr/2006/novisimaRecopilacionT1.pdf [Accessed 1 Mar. 2017].

68 See for example, the appointment of Lic. Hernando de Sepúlveda in 1537 (*Libros de cabildos de Lima,* 1: 141–42 Licenica al dotor [sic] Sepúlveda 27 Apr. 1537), Lic. Alvaro de Torres in 1551 (*Libros de cabildos de Lima,* 4: 371 Médicos 17 Apr. 1551), and Lic. Francisco Gutiérrez in 1556 (*Libros de cabildos de Lima*, 5: 520 El hospital despañoles médico y concierto con el 4 Sep. 1556).

69 Lastres, *Historia de la medicina peruana*, 2: 29–38, 57–58; Lanning, *Royal Protomedicato*, 29–30, 62. The municipality exerted control over the appointment of *protomédicos* for longer in Mexico City where a Tribunal was not established until 1646.

had not presented his title to the *cabildo* and should do so on pain of payment of the considerable fine of 1,000 pesos.[70] The *cabildo* was still complaining in 1576 when it suggested that even though Sánchez Renedo had inspected the *boticas, visitas* should follow the "custom [costumbre]" and be undertaken by members of the *cabildo* (*alcalde ordinario, regidor* and *fiel ejecutor*) with a physician of their choice.[71] Following the death of Sánchez Renedo in 1579, the *cabildo* reasserted its authority commissioning the *fiel ejecutor* (its officer for weights and measures) to conduct pharmacy inspections with a physician and *boticario*.[72]

Inspections in Lima do not appear to have been undertaken on the regular and systematic basis that was specified in the Pragmatics, but rather intermittently in response to complaints.[73] In 1581 it was said that the Lima's *boticas* had not been visited for "a long time" and that there were "great irregularities because the medicines that were being sold were corrupted and damaged and many of them [the *boticas*] lacked the principal ingredients from which complex medicines had to be made…"[74] In response, the *cabildo* appointed the interim *protomédicos* Doctors Alvaro de Torres and Enríque Méndez to conduct an inspection. Again in 1598 the *cabildo* observed that because it had been a long time since the city's pharmacies had been inspected, they contained many "corrupt medicines" which were "of no value." On this occasion it ordered all *boticas* to be closed and for them to be inspected by the *alcalde* and *fiel ejector* in company with the then royal *protomédico*, Iñigo de Hormero, who was ordered to take remedial action.[75] The exercise was repeated for the same reason five years later.[76]

---

70 *Libros de cabildos de Lima*, 7: 9 Que el protomédico mostrase en el cabildo el título, en ínterin no usase el dicho oficio 5 May 1570.

71 *Libros de cabildos de Lima*, 8: 226 Lo que toca a la visita de las boticas que le guarda la costumbre 9 Apr. 1576.

72 *Libros de cabildos de Lima*, 9: 73 Que se visiten las boticas 3 Aug. 1579.

73 Lanning, *Royal Protomedicato*, 52. See for example, the reaction of the *cabildo* to criticism of the *botica* belonging to Alonso de Alemán (*Libros de cabildos de Lima*, 2: 115 El bachiller Pacheco malquerencia 1 Dec. 1545).

74 AHML Libro 2 de cédulas y provisiones fol. 15v–16 and *Libros de cabildos de Lima*, 9: 408 Al señor licenciado Torres para lo de las boticas 14 Jul. 1581.

75 *Libros de cabildos de Lima* 13: 185–86 Que ysiban [exhiban] los títulos de cirujanos y barberos y boticarios 6 Nov 1598. Lanning (*Royal Protomedicato*, 53) suggests that overnight closure of all *boticas* was to prevent spoiled medicines being hidden or transferred between *boticas*.

76 *Libros de cabildos de Lima*, 14: 561–62 Sobre visitar los boticas 10 Nov. 1603.

From the earliest days, there were complaints about the poor quality of medicines available and the high prices that *boticarios* charged. The *cabildo* generally responded to these complaints by sending a doctor to investigate. In 1538, when Lima possessed only one *boticario*, Doctor Hernando de Sepúlveda was appointed to carry out an inspection and establish a list of prices.[77] However, less than a year later this *botica* still contained many medicines that were described as "muy dañadas," which the physician, Bachiller Juan de Castro was ordered to inspect and destroy.[78] Commonly inspections involved the posting of a list of prices.[79] This was regarded as particularly important for the poor, who often resorted to pharmacies and paid with "silver in hand" rather than getting a prescription from a physician.[80]

The regulation and inspection of pharmacies extended to hospitals, convents, and prisons. At the time the pharmacy of the Hospital of Santa Ana was established in 1551,[81] its *boticario* Francisco de Bilbao was contracted to supply it with medicines for a period of six years. However, this contract was dissolved two years later and in 1554 a new one was made with Pedro López to supply the Hospital, which at that time operated in conjunction with the hospital for Spaniards; two years later he was failing to meet his obligations.[82] Subsequently responsibility for ensuring that hospitals were supplied with medicines lay with their *mayordomos*.[83] The purchase of medicines constituted a significant proportion of the limited expenditure of hospitals, which depended largely on Crown disbursements, income from lands, donations, or charity. Typically the expenditure of hospitals exceeded their income. In 1570 the cost of running the Hospital of Santa Ana was estimated to be 8,000 pesos when its income was only 2,500 pesos; the cost of the medicines alone was estimated to be 3,000

77 *Libros de cabildos de Lima*, 1: 254 Visita de la botica desta ciudad 6 Dec. 1538. Unfortunately no records of the *arancel* have been found.

78 *Libros de cabildos de Lima*, 1: 387 [Curas con e[n]salmos] 16 Oct. 1539.

79 *Libros de cabildos de Lima*, 5: 108–09 Arancel de boticarios 17 Nov. 1553; *Libros de cabildos de Lima*, 5: 393 Para que hiciese arancel de boticarios 7 Feb. 1556.

80 *Libros de cabildos de Lima*, 5 108–09 Arancel de boticarios 17 Nov. 1553; *Libros de cabildos de Lima*, 5: 425 Boticarios 28 Mar. 1556.

81 AGNP Protocolos Siglo XVI 63 Diego Gutiérrez fols. 167–88 Hospital de los españoles y naturales compañía con Francisco de Bilbao de la botica 20 Oct. 1551; Cobo, *Obras*, 2:444; Rabí Chara, "Primera botica," 273–77. The initial construction of the Hospital of Santa Ana was made possible by the assignment of land by the Dominican, Archbishop, Jerónimo de Loayza, and several donations from the Crown (Guerra, *El hospital*, 433–34).

82 *Libros de cabildos de Lima*, 4: 511 El hospital que es tome la cuenta 18 Aug. 1556.

83 See Chapter 4.

pesos.[84] Similarly, in 1602 the *mayordomo* of San Andrés complained that the building was in ruins and that there were insufficient funds to pay the salaries of its employees or buy items for the pharmacy. At that time the Hospital's income was 12,000 pesos, while its running costs were estimated at 26,000 pesos a year.[85] The high cost of medicines reflected in part the large number of people being treated in the hospitals; the two main hospitals of Santa Ana and San Andrés were caring for about 2,000 patients each a year.[86] Meanwhile the basic income of the hospitals, which had initially derived from the tribute income from *encomiendas*, had declined significantly with the demographic collapse of the native population.[87] Not surprisingly, the archival sources are replete with petitions to the Crown for donations.

The *mayordomos* of the Hospitals of Santa Ana and San Andrés, who were appointed by the archbishop and *cabildo*, were required by hospital ordinances to ensure that their *boticas* were inspected by a physician every three to four months to ensure they were well stocked and that the medicines were not damaged.[88] Inspections of hospitals as a whole occurred less regularly, though they also included the pharmacy. Some, inspections such as those conducted of the Hospital of Santa Ana in 1606 and 1620, were conducted by the *protomédico* at the time, and the cost of these constituted an added burden on already stretched hospital finances.[89] The *visita* by the *protomédico* Melchor de Amusco in 1620 cost 52 pesos.[90] More often, however, inspections were undertaken by a member of a religious order and representatives of the *cabildo*.[91]

84 AGI Lima 122 Del hospital de los naturales de la ciudad de los reyes 16 Jan. 1570.

85 AGI Lima 214 no. 19 doc. 1 Don Fernando de Cordóva y Figueroa, mayordomo of San Andrés 6 Oct. 1602.

86 AGI Lima 301 Relación de las ciudades, villas y lugares, parrochias y doctrinas que ay en este Arzobispado de Lima 20 Apr. 1619.

87 AGI Lima 125 Juan Manuel Carrasco, mayordomo del hospital de Santa Ana 15 May 1604.

88 ABPL 9086 Cargos que resultan contra el bachiller Juan Manuel Carrasco, mayordomo del hospital de los naturales 17 Jan 1607; *Constituciones y ordenanzas del hospital de Santa Ana de Lima* art. 27 and 45 28 Feb. 1609; Lohmann Villena and Sarabia Viejo, *Francisco de Toledo*, 318 Ordenanzas para el hospital de San Andrés de Lima, 9 Oct. 1577.

89 The *visita* by the *protomédico* Melchor de Amusco in 1620 cost 52 pesos (ABPL 9097 Libro de cuentas de gastos del hospital de Santa Ana 1620 y 1621).

90 ABPL 9097 Libro de cuentas de gastos del hospital de Santa Ana 1620 y 1621.

91 For visitas see: ABPL 9086 fols. 65–92 Visita al hospital de Santa Ana 4 Dec. 1590, fols. 271–74 1588, fols. 118–23 1606, fols. 320–23 1607; ABPL 9097 Libro de cuentas de gastos del hospital de Santa Ana 1620 y 1621.

Formal responsibility for the supply of medicines to hospitals lay with their *mayordomos*. In 1607 when the mayordomo of the Hospital of Santa Ana, Bachiller Juan Manuel Carrasco, was accused with not having ensured that the *botica* was supplied with medicines, he claimed that neither the hospital's physician nor *boticario* had informed him that they were required.[92] Nevertheless, the medicines inspected during the 1606 *visita* were generally found to be in good condition. Only one item – bladder cherry (*trociscos de alquequenje*) – was to be destroyed, while the violets and polypody (*trociscos de polipodio*) were described as damp and were to be dried in the sun. An ointment with litharge (*unguento de almartaga*) was to be softened with Castilian oil (*aceite de Castilla*) and four of the syrups (*jarabes*) were to be perfected (*suba al punto*). Finally the senna was to be cleaned of dust and stems (*palillos*). The *visita* also noted that the pharmacy lacked chebulic myrobalan (*mirabolanos chebulos*). There is no evidence that *boticarios* were fined for not possessing particular types of medicines, though as will be shown in the next chapter, the general approach encouraged the use of Old World plant materials rather than native ones.

The Impact of the Counter Reformation and Inquisition

Just as it is difficult to be precise about the impact of the Counter Reformation and Inquisition on medical practice in Spain, so it is for Peru. In Spain the actions of the Inquisition served to create a conservative intellectual climate which, even if they did not result in the imprisonment or confiscation of goods of offenders, may have encouraged medical practitioners of a more progressive persuasion to migrate to Spanish America. One clear example, though for Cartagena rather than Lima, is Pedro López de León who arrived in the city in 1590 as surgeon to its hospital, garrison and the galleons that visited the port.[93] López de León spent more than twenty-five years in Cartagena and published a notable manual entitled *Práctica y teórica de las apostemas en general y particular. Questión y prácticas de cirugía, de heridas, llagas y otras cosas nuevos y particulares* (1628). He was a student of the progressive surgeon, Bartolomé Hidalgo de Agüero at the Hospital del Cardenal in Seville, who as a result of

92 ABPL 9086 Cargos que resultan contra el bachiller Juan Manuel Carrasco, mayordomo del hospital de los naturales 17 Jan. 1607.

93 Jairo Solano Alonso, *Salud, cultura y sociedad en Cartagena de Indias siglos XVI y XVII* (Universidad del Atlántico: Barranquilla, 1998), 127–29.

experimentation had come to oppose trepanation and periosteotomy and favour the drying and closing wounds to prevent contamination.[94]

In theory the greater distance of Peru from the metropolis should have made it an attractive destination for medical practitioners seeking to escape the surveillance of the Inquisition, even though a tribunal of the Inquisition was established in Lima in 1570.[95] *Conversos* were particularly vulnerable to being charged with heresy, as were those who employed chemical methods who might be accused of witchcraft. However, from the limited evidence available, it seems that the licenced medical practitioners that arrived in Peru generally had a conservative training at a Spanish university, very often at the University of Salamanca. One progressive physician and surgeon was Pedro Gago de Vadillo. He worked in hospitals in Huamanga and the mining district of Castrovirreyna, and ultimately in the Hospital of San Andrés in Lima. Although not a student of Hidalgo de Agüero, he similarly advocated the drying of wounds, a procedure that challenged traditional Galenic methods that used unguents and kept wounds open.[96] In his book, *Discurso de la verdadera cirujía y discursos de censura de ambas vías* (1632) he described various cures he had undertaken in the Hospital of San Andrés. Prior to 1640 only two physicians and two surgeons were brought before the Inquisition in Lima and it was for judaizing or promoting heretical beliefs rather than unorthodox medical practices.[97] One surgeon was Alexandre Pérez who came from Flanders. When his possessions were seized by the Inquisition in 1612, his library contained a mixture of traditional and progressive medical texts, including books on surgery by Doctor León,[98] Doctor Hidalgo,[99] Doctor Francisco Díaz,[100] and

94 Ibid., 110–11. On Bartolomé Hidalgo de Agüero see Kristy Wilson Bowers, "Tradition and Innovation in Spanish Medicine: Bartolomé Hidalgo de Agüero and the Vía Particular," *The Sixteenth Century Journal* 41(1)(2010): 29–47.

95 Kamen, *Spanish Inquisition*, 103–34; José T. Medina, *Historia del tribunal del Santo Oficio de la Inquisición de Lima* (Santiago, Chile: 1956), 1: 1–18.

96 Bowers, "Tradition and Innovation," 46–47. The full title is *Discursos de verdadera cirugía, y censura de ambas vías, y elección de la primera intención curativa, y unicion de las heridas* (Madrid: Juan Gonçalez, 1632).

97 Medina, *Historia del tribunal del Santo Oficio*, 1: 39, 310, 319; 2: 137. The two physicians, Juan Álvarez and Álvaro Nuñez, and the surgeon Tomé Cuaresma were all accused of judaizing. On Tomé Cuaresma see Newson and Minchin, *From Capture to Sale*, 264.

98 This would not have been the book by Pedro López de León which was not published until 1628. Possibly it was Andrés de León's *Tratados de medicina, cirugía, y anatomía* (Valladolid: Luis Sánchez, 1605).

99 Probably Bartolomé Hidalgo de Agüero, *Thesoro de la verdadera cirugía* (Seville, 1604).

100 Possibly Francisco Díaz, *Tratado nuevamente impresso, de todas las enfermedades de los riñones, vexiga, y carnosidades de la verga, y urina* (Madrid. F. Sánchez, 1588).

Doctor Murillo.[101] Other healers brought before the Inquisition included a priest and an *oidor* who were using herbs.[102] Another priest from Genoa was deported as an illegal immigrant, but not without some of his medical books being seized.[103]

Even though there may be little evidence of progressive medical practitioners migrating to Peru, this does not mean that all those who arrived were ignorant of modern methods. It is significant that in his manuscript Antonio de Robles Cornejo while referring most frequently to traditional texts such as those by Galen, Dioscorides, Avicenna, Mesüe and others, in discussing the qualities of minerals particularly also revealed knowledge of the methods that Paracelsus was advocating.[104] Neither does the limited evidence for progressive texts mean that medical practice lacked innovation. The absence of facilities for training and examining doctors in Spanish America meant that during the sixteenth century Creole practitioners often had little option but to work by trial and error, the experience of which led some to the conclusion that formal training was unnecessary. Moreover, they might be preferred as doctors. One surgeon in Santa Fe de Bogotá in the 1620s noted how new medical graduates arriving from Spain made many mistakes because the country was "a different region [with] a different climate, different medicines, different complexions, different foods, and as such different subjects."[105]

The Circulation of Medical Texts

Very little is known about the educational background of *boticarios* practising in Lima, but it seems likely that most acquired knowledge of the properties of different products and how to prepare them through a combination of on-the-job training, as described in Chapter 2, and consulting medical tracts and manuals that were available locally.[106] In this context the Inquisition through its scrutiny of books sent to the New World as well as those published

101 AGNP SO CO ca. 16 doc. 194 fols. 1–8 Secuestro de Alexandre Pérez, cirujano, de Flandes, en el pueblo de Yllimo (Saña) 1612. Other items confiscated were two books of remedies, part in Spanish and in Flemish, several boxes containing drugs wrapped in paper, a small chest with pharmacists' weights, two blistering jars and their covers, a bag of surgical instruments and some jars of medicinal cream. Murillo probably referred to Jerónimo Murillo, *Therapeutica, método de Galeno en lo que toca a cirugía* (Zaragoza, 1561).

102 Medina, *Historia del tribunal del Santo Oficio*, 109–10, 257–59.

103 Ibid., 138.

104 ARJBM División 1, legajo 17 fols. 588–91 Libro de examen de los simples medicinales Antonio de Robles Cornejo 1617.

105 AGNC Colonia Médicos y Abogados 11 fol. 853 Miguel de Çepeda Santa Cruz [1626].

106 See Chapter 6 on how medicines were prepared.

or circulating in Lima had the potential to influence the types of medicines prescribed. It is recognised, however, that this was not the only way in which knowledge of medicine and medical practice was diffused.

In general medical texts and manuals were not easy to acquire. At first the Crown banned the establishment of a printing press in Lima fearing that it might facilitate the spread of heretical literature. However, this policy became untenable in the face of the demand for catechisms to facilitate native conversion and the growing needs of the university.[107] In 1580 an Italian printer, Antonio Ricardo, having set up a printing press in Mexico City, was encouraged to move to Lima and four years later was given permission to establish one under the supervision of the Jesuits at the College of San Pablo. The first book to be printed there was the *Doctrina cristiana y catecismo para la instrucción de los indios* (1584).[108]

Any text to be printed in Lima had to be licensed by the viceroy or *audiencia*; ecclesiastical authors in addition had to obtain permission from their superiors. These bureaucratic obstacles were probably less significant barriers to publication than the practical difficulties that authors and printers faced. These included the cost of printing, shortages of paper, and poor quality typefaces that generally resulted in physically inferior books. Moreover, before 1630 there were only three printing presses in the city. These limitations meant that relatively few texts were produced in Lima, print runs were short, and authors commonly sought publication in Europe.[109] Those books that were published were predominantly theological, though legal texts also figured significantly, while some books on poetry and the history and culture of Peru and Europe were also printed.[110] Pedro Guibovich has calculated that only seven percent of over a thousand books published in Lima before 1700 were in the field of

107 AGI Lima 126 Vecinos of Lima 12 Aug. 1581, Doctores de la universidad 13 Aug. 1581.

108 Rubén Vargas Ugarte, *Impresos peruanos (1584–1650)* (Lima: Universidad Nacional San Marcos, 1953), 1: 6–10; José Toribio de Medina, *La imprenta en Lima,* 1584–1824 (Santiago: Impreso y grabado en casa del autor, 1904), xx–xxx; Pedro M. Guibovich Pérez, "Printing Press in Colonial Peru: Production Process and Literary Categories in Lima," *Colonial Latin American Review* 10 (2) (2001):168–69. For the testament of Antonio Ricardo see: "Testamento de Don Antonio Ricardo," *Revista del Archivo Nacional del Perú* 19 (1955): 293–306.

109 Salinas y Córdova, *Memorial,* 1: 257; Guibovich Pérez, "Printing Press," 168–73; Catalina Romero Romero, "Libros, cultura y sociedad en el Perú virreinal: Las bibliotecas de los colegios jesuitas de Arequipa, Potosí y las misiones de Moxos en el siglo XVIII." PhD diss., Universidad de Complutense de Madrid, 1992. Vol 1: 4–21.

110 Antonio Rodríguez-Buckingham, "Change and the Printing Press in Sixteenth-Century Spanish America," in *Agent of Change: Print Culture Studies after Elizabeth L. Einstein,* eds. Sabrina Alcorn Baron, Eric N Lindquist and Eleanor F. Shevlin (Amherst and Boston: University of Massachusetts Press, 2007), 228–30.

applied sciences, and of those only ten were on medicine.[111] The first medical text to be published in Lima was Matías de Porres's book, *Breve advertencies para beber frío con nieve*, which was published in 1621.[112]

With few books being printed locally, the majority of those circulating in Lima either accompanied their immigrant owners to Peru or were brought by merchants for booksellers on commission or for sale on their own account. Many professionals, including medical practitioners, who migrated to the New World took books with them, no doubt forewarned that they were harder to come by and more expensive than in Spain. The high cost of books derived in part from transport costs, taxes, and the higher cost of living,[113] but also from the significant profits expected by merchants involved in the book trade.[114] Cheaper second-hand books might be sold in the marketplace when their owners died and undoubtedly others were inherited.[115]

In theory all books that arrived in the New World had been inspected and approved by the Inquisition. In 1558 the Spanish Inquisition was ordered to compile an Index of prohibited books and between 1559 and 1784 the Spanish crown published eight Indexes of banned books. All books written by Paracelsus were banned.[116] Other texts were permitted but with certain sections expurgated. These included the writings of the physician and alchemist Arnau de Vilanova, who employed experimental methods that included distillation and the preparation of alcohol and mineral acids.[117] Books on the banned list

111 Guibovich Pérez, "Printing Press," 179.

112 Vargas Ugarte, *Impresos peruanos*, 1: 90. For a discussion of Porres and his work see: Mar Rey Bueno, "Concordias," 347–62.

113 Guillermo Aulet Sastre, "Precios autorizados de libros españoles en Indias," *Revista de Indias* 24 (1946): 311–12; Teodoro Hampe-Martínez, "The Diffusion of Books and Ideas in Colonial Peru: A Study of Private Libraries in the Sixteenth and Seventeenth Centuries," *Hispanic American Historical Review* 273(2)(1993): 220–21; Romero Romero, "Libros, cultura y sociedad," 20; Pedro Guibovich Pérez, "Bibliotecas de médicos en Lima colonial," in *Del autor al lector*, ed. Carmen Castañeda (Mexico City: CIESAS, 2002), 296. Prices were even higher in Cuzco (Leonard, *Books of the Brave*, 299–300).

114 AGI Lima 143 fol. 35 Memorial y apuntamiento sumario de las cosas muy importantes 1612. For the import taxes paid on books see Chapter 4.

115 Guibvoich, "Bibliotecas de médicos," 303.

116 *Index et catalogus librorum prohibitorum* (Madrid: Apud Alphonsum Gomezium, 1583), 60; José Pardo-Tomás, *Ciencia y censura: la Inquisición española y los libros científicos en los siglos XVI y XVII* (Madrid: CSIC, 1991), 220–27.

117 *Index et catalogus librorum prohibitorum*, 10; Debus, *Chemical Philosophy*, 20–21, 24. For other medical texts banned by the Inquisition in the sixteenth century see: José Pardo-Tomás, "Autores médicos en los índices inquisitoriales españoles del siglo XVI," *Dynamis* 5–6 (1985–1986): 201–14.

did not remain constant over time. Books that had been previously sanctioned might later be expurgated or banned. For example, Andrés Laguna's interpretation of Dioscorides, which had been widely used in the sixteenth century, was expurgated in 1632 to make it more faithful to the original classical text and to remove any hypothetical superstitions, for example the supposed abortifacient properties of cardamom. From 1612 some of the philosophical writings of Francisco Valles who had been *protomédico* in Spain, were also expurgated.[118] In general, censorship was most intense in the first part of the seventeenth century.[119]

All books that were imported, whether they were carried by individuals or merchants, were examined when they were loaded onto ships in Seville and again when they arrived in Callao, whilst in Lima printing presses, bookshops and public and private libraries, including the extensive ones owned by the religious orders, were all subject to inspection.[120] While the focus of the Inquisition was on books of a heretical nature, there were areas where medical texts might be interpreted to fall in this category. These were the fields of alchemy and astrology, especially judicial astrology, where practices were often associated with magic and witchcraft.[121] It was in these areas that the inquisitors focussed their attention.[122]

Pedro Guibovich argues that in Lima the impact of Inquisition was greater on the circulation of books than on their production, since authors often self-censured to ensure that their works did not attract Inquisitorial attention. Hence, when Juan de Figueroa, himself apothecary to the Inquisition, argued in his *Opúsculo de astrología de medicina* (1660) that astrology and knowledge of the heavens was essential for curing sickness, he was at pains to point out that the text did not contradict orthodox Christian beliefs and indeed he

118 Pardo-Tómas, *Ciencia y censura*, 217–19, 252.

119 Guibovich Pérez, *Censura*, 346.

120 Eguiguren, *Diccionario histórico cronológico*, 1: 664; Guibovich, *Censura*, 99–132. For the process in Spain see: Pardo-Tomás, *Ciencia y censura*, 29–45; Gonzalez Sánchez, *New World Literacy*, 50–54. The AGI contains the records of all cargoes sent to the Americas in individual ships. The titles of all books are included, whether or not they belonged to private individuals or were being shipped to booksellers. For the period under study see AGI Contratación 1080–1184 Registros de navíos 1583–1640.

121 Pardo-Tomás, *Ciencia y censura*, 154–82, 248–64; Guibovich, *Censura*, 219–20.

122 Medina, *Historia del tribunal del Santo Oficio*, 2: 35–41. A decree issued by the Inquisition in Lima in 1629 specified the activities that fell under "astrología judiciaria" and "arte mágica" were to be regarded as heretical, including the use of hallucinogenic drinks made of achuma (*Trichocereus bridgesii* Britton & Rose), chamico (*Datura ferox* L.) and coca.

dedicated the volume to the Viceroy.[123] Others avoided censure by circulating their writings in manuscript form rather than through publication.[124] Pedro Guibovich has also argued that Juan Gerónimo Navarro's *Sangrar y purgar en días de convención* (1645) which employed empirical methods to challenge prevailing ideas about bloodletting on particular days, escaped censure because he was part of Lima's small elite that included the Jesuits who scrutinised his work.[125]

As was recognised at the time, the inquisitorial process was imperfect. Books might be hidden in other cargoes of merchandise or bound in with other works, while often there were insufficient personnel to conduct inspections, which were often cursory.[126] In 1583 Inquisitor Ulloa was concerned that cargoes destined for Lima were being unloaded in the ports of Guayaquil, Manta, and Paita where there were no commissioners to carry out inspections. He recommended that priests who lived in the vicinity should arrange for all books to be returned to the ships and brought to Lima.[127] Even where books were inspected, it has been argued in the case Mexico that most examinations were conducted by theologians with little knowledge of science and medicine, so that unless books were obviously related to magic or witchcraft banned items might easily slip through the inquisitorial net.[128] Among the belongings of one Pedro Durango de Espinosa, a deceased Spanish bookseller who became domiciled in Lima, were two copies of the Koran![129] One shipment of books to two booksellers of Cuzco, Rodrigo de Fajardo and Bartolomé Arias contained a text entitled "fisonomía natural."[130] This might be *Libro de phisonomia natural y varios secretos de naturaleza* by Jerónimo Cortés, a scientist and astrologist, which was published in Valencia in 1598. Although this volume was not

123 Guibovich, *Censura*, 249, 258–60, 268. He comments that the three copies he consulted had a lot of marginalia indicating that it was widely consulted.

124 Guibovich, *Censura*, 274.

125 Pedro Guibovich, "Autores, censores y producción de libros en el virreinato peruano" (2015), 7–8. Unpublished manuscript at: https://escriturasvirreinales.files.wordpress.com/2015/05/guibovich-autores_y_censores-u-c.pdf [Accessed 11 Nov. 2015].

126 Leonard, *Books of the Brave*, 174–79; Guibovich, *Censura*, 266–67.

127 Archivo Histórico Nacional, Madrid. Inquisición 1034 fol. 355 Licenciado Antonio Gutiérrez de Ulloa 1 Feb. 1583.

128 Guibovich, *Censura*, 263.

129 AGI Contratación 273 R3 N73 (fol.37) Bienes de difuntos de Pedro Durango de Espinosa 1606; Carlos A. González Sánchez, "Consideraciones sobre el comercio de libros en Lima a principios del siglo XVII," *Anuario de Estudios Americanos* 54 (2) (1997): 681.

130 AGI Contratación 1138B N8: 366 Registro del navío Nuestra Señora del Rosario 1601.

censured, a popular almanac by the same author entitled *Lunario y pronostico perpetuo*, first published in Valencia in 1594, was expurgated in 1632.[131]

A wealth of evidence exists for the texts that were available for use by medical practitioners in early colonial Lima. These take the form of inventories of books accompanying imported cargoes and handled by booksellers, as well as lists included in the testaments of apothecaries. Studies using these sources have revealed that they were dominated by theological and legal titles, though chronicles, histories, novels, and poetry also figured.[132] Booksellers generally possessed only a few titles related to medicine and pharmacy, but this might be expected given the small size of the market for books of this kind.[133] The titles in circulation were overwhelmingly traditional texts, such as those by Galen, Hippocrates, Dioscorides, Mesuë, or their interpreters. The field of surgery was dominated by Juan Fragoso's *Chirugía universal* (1581) and that of pharmacy by Bernardino de Laredo's *Modus faciendi* (1527). There were a number of other texts, however, which figured quite often and reveal an interest in the medicinal use of plants recently encountered in both the Americas and Far East.[134] These included the Seville-based physician-trader-botanist, Nicólas Monardes', *Dos libros, el uno que trata de todas las cosas que se traen de nuestras Indias Occidentales que sirven al uso de la medicina* (Seville, 1565) and the Sevillan physician, Simón de Tovar's *De compositorum medicamentorum examine, nova methodus* (Antwerp, 1586). Also represented was Cristóbal de Acosta's, *Tratado*

131 López Piñero, *Ciencia y técnica*, 195; Pardo-Tomás, *Ciencia y censura*, 282.

132 Irving A. Leonard, "Best Sellers of the Lima Book Trade," *Hispanic American Historical Review* 22(1) (1942): 5–33; Irving A. Leonard, "On the Lima Book Trade, 1591," *Hispanic American Historical Review* 33(4)(1953): 515–19; Carlos A. González Sánchez, "El libro y la carrera de Indias: registro de ida de navíos," *Archivo hispalense* 72 (1989): 96–97; Hampe-Martínez, "Diffusion of books"; Carlos A. González Sánchez, "Los libros de los españoles el virreinato del Perú. Siglos XVI y XVI," *Revista de Indias* 56 (no. 206) (1996), 7–47; González Sánchez, "Consideraciones sobre el comercio," 665–92. Several of Leonard's articles have been revised and published as, *Books of the Brave*. Even though these are scholarly studies, they have only used a fraction of the documentary sources available.

133 Leonard, "Best Sellers," 5–33; Eguiguren, *Diccionario histórico cronológico*, 697–722; Guibovich, "Bibliotecas de médicos," 302. For the development of the Lima book trade see Leonard, "On the Lima Book Trade," 515–19; Guibovich, *Censura*, 265–67.

134 For books published that showed an interest in newly-discovered plants see: Puerto Sarmiento, "Farmacia renacentista," 105–11, 116–18. For Nicolás Monardes see: Barrera-Osorio, *Experiencing Nature*, 122–23; J. López Piñero, "Las 'nuevas medicinas' americanos en la obra (1565–1574) de Nicolás Monardes," *Asclepio* 42 (1)(1990): 3–67; for the biography of Monardes see: Francisco Guerra, *Nicolás Bautista Monardes: Su vida y su obra (ca. 1493–1588)* (Mexico: Fundidora, 1961).

de las drogas y medicinas de la indias orientales (Burgos, 1578), which contained descriptions of plants found in the Far East, based on Garcia de Orta's *Colóquios dos simples e drogas da India* (Goa, 1563).[135]

Experimental tendencies are suggested by the work entitled "esperimentos médicos" which was imported by the booksellers of Cuzco, Rodrigo de Fajardo and Bartolomé Arias.[136] This was probably Jerónimo Soriano's, *Libro de experimentos médicos, fáciles y verdaderos* (Zaragoza, 1598). It was a collection of "experiments" or remedies assembled for use by the public at large. It included those that had been proposed by eminent medical experts to which the author suggested modifications that he had gleaned from his own experience.[137] From a different, but potentially experimental perspective, there is evidence for the possession of texts dealing with minerals. This is perhaps not surprising given the abundance of minerals that the Spanish encountered in Peru. One text found in many cargoes of books was Gaspar de Morales's, *Libro de las virtudes y propiedades maravillosas de las piedras preciosas* (1605), which was eventually banned by the Inquisition.[138] In fact in 1606, one Juan de Sarria, a book seller in Lima, certified that he had received nineteen copies of the book from Spain.[139]

In understanding what the Spanish authorities perceived to be best pharmacy practice, the list of books imported for the establishment of the first *botica* for the Hospital of Santa Ana in 1551 is instructive (Appendix B). It is worth noting that it pre-dates policies emanating from the Council of Trent.[140] Of the eleven books shipped for the pharmacy, the overwhelming majority were written in the Galenic tradition.[141] There was no copy of Galen's *De Simplicibus*,

135 AGI Contratación 1127 N2 fol. 94 Registro de Francisco Marroquín 1598. For Simón de Tovar see: Miguel López Pérez y Mar Rey Bueno, "Simón de Tovar (1528–1596): redes familiares, naturaleza Americana y comercio de maravillas en la Sevilla del XVI," *Dynamis* 26 (2006): 69–91.

136 AGI Contratación 1138B N8: 366 Registro del navío Nuestra Señora del Rosario 1601.

137 Michael Solomon, *Fictions of Well-Being: Sickly Readers and Vernacular Medical Writing in Late Medieval and Early Modern Spain* (Philadelphia: University of Pennsylvania Press, 2010), 54–55, 58–60.

138 González Sánchez, "Consideraciones sobre el comercio," 683, 689; Pardo-Tomás, *Ciencia y censura*, 361; Guibovich, *Censura*, 339. Guibovich gives the date of the ban as 1649, whereas Pardo-Tomás gives it as 1707.

139 Leonard, *Libros del conquistador*, 343.

140 AGNP Protocolos Siglo XVI 63 Diego Gutiérrez fols. 177v.–178 Quenta de la botica 20 Oct. 1551. See also: Rabí Chara, "Primera botica," 276–77.

141 For an account of medical texts produced and circulating in Spain in the sixteenth century see López-Piñero, *Ciencia y técnica*, 339–70.

but various Arab treatises based on his works, such as those by Avicenna, Mesuë, and Serapion, the last being the author of a twelfth century herbal.[142] It also included Luis Lobera Ávila's *Banquete de los nobles caballeros* (Augsburg, 1530). Lobera de Ávila was *protomedico* to Charles I and his text focussed on ailments commonly found among the privileged. Books on pharmacy included Bernardino de Laredo's *Modus faciendi* (Seville, 1527), which followed the Galenic-Arab tradition, but incorporated his own experiences; it was the first to be written in the Spanish language.[143] On surgery, there was a volume by the Italian Juanes de Vigo (Giovanni da Vigo), possibly *Practica in arte chirurgica copiosa* (Rome, 1514). He was noted for his powders of red precipitate of mercury which were used to treat skin diseases and syphilitic ulcers.[144] Noteworthy for this discussion was the inclusion of "un arnaldo," which almost certainly referred to a text written by the physician and alchemist, Arnau de Vilanova, based in Montpellier. Although Vilanova wrote within the Galenic tradition, his works contained accounts of distillation and the preparation of alcohol and mineral acids.[145] He has traditionally been described as an alchemist, but more recent scholarship suggests that such methods were widely known at the time and has questioned whether Vilanova was the innovator in the field that he is often held up to be.[146] There is evidence that his work was also owned by the two *boticarios* in Lima, Bartolomé Díaz Cabeza de Vaca[147] and Francisco Martín Reyna.[148] For the aid of pharmacists there was also a copy of the "vocabulario de Antonio," the first Latin vocabulary in Spanish published in Salamanca in 1492. Other unspecified titles included "un guido,"[149] probably one of the texts of the fourteenth-century French physician and surgeon,

142 These works were recommended for *boticarios* by the fifteenth century Italian pharmacist Saladino d'Ascoli, whose *Compendio de los boticarios* was translated into Spanish by Alonso Rodríguez de Tudela in 1515 (Saladino de Ascoli, *Compendio de los boticarios* (Valladolid: Arnao Guillén de Brocar, 1515)). See also Pastor Frechoso, *Boticas,* 38 n. 3.

143 López Piñero and Bujosa Homar, "Tradición y renovación," 293.

144 Laval, *Botica*, 139.

145 Debus, *Chemical Philosophy,* 20–21, 24.

146 See Michael McVaugh, "Chemical Medicine in the Medical Writings of Arnau de Vilanova." II *Trobada Internacional d'Estudis sobre Arnau de Vilanova. Arxiu de Textos Catalans Antics* 23–24: (2004–2005): 239–264.

147 AGNP Protocolos Siglo XVI 819 Hernández Francisco fol. 176v. Testamento de Bartolomé Díaz Cabeza de Vaca boticario 1608.

148 AGI Contratación 1097 N5 fol. 197 Registro de Santa Catalina 1592.

149 Probably a reference to *Cirugía magna* by the fourteenth-century French physician and surgeon Guy de Chauliac (Guido de Cauliaco), whose treatise relied heavily on Galen and Avicenna (Garciá Ballester, *Búsqueda de la salud,* 550).

Guido de Cauliaco (Guy de Chauliac); "un pandeta,"[150] an encyclopaedia, probably that compiled by Matthaeus Silvaticus; and "un luminaria mayor." The last was probably *Luminare maius, seu Interpretatio antidotarii et practicae Johannis Mesue* (Venice, 1496). This interpretation of Mesuë by Giovanni Giacomo Manlio was one of the earliest books on pharmacy to be written by an apothecary.

The largest library relating to pharmacy in Lima was in the hands of the Jesuit College of San Pablo.[151] In the early seventeenth century Bernabé Cobo recorded that it possessed 40,000 books and that "it is rare that there is a book it doesn't possess."[152] At that time its library of medical books was dominated by those from Spain, but it also included a significant number of texts on different branches of medicine published in Italy and France. On pharmacy it included the great pharmacopoeias of Luis de Oviedo and Juan de Castillo that were widely used in Spain at the time,[153] and Girolamo Mercuriale's *De compositione medicamentorum* published in Venice in 1590. The last, and maybe many other books published in Italy, may have been brought by the Italian Jesuit pharmacist, Augustino Salumbrino, who arrived at the College in 1605 with the intention of establishing a pharmacy that would supply Jesuit colleges and haciendas throughout the Viceroyalty.[154] Other medical texts included books on surgery such as *Thesoro de la verdadera cirugía y via particular contra la común* (Seville, 1604) written by the progressive surgeon Bartolomé Hidalgo de Agüero, referred to above, as well as Juan Calvo's *Primera y segunda parte de la cirugía universal del cuerpo humano* (Seville, 1580). Italian authors included Giovanni Battista Cortesi, an anatomist from Bologna, Gabrielle Falopio from Padua, and Michele Mercati from Milan, and there were also writings by the Parisian anatomist, Jean Riolan, an ardent defender of Galen, and the sixteenth century Italian mathematician, physician, and astrologer, Girolamo

150 Probably *Liber pandectarum medicinae*, completed by Matthaeus Silvaticus in 1317 and published in number of editions from the fourteenth century.

151 Martín, *Intellectual Conquest*, 106–08; Luis Martín, La biblioteca del Colegio de San Pablo (1568–1767), antecedente de la Biblioteca Nacional pp. 24–29 http://www.comunidadandina.org/bda/docs/PE-CA-0015.pdf [Accessed 16 Nov. 2015]; Guibovich, "Bibliotecas de médicos," 302.

152 Cobo, *Obras*, 2: 425.

153 Oviedo, *Methodo* and Juan de Castillo, *Pharmacopoea, universa medicamenta in officinis pharmaceuticis usitata complectens, et explicans* (Cádiz: Gadibus: Apud Joannem de Borja, 1622).

154 Martín, *Intellectual Conquest*, 99–100, 106; Martín, "Biblioteca del Colegio de San Pablo," 29; Rocco, *Miraculous Fever-Tree*, 69–71.

Cardano, who defended the Hippocratic tradition.[155] Many of these volumes may have been brought by members of the clergy who came from different parts of Europe. Medical practitioners in Lima may have been aware of these texts since they were allowed to borrow books from the College's library.[156] In 1647 the physician Juan Gerónimo Navarro had in his possession some eighty books belonging to the library which he ordered to be returned on his death.[157] Progressive texts may not have been exclusive to the College, either. While Navarro's personal library of some 400 books contained many medical volumes by interpreters of Galen and Hippocrates, including Laguna, Avicenna, and Mesuë, and even duplicate copies of some of them, he also owned a significant number of volumes on meteorology and astronomy in which he had a particular interest. These were no doubt sources for his book *Sangrar y purgar en días de conjunción* (Lima, 1645), which challenged some prevailing humoral beliefs.[158]

Focussing on the volumes possessed by private individuals one of the largest cargoes of medical books was sent from one Licenciado Pedro Colorado in Spain to the physician Doctor Melchor de Amusco in 1584. At that time Amusco was resident in Nombre de Dios, but he was to become *protomédico* in Peru.[159] The whole consignment contained over seventy books (See Appendix C). It is not possible to identify all the titles and not all were medical texts, but traditional Galenic medical titles figured prominently.[160] Most cargoes of medical books, as well as those owned by *boticarios* themselves, contained far fewer titles, generally only ten or twelve. Commonly they included copies of Mesuë, Dioscorides, Arnau de Vilanova, Serapion, and Laredo, all practical manuals rather than medical texts, but occasionally books by Italian doctors, all of a conservative persuasion, such as Antonio Musa[161] and Giovanni

155 Martín, *Intellectual Conquest*, 107. On Girolamo Cardano see: Girolamo Cardano, *Book of My Life*, trans. Jean Stoner (New York: New York Review Books, 2002) and Guido Giglioni, "Girolamo Cardano: University Student and Professor," *Renaissance Studies* 27 (4) (2013): 517–32.

156 Guibovich, "Bibliotecas de médicos," 302.

157 AAL Testamentos 28A-1 fols. 81v.–82, 207v.–208 Testamento de Juan Gerónimo Navarro 1648.

158 AAL Testamentos 28A-1 fols. 127–47 Testamento de Juan Gerónimo Navarro 1648.

159 According to the Jesuit annual letter of 1632 Melchor de Amusco left a thorn from the Christ's crown of thorns and other holy relics in his will (ARSI Peru 15 fol. 32v. Carta annua [1632 sic] Antonio Vázquez 28 May 1635).

160 AGI Contratación 1081 N2 R1 fols. 67–68 Registro de Juan Bautista de Olarte 1584.

161 He published a number of texts on pharmacy: *Examen omnium simplicium medicamentorum, quorum in officinis usus est. Jean & François Frellon* (Lyon, 1537); *In octo libros*

Battista Susio,[162] and the orthodox Italian Franciscan, Bernardino de Siena.[163] Apart from using the texts themselves, it seems that *boticarios* may also have been selling books of a medical nature. The testament of Bartolomé Díaz Cabeza de Vaca indicated that one Licenciado Alvaro owed him for an "Arnaldo" and five large copies of "Galen,"[164] while in 1610 the *boticario* Pedro de Bilbao imported three copies of "Oviedo," one "Laguna sobre Dioscórides," and one of "Fragoso de botica" from Spain.[165] This emphasis on traditional Galenic texts parallels the findings of Germán Somolinos for Mexico where in the sixteenth and first half of the seventeenth century the libraries of physicians were dominated by traditional texts.[166]

Conclusion

The Spanish authorities clearly took steps to encourage orthodox medical practices, among other things through the regulation of pharmacies and by restricting the medical texts available. Specifying the types of products pharmacists were to stock and instituting inspections clearly encouraged the adherence to humoral medical practice, even if the inspections were not conducted regularly or systematically. Also influential was the fact that medicines

---

aphorismorum Hippocratis & Galeni commentaria & annotationes (Basel, 1541); *In libros de ratione victus in morbis acutis Hippocratis & Galeni commentaria & annotationes* (Venice, 1546); *Examen omnium electuariorum* (Venice, 1548).

162 Possibly *Liber de sanguinis mittendi ratione* (Basel, 1558) or *Libro del conoscere la pestilenza* (Mantova, 1576).

163 For examples of the books imported and contained in pharmacies see: John Carter Brown Library Mss Codex Sp. 136 Francisco Martínez y compañía, obligación 1555; AGI Contratación 1097 N5 fol. 197 Registro del navío Santa Catalina 1592; AGI Contratación 1127 lib. 2: 94 Registro del navío San Cristóbal 1598; AGI Contratación 1155B N8: 463, 703 Registro del navío Nuestra Señora de la Concepción 1610; AGNP Protocolos Siglo XVII 819 Francisco Hernández fol. 176v. Testamento de Bartolomé Díaz Cabeza de Vaca boticario 1608.

164 AGNP Protocolos Siglo XVI 819 Francisco Hernández fol. 176v. Testamento de Bartolomé Díaz Cabeza de Vaca boticario 1608.

165 AGI Contratación 1155B N8 463, 703 Registro del navío Nuestra Señora de la Concepción. This could be *Discursos de las cosas aromáticas, árboles y frutales, y de otras muchas medicinas simples, que se traen de la India oriental y sirven al uso de medicina* (Madrid, 1572), which was based on the works of Monardes and Garcia da Orta. *See* José Luis Fresquet Febrer, ed., *Juan Fragoso y los "Discursos de las cosas aromáticas, árboles y frutales…"(1572)* (CSIC – Universidad de Valencia: Valencia, 2001), 10–111. Alternatively it could have been *Catalogus simplicium medicamentorum* (Alcalá, 1566).

166 Cited in Guibovich, "Bibliotecas de médicos," 303–04.

were being prescribed by trained physicians who had received a conservative medical education and who were discouraged from acquiring knowledge of progressive practices by the actions of Inquisition. There is no evidence for "progressive" medical practitioners seeking refuge in Lima. At the same time, the range of books imported and used in *boticas* suggests that *boticarios* were working within a humoral tradition. Whether or not this reflected the success of Inquisition in suppressing "progressive" texts or the general scarcity of books of this kind in Europe at the time is less clear. In any case, this did not necessarily mean that *boticarios* did not develop innovative methods or experiment with new *materia medica.* In fact knowledge about new preparations and treatments might be circulated in other ways, in manuscript form or by word of mouth. Further insight into *boticario* practices can be gained from an analysis of the medicines they actually prepared and employed, which is explored in the following chapter.

CHAPTER 6

Making Medicines

> Having medicines as safe as is known and with good effects, why is it necessary to use exotic foreign medicines?[1]

∴

The *materia medica* that were actually employed by *boticarios* reflected a wide range of interwoven factors that among other things included the prevailing natural philosophy, healing traditions, the knowledge, experience and preferences of particular *boticarios*, and the availability and demand for particular medicinal products. Spain's attempt to regulate medical practice extended only partially into the realm of *materia medica*, where the primary aim was to ensure the *quality* of medicines being employed rather than their *type*. Nevertheless, as the previous chapter has shown, *boticarios* and inspectors were given guidance on the types of products that a well-stocked *botica* should possess and, in Spain at least, *boticarios* whose pharmacies lacked certain items might be fined and required to acquire them. However, medicinal imports from Spain failed to meet the demand in Peru and as such at least some *materia medica* had to be acquired locally. At the same time, changes in medical practice in Europe were beginning to influence the types of medicines being employed. In the sixteenth century humoral beliefs dominated, such that the treatments prescribed consisted primarily of purgatives and emetics, based largely on plant materials, which were aimed at correcting imbalances in the humours. Overtime, however, with the development of chemical methods, minerals and chemicals began to feature more frequently in prescriptions. The widespread availability of minerals in the Andes should, in theory, have facilitated their adoption in Peru at an early date.

1 "Teniendo pues medicamentos tan seguros como conocidos, y experimentados felizmente, porque avemos de usar medicamentos tan exóticos, y tan peregrinos…" José Colmenero, *Reprobación de pernicioso abuso de los polvos de la corteza de quarango o China china* (Salamanca: Eugenio Antonio Garcia, 1697) cited in Slater, "Green Gold Fallacies," 107. This view was expressed by a Spanish physician about the adoption of plants from the New World, but might apply equally to those practising in the Americas.

 | DOI 10.1163/9789004351271_007

It might be expected that these new influences on medical practice might be reflected in the medicines that were prepared by apothecaries in Lima. However, this chapter will demonstrate that contrary to what might be expected, apothecaries in Lima prepared medicines according to humoral principles and focussed on Old World botanical materials, while making little use of native plants and minerals. Even where native plants were adopted they were placed and applied within a Galenic framework. Thus the chapter shows that there was little cross-over in Old and New World healing traditions and practices and that, despite the putative pioneering role of apothecaries in developing new practices, they remained largely separate from other healers. The chapter explores the multiple reasons why the medical marketplace in Lima was segmented and why professional *boticarios* adopted so few native products.

Types of Medicines

Most medicines were prepared from the dried parts of plants – roots, seeds, and flower heads – and less commonly incorporated minerals, as well as precious and semi-precious stones. These simple raw ingredients were mixed with water, oil, sugar, honey, vinegar or wax, or were dried and ground to a powder; in other cases they were pressed to extract an oil, pulp or juice.[2] These *materia medica* would be incorporated into syrups, waters or electuaries that were taken internally or they would be applied externally in ointments or plasters. Not surprisingly medicinal imports were dominated by unprocessed simples from which medicines might be composed. Complex medicines or *compuestos* that combined ingredients were mainly prepared in the New World and as such they figured more commonly in pharmacy inventories and prescriptions. The cargoes destined for hospitals and *boticarios* in Lima were dominated by three broad types of simples: plant and animal simples; resins, gums and oils; and minerals. These simple products accounted for over eighty percent of medicinal imports in terms of their total value and the total number of entries (Table 8). Resins, gums and oils were important as binders for ointments and plasters,[3] and their importance among imports (about 30 percent) can probably be explained by their relative absence on the treeless coast of Peru. Minerals accounted for only a small proportion of the imports (about 10 percent by

2 De Vos, "Art of Pharmacy," 125.

3 Paula S. De Vos, "Apothecaries, Artists and Artisans: Early Industrial Material Culture in the Biological Old Regime," *Journal of Interdisciplinary History* 45 (3) (2015): 307–08.

TABLE 8 *Percentage of types of medicines in cargoes destined for hospitals and private apothecaries, 1596–1622.*

	Apothecaries percent	Hospitals percent	Percentage di…
Plant simples (excluding seeds and roots)	32.9	36.1	3.20
Seeds	0.8	1.8	1.0
Roots	1	2.2	−1.20
Spices	0.6	0.3	0.30
Minerals	11.4	8.7	2.70
Stones	0.4	1	−0.60
Oils	16.7	20.9	−4.20
Gums and resins	11.5	13.1	−1.60
Animal	1.6	0.1	1.50
Wax	6.3	0.6	5.70
Syrups	1.2	0.2	1.00
Conserves	2.1	3.4	−1.30
Electuaries	5.5	1.5	4.00
Plasters	4	0.8	3.20
Ointments	2	5.8	−3.80
Chemicals	1	0	1.00
Miscellaneous	1.1	3.6	−2.50
Total percentage	100.0	100.0	
Number cargoes	15	8	
Total number entries	1,062	348	
Total value in maravedís	870,116.5	212,159	
Average value of cargo in maravedís	58,007.5	26,519.9	
Average number of entries per cargo	70.8	43.5	

Sources: For hospitals: AGI Contratación 1115 N4 676–77 Registro del navío La Magdalena and AGI Contratación 1116 N6 349–50 Registro del navío San Gregorio, Gaspar de Perales, 1596; AGI Contratación 1145A N2 341 Registro del navío San Vicente y San Juan Bautista and AGI Contratación 1145B N8 403–05 Registro del navío San Pedro, Rodrigo Arias de Buiça, 1605; AGI Contratación 1158 N1 R2 423–25 Registro del navío Nuestra Señora de Gracia, Ambrosio Villaza, 1612; AGI Contratación 1166 N4 277–80 Registro del navío San Francisco, Alonso González de la Canal, 1618; AGI Contratación 1166 N10 353–56 Registro del navío San Pedro, Adriano Legasso, 1618; AGI Contratación 1166 N4 297–98 Registro del navío San Francisco, Alonso González de la Canal, 1618. For *boticarios*: For Juan de Bilbao: AGI Contratación 1094 N4: 332–35 1591; 1107 N6: 197–99 1593–94; 1138A 3: 113–16 1601; 1138B 6: 101–04 1601; for Francisco Crespo: AGI Contratación 1138B N9: 137–40 Registro del navío San Ignacio 1601; for Gerónimo Pujadas: AGI Contratación 1166 N10: 425–30 1618; 1168 N1: 191–96 1619; for Francisco Martín Reyna: AGI Contratación 1162 N4: 43–46 1615; 1166 N1: 403–06 1618; for Pedro de Bilbao: AGI Contratación 1151B N1: 683–84 Registro del navío La Magdalena 1607; AGI Contratación 1151B N6: 429–30 Registro del navío Santa Cruz 1607; AGI Contratación 1155B N8: 445–48 Registro del navío Nuestra Señora de la Concepción 1609; AGI Contratación 1163 N1 R2: 81–83 Registro del navío La Visitación de Nuestra Señora 1615; AGI Contratación 1164 N4: 295–96 Registro del navío San Francisco 1618; AGI Contratación 1172 N2 R7: 281–82 Registro del navío Nuestra Señora de la Candelaria 1622.

value), but even this low figure is rather larger than might be expected given their local availability. Meanwhile, prepared medicines, including ointments, electuaries, plasters, and syrups, figured less prominently among imports since many would have lost their potency during transport. These products, which were often more expensive, were more common among cargoes destined for *boticarios* than for hospital pharmacies.

The hospitals of Santa Ana, San Andrés and La Caridad treated patients from largely distinct social and ethnic groups. The Hospital of Santa Ana for Indians was founded in 1549.[4] It possessed two wards for men and women, and another for contagious diseases. It was described by Bernabé Cobo as the richest in the kingdom.[5] African slaves were not supposed to be treated in there, but it seems that the Viceroy permitted this if their owners were poor.[6] The Hospital of San Andrés was founded in 1538 and treated about two thousand people a year, most of whom were Spaniards but some were free Blacks and Mulattoes, in addition to which it ordinarily housed about sixteen to twenty mentally ill patients.[7] The Hospital of La Caridad cared for fifty to sixty poor women and housed female orphans who were provided with dowries. Although it catered for women of all ethnic backgrounds, Spanish women lived in a separate section of the hospital.[8] Despite these broad differences in the ethnicity of the patients cared for in these three hospitals, there is little evidence to suggest that different types of medicinal products were being used in their treatments (Table 9).[9]

4 For the establishment of hospitals in Lima see Guerra, *El hospital*, 429–75.

5 AGI Lima 301 Relación de los hospitales que ay en esta ciudad de los Reies…20 Apr. 1619; Cobo, *Obras*, 2: 445–47.

6 ABPL 9806 fols.313–16, fol. 345 Visita al Hospital de Santa Ana [1588], ABPL 9085 fols.10–11 Libro de cuentas del hospital de Santa Ana 1595–1597.

7 AGI Lima 301 Relación de los hospitales que ay en esta ciudad de los Reies…20 Apr.1619; BNP Manuscritos B1563 Libro en que se escriben y asientan los enfermos…Hospital Real de San Andrés…desde 1 Mayo 1609.

8 AGI Lima 301 Relación de los hospitales que ay en esta ciudad de los Reies…20 Apr. 1619; AGI Lima 154 Autos sobre el Hospital de la Caridad de Lima 1622; Cobo, *Obras*, 2: 449.

9 AGI Contratación 1115 N4 676–77 Cargo of La Magdalena 1596; AGI Contratación 1116 N6 348–50 Registro del navío San Gregorio 1596; AGI Contratación 1145B N8 403–05 Registro del navío San Pedro 1605; AGI Contratación 1145A N2 341 Registro del navío San Vicente y San Juan Bautista 1605; AGI Contratación 1145B N8 403–05 Registro del navío San Pedro 1605; AGI Contratación 1158 N1 R2 423–25 Registro del navío Nuestra Señora de la Gracia 1612; AGI Contratación 1166 N4 277–80, 297–98 Registro del navío San Francisco 1618; AGI Contratación 1166 N10 353–56 Registro del navío San Pedro 1618; AGI Contratación 1166 N 1 459–65 Registro del navío El Espíritu Santo 1618.

In general pharmacies contained between 200 and 300 types of *materia medica*.[10] When the *botica* of the Hospital of Santa Ana was installed in 1552 it contained 251 types of medicinal product, but over time the range it possessed declined such that when it was inspected by the physician Doctor Melchor Romero in 1606 it possessed only 146 different varieties.[11] This decline probably reflected the limited number and range of products being shipped for hospital pharmacies in general, deficiencies that were not made good by the adoption of local products. The accounts kept by the Hospital of Santa Ana indicate that between 1595 and 1600 only 62 types of *materia medica* were acquired. It is worth noting that not only do these figures fall far short of the 429 products that were recommended by the *protomédico general,* Andrés Zamudio de Alfaro, in Spain, but only about half of them were those specified in his list (Table 10 and Appendix B).[12]

Preparing Medicines

How then were medicines prepared? Evidence from the books used by *boticarios* in early colonial Lima suggests that they were guided by the classical writings of Galen and Arabic treatises written in the Galenic tradition, while for practical advice they referred to the first-century herbal of Dioscorides interpreted by Andrés Laguna or to manuals based on the writings of the Arabic scholar Mesuë (Yūhannā ibn Māsawayh).[13] That Dioscorides was widely used is evident from the large number of copies of his work imported in comparison to other medical texts. In 1591 the merchant Francisco de Butrón acquired 22 copies of Andrés Laguna's interpretation of Dioscorides.[14] However, it was a

10 The private *botica* belonging to Pedro de Bilbao contained only 94 items, the majority being simples, but this was probably an underestimate since it was visited and assessed nearly two years after his death when many of the prepared and complex medicines would have been out of date and worthless. Pedro de Bilbao died on 4 September 1634 (AGNP Protocolos Siglo XVII 1791 Diego Sánchez Vadillo fol. 3519), but the assessment of this *botica* by the *boticario*, Alonso de Carrión, was not conducted until 28 Jul. 1636 (AGNP Protocolos Siglo XVII 1789 Diego Sanchez Vadillo fol. 1312v.).

11 ABPL 9086 fols.118–23 Visita al Hospital de Santa Ana por Doctor Melchor Romero Oct. 1606.

12 De Vos, "From Herbs to Alchemy," 140–48 notes that the majority of products found in Mexican pharmacies were those specified in traditional texts.

13 For a study and re-evaluation of the importance of Mesuë see: De Vos, "'Prince of Medicine'."

14 Leonard, "On the Lima Book Trade," 514, 521.

volume by Mesuë that was used in apothecary examinations in Lima.[15] Copies of Bernardino de Laredo's, *Modus faciendi* (1527) based on Mesuë were inventoried in a number of *boticas* and cargoes of medicines, while Antonio de Aguilera's *Exposición sobre las preparaciones de Mesuë* was owned by the *boticario*, Juan Sánchez.[16] In fact up to the end of the seventeenth century in Europe, there was no pharmaceutical book of formulas that did not rely on Mesuë.[17] All these authors were Galenists and it was not until the seventeenth century that European pharmacopoeias included chemicals.

Initially the use of published texts was limited to *boticarios* who could understand Latin. Many of these texts were later translated into Spanish, but in the sixteenth century very few translations were available. In the early seventeenth century Antonio de Robles Cornejo observed that many *boticarios* in Peru did not know Latin and were therefore preparing medicines on the basis of training and experience, producing medicines that were often ineffective and relying on poor quality and expensive imports.[18] It was these circumstances that persuaded him to produce a manual on how to prepare medicines in Spanish.[19] In contrast to scholastic medical works that were structured around ailments, entries in Robles Cornejo's work in common with the writings of Mesuë and his followers were ordered alphabetically by medicinal product or recipe.[20] These included Bernardino Laredo's *Modus faciendi*, which was that most commonly used in Lima at the time. Luis de Oviedo's *Methodo de la colección y reposición de las medicinas simples* (1581), which was widely used in Spain and the Americas, possessed a similar format.

The format used in pharmacopeias reflected the growing interest of apothecaries in the qualities and therapeutic value of individual medicinal products and remedies that were often contested. While there might be a general consensus over whether a medicinal product was hot or cold, there were often differences of view about its healing properties. There was also growing concern that the life of patients was being endangered by the employment of medicines that had been prepared inappropriately or applied in the wrong

15 AHML Libro 4 de cédulas y provisiones. Título de Bernardo Gil, boticario fol. 161 7 Nov. 1609. See Chapter 2.

16 Aguilera's *Exposición*, fols. 16v–25. For the discussion of books used in Lima see Chapter 5.

17 George Urdang, "The Development of Pharmacopoeias," *Bulletin of the World Health Organisation* 4 (4) (1951): 580, 588.

18 ARJBM División 1, leg. 17 Prologo. Libro de examen de los simples medicinales Antonio de Robles Cornejo 1617.

19 See Chapter 5 for a more detailed discussion of the history of Antonio de Robles Cornejo's texts.

20 Urdang, "Development of Pharmacopoeias," 581; De Vos, "'Prince of Medicine'," 691.

TABLE 9 *Cargoes of medicines for hospitals in Lima by value and product type, 1596–1618.*

Hospital	San Andrés[a]	San Andrés[b]	San Andrés[c]
Trader	Gaspar de Perales	Rodrigo Arias de Buiça	Ambrosio Villaza
	1596	1605	1612
Plant simples	26.0	46.9	40.9
Seeds	0.2	1.6	4.5
Roots	0.5	1.5	3.0
Spices	0.2	0.0	1.7
Minerals	13.0	4.9	4.6
Stones	0.1	2.5	1.1
Oils	23.2	7.6	24.1
Gums and resins	19.0	12.0	6.8
Animal	0.2	0.0	0.0
Wax	0.0	2.2	1.3
Syrups	0.0	0.0	0.0
Conserves	8.1	3.0	4.9
Electuaries	4.5	3.0	1.5
Plasters	0.0	0.0	0.0
Ointments	3.7	14.8	2.9
Miscellaneous	1.4	0.0	2.6
Value in maravedís	88,947	55,182	30,791
Number of entries	90	62	71
	100.0	100.0	100.0

a AGI Contratación 1115 N4 676–77 Registro del navío La Magdalena and AGI Contratación 1116 N6 349–50 Registro del navío San Gregorio, Gaspar de Perales, 1596.

b AGI Contratación 1145A N2 341 Registro del navío San Vicente y San Juan Bautista and AGI Contratación 1145B N8 403–05 Registro del navío San Pedro, Rodrigo Arias de Buiça, 1605.

c AGI Contratación 1158 N1 R2 423–25 Registro del navío Nuestra Señora de Gracia, Ambrosio Villaza, 1612.

d AGI Contratación 1166 N4 277–80 Registro del navío San Francisco, Alonso González de la Canal, 1618.

e AGI Contratación 1166 N10 353–56 Registro del navío San Pedro, Adriano Legasso, 1618.

f AGI Contratación 1166 N4 297–98 Registro del navío San Francisco, Alonso González de la Canal, 1618.

San Andrés[d]	La Caridad[e]	Santa Ana[f]	Average percent by value	Total number of entries
Alonso González de la Canal	Adriano Legasso	Alonso González de la Canal		
1618	1618	1618		
22.7	30.5	49.6	36.1	104
1.1	3.3	0.0	1.8	18
2.7	0.3	4.8	2.2	16
0.0	0.0	0.0	0.3	3
9.8	12.3	7.3	8.7	25
0.5	1.6	0.0	1.0	6
12.3	38.6	19.4	20.9	71
25.8	7.7	7.1	13.1	42
0.0	0.4	0.0	0.1	2
0.0	0.0	0.0	0.6	2
0.0	1.3	0.0	0.2	4
1.9	2.7	0.0	3.4	11
0.0	0.0	0.0	1.5	3
4.4	0.3	0.0	0.8	6
7.7	1.0	4.8	5.8	20
10.9	0.0	6.9	3.6	15
24,866	25,959	8,414	35,360	
37	67	21		348
100.0	100.0	100.0		

TABLE 10 *Types of* materia medica *recommended by Andrés Zamudio de Alfaro and found in pharmacies in Lima.*

Apothecary or pharmacy	**Andrés Zamudio de Alfaro 1590s**	**First *botica* of Hospital of Santa Ana 1551**	**Diego de Tineo 1555**
Plant simples	28	63	91
Seeds	46	18	34
Roots	30	11	13
Flowers	11	10	8
Herbs	24	6	15
Pieces/fragments	16	5	1
Minerals	10	11	19
Stones	11	6	14
Oils	35	29	42
Gums	19	25	27
Animal	11	5	8
Cordials	16	1	4
Syrups	47	4	22
Juices	14	4	10
Waters	12	15	9
Conserves	10	4	11
Electuaries	12	3	17
Opiates	11	4	5
Pills	17	0	2
Powders	0	0	2
Plasters	19	8	16
Ointments	26	10	16
Flour	4	2	2
Chemicals	0	4	1
Miscellaneous	0	3	5
Total	429	251	394
Number and percent in Alfaro's list		125 (49.8)	192 (48.7)

Sources: Charles Davis and María Luz López Terrada, "Protomedicato y farmacia en Castilla a finales del siglo XVI: edición crítica del Catálogo de las cosas que los boticarios han de tener en sus boticas, de Andrés Zamudio de Alfaro, protomedicato general (1592–1599)," *Asclepio* 53(2)(2010): 579–626; JCB Mss Codex Sp. 136 Memoria de la botica 1555; AGNP Protocolos Siglo XVI 63 Diego Gutiérrez fols. 167–88 Hospital de los españoles y naturales compañía con Francisco de Bilbao de la botica 20 Oct. 1551; AGNP Real Audiencia. Causas Civiles legajo 16 cuad. 81 fols. 119–26, 687–715 Pleito… contra los bienes y herederos del bachiller Francisco de Alba, boticario morador en la Ciudad de los Reyes. 1576; AAL Testamentos 5-1 fols. 44–49v. Testamento de Bartolomé Díaz Cabeza de Vaca 1608; AGNP Protocolos Siglo XVII 1789 Sánchez Vadillo fols. 1372v.-1374v. Inventario de bienes de Pedro de Bilbao 25 Aug. 1636.

Francisco de Alva 1576	Bartolomé Díaz Cabeza de Vaca 1608	Pedro de Bilbao 1636
42	59	30
12	30	11
10	11	5
3	4	1
0	5	3
0	7	0
8	19	4
11	9	5
22	19	8
19	18	13
5	6	4
6	9	2
22	33	4
10	10	0
13	11	0
1	1	1
7	7	0
3	5	1
1	9	0
0	14	0
2	18	0
14	15	1
0	0	0
0	2	0
3	2	1
214	323	94
127(59.3)	170 (53.6)	52 (55.3)

Note: To make the figures comparable medicinal products are classified according to the list drawn up Andrés Zamudio de Alfaro. They are classified differently in different inventories.

quantities.[21] It was argued that the only way that the correct treatments could be applied was through trial and error, or as the Jesuit priest in Peru, Martín Vitillejos stated succinctly by referring to the Latin aphorism – *juvantibus et nocentibus sumi indicia* (evidence comes from help and harm).[22] The focus on the medicines themselves and the use of experimental methods, which began to emerge in the thirteenth century, signalled the increasing specialisation and professionalisation of pharmacy within medical practice.[23]

Even though *boticarios* might be guided by particular manuals, they often annotated them in the margins, perhaps varying the quantity or combination of ingredients specified in the text according to their own experience. They might also compose their own hand-written recipes, as might members of religious orders who also compiled *recetarios*, some for use by the population at large. The latter included those by Agustín Farfán, *Tratado breve de anothomía y chirugía* (1579), Juan de Esteyneffer's popular *Florilegio medicinal* (1712), and the hermit Gregorio López's *Tesoro de medicinas* (1672, but compiled in the 1580s).[24] These texts all followed the Galenic tradition but occasionally incorporated some local medicinal products.[25] These books were published in Mexico and as yet there is no evidence that Farfán's work or a similar volume was either used or published in Peru before the eighteenth century. However, the Jesuits in sending *materia medica* to outlying colleges in Peru might include brief notes on how to utilise particular products. For example, one despatch of medicines from the Jesuit College of San Pablo to its College at Nazca included a handwritten note as follows:

> For rose sugar it is sufficient for a light purgative to use one ounce and if you want a strong purgative one and a half ounces
> The plaster for treating dysentery it is necessary to place it on the stomach
> The rest of the ointments you already know how to use them.[26]

---

21 Michael McVaugh, "The 'Experience-Based Medicine' of the Thirteenth Century," *Early Science and Medicine* 14 (2009): 105–30.

22 ARSI Perú 1a Epistolae Generalium cartas originales fol. 170 Martín Vitillejos 20 Sep. 1620.

23 Urdang, "Development of Pharmacopoeias," 581; De Vos, "'Prince of Medicine'," 694.

24 Francisco Guerra, *El Tesoro de Medicinas de Gregorio López* 1542–1596 (Madrid: Ediciones Cultura Hispánica del Instituto de Cooperación Iberoamericana, 1982).

25 Saul Jarcho, "Medicine in Sixteenth Century New Spain as Illustrated by the Writings of Bravo, Farfán and Vargas Machuca," *Bulletin of the History of Medicine* 31 (1957): 431–39; Foster, *Hippocrates' Latin American Legacy*, 155–56.

26 AGNP Compañía de Jesús, Colegio San Pablo, Cuentas leg. 119 fol. 136v. Libro de viáticos y almacén 1628–1631. The medicines sent were rose sugar, *polvos de juanes*, Zacarias

Pharmacy Methods and Equipment

Prior to the late seventeenth century most pharmacy processes were fairly simple. They involved mixing and/or heating on a stove or in an oven, and the employment of skimmers, sieves, funnels, or presses to remove impurities.[27] Paula De Vos argues that the preparation of medicines at this time did not require a distillation laboratory.[28] However, evidence from the inventories of equipment found in pharmacies in Lima suggests that there distillation methods were widely employed. This does not necessarily mean, however, that they were being used for the processing of minerals or the production of mineral salts. Distillation methods were used to produce purer forms or "fifth essences" or quintessences of other *materia medica*, particularly herbs. Indeed Mesuë was familiar with distillation methods.[29] Nevertheless, the fact that texts by Arnau de Vilanova, which described some chemical processes, were commonly found in pharmacies in Lima suggests that distillation methods may have also been employed to prepare chemical medicines. (Figure 9)

The consignment of equipment destined for the establishment of the Hospital of Santa Ana in 1552 is indicative of the type and range of apparatus employed in hospital pharmacies of the time.[30] The shipment consisted of two *alquitaras* (metal stills) and two tin-plated boilers with four ladles and two skimmers. Other items included a measuring bowl, a balance, various weights and measures, brass funnels, golden spatulas, sieves, baskets, syrup jars and a large numbers of jars, pots, vases and boxes of different sizes.[31] *Alquitaras* were essential for the functioning of a hospital pharmacy and as such were

ointment, *basilicón* ointment, *sanalatodo* (*Gnaphalium attenuatum* DC) (an American herb), *diapalma* and *diafinicón* (plasters), some pigeons and turpentine.

27 For an account of the processes used to make medicines, see Oviedo, *Methodo*, and for a brief summary based on Oviedo see: De Vos, "Art of Pharmacy," 111–55.

28 De Vos, "Art of Pharmacy," 128 and "From Herbs to Alchemy," 145–48.

29 Palmer, "Pharmacy in the Republic of Venice," 115–17.

30 AGNP Protocolos Siglo XVI 63 Diego Gutiérrez fols. 167–88 El hospital de los españoles y naturales con Francisco de Bilbao de la botica 9 Mar. 1552. The detailed cost of individual items in the *botica* suggests it was worth 289,110 *maravedís*, of which 61,750 was spent on equipment. As noted above, the figure of 289,110 *maravedís* is difficult to reconcile with the stated total cost of 3,400 *pesos de oro en plata ensayado*.

31 AGNP Protocolos Siglo XVI 63 Diego Gutiérrez fols. 167–88 El hospital de los españoles [sic] compañía con Francisco de Bilbao 9 Mar. 1552. For the use of *alquitaras* in the Hospital of San Andrés see AGI Lima 131 Petition from Francisco Molina, mayordomo of the hospital de los españoles [San Andrés] 9 Mar. 1563.

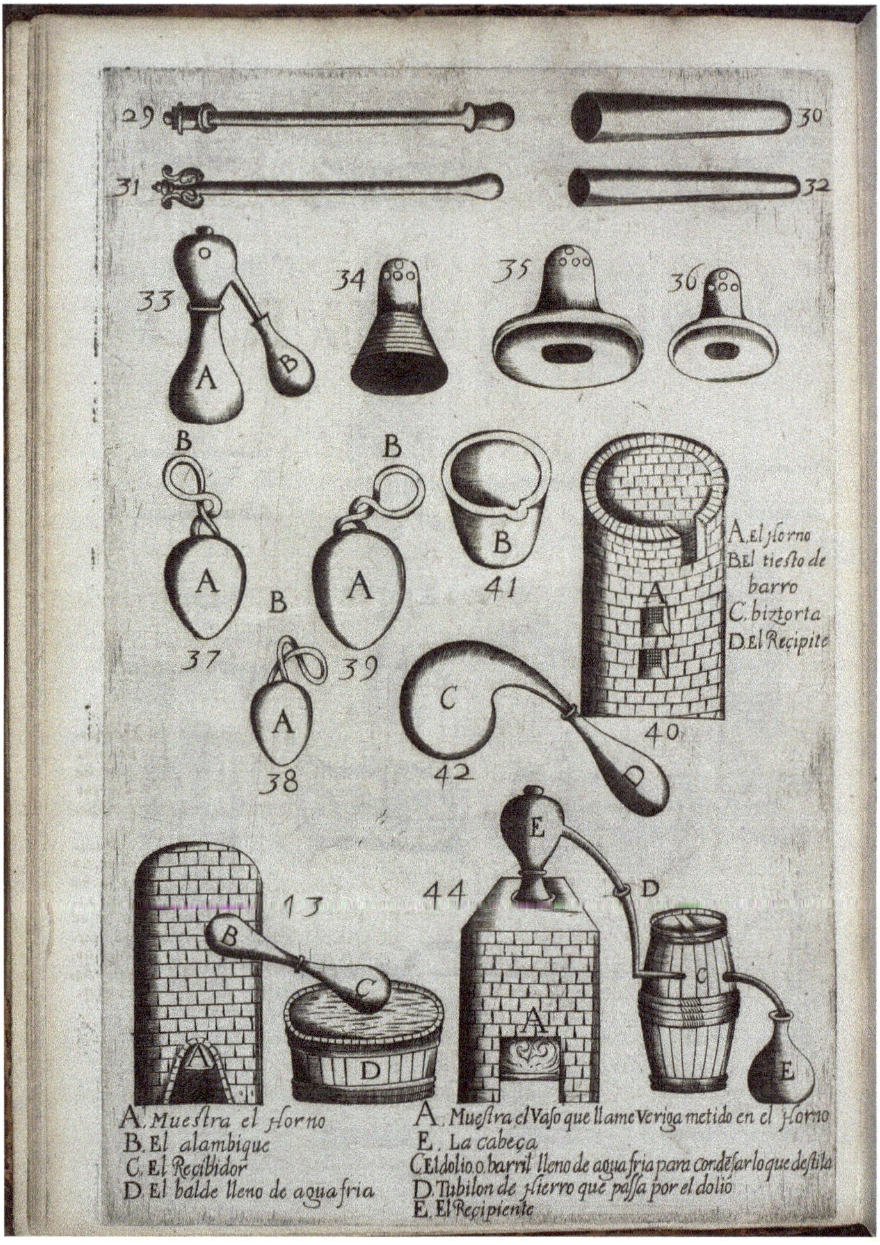

FIGURE 9 *Distillation equipment depicted by Licenciado Pedro López de León, 1628.*
Source: Licenciado Pedro López de León, Prática [sic] y teórica de las apostemas en general y particular: questiones y prácticas de cirugía, de heridas, llagas, y otras cosas nuevas, y particulares. *Seville: Oficina de Luys Estupiñan, 1628.*
COURTESY BRITISH LIBRARY.

regularly repaired or replaced.[32] Some mortars were imported but others were manufactured locally.[33] The emphasis on bloodletting in treatments meant that smaller pieces of equipment such as cupping-glasses and lancets were also purchased by hospitals on a regular basis, but for use by barber-surgeons rather than *boticarios*.

Distillation equipment was not confined to hospitals. The inventory of the well-endowed *botica* of Francisco de Alva drawn up in 1576 listed four *alquitaras*, six mortars, including one each of lead and marble, three boiling pans, three ladles, two funnels, a press, and eight spatulas, of which three were of silver.[34] Similarly, in 1608 the private *botica* belonging to Bartolomé Díaz Cabeza de Vaca possessed an unspecified number of *alquitaras*, along with mortars, spatulas, sieves, ladles, a variety of boxes and phials, an oven and other essential tools.[35] The range of equipment found in pharmacies in Lima was not dissimilar to that found in *boticas* in Spain.[36]

It is worth noting that the stills were *alquitaras* rather than *alambiques*. In 1589 the *Protomédico* in Madrid, Francisco Valles, ordered that all distilled waters that were to be drunk, including syrups, should be produced using a glass *alambique* in a *bain marie* rather than a copper *alquitara*.[37] This view was also held by Antonio de Robles Cornejo, who advised *boticarios* in Peru that the best essences were distilled using a double *bain marie*.[38] *Alambiques* were said to produce better quality extracts, while *alquitaras* were believed to result in contamination by verdigris. However, not everyone agreed. *Boticarios* in Madrid argued that distillation using a *bain marie* was suitable for delicate plants, such as violets, but not for extracting the essences of other plants which

---

32 AGI Lima 131 Hospital de los españoles de la ciudad de los Reyes 13 Jan. 1564; ABPL 9085 Libro de la razón que toma Bartolomé de la Cueva 1595–1597 Accounts of the botica of the Hospital of Santa Ana, include expenditure on the repair of *alquitaras*.

33 ABPL 9084 Libro de cuentas de gastos del hospital de Santa Ana Ano 1598 fols. 169–70. The hospital of Santa Ana was buying mortars from one Diego Rajel, a *calderero* (a boiler maker or coppersmith).

34 AGNP Real Audiencia leg. 16 cuad. 81 fols. 14v–15 Inventario de los bienes del bachiller Francisco de Alva 9 Feb. 1576.

35 AAL Testamentos 5 leg 1 fol. 165v. Memoria de las medicinas que tenía esta botica cuando la compraron… Bartolomé Díaz Cabeza de Vaca 1608.

36 Pastor Frechoso, *Boticas*, 171–201.

37 Francisco de Valles, *Tratado de las aguas destilladas, pesos y medidas* (Madrid: Luis Sánchez 1592); Mar Rey Bueno, "El informes Valles: Los desdibujados límites del arte de boticarios a finales del siglo XVI (1589–1594)," *Asclepio* 56 (2) (2004): 245–68.

38 ARJBM División 1, leg. 17 cap. 2 del arte del boticario, Libro de examen de los simples medicinales Antonio de Robles Cornejo 1617.

required more heat.[39] It was also claimed that the waters (*aguas*) produced in a *bain marie* were not as effective for as long and that glass was more expensive than copper, a factor which may have been of greater consideration in Peru than in Spain. The belief that metal stills resulted in the contamination of distilled waters ultimately prevailed and the requirement to use glass *alambiques* for waters that were to be drunk was incorporated into legislation in Spain.[40] However, there is little evidence that *alambiques* were used in Lima at this time, whereas *alquitaras* were widely employed.[41] It is worth noting in the context of the running costs of a *botica* that the use of stills and the preparation of medicines in general required some form of heating or cooking, which consumed significant quantities of charcoal that was in particularly short supply on the coast of Lima.

Categories of Medicines

The range of equipment from simple to sophisticated was used to prepare an equally wide variety of simple and complex medicines ranging from simple waters and syrups to complex compound medicines that might be composed of more than twenty ingredients.[42] Simple syrups, as well as plasters and ointments, cost between two and three reals an application, whereas the fee for more complex purgatives could be between sixteen and twenty-four reals.

Simple syrups were made from honey or sugar with a plant ingredient of some kind. No less than sixty-six types of syrups were found in the four *boticas* examined here, the most prominent being violet, rose, verjuice (*agraz*), quince, borage and myrtle. Most of these syrups were used as general tonics and to treat diarrhoea and vomiting.[43] Sometimes similar ingredients were employed in very thick syrups called *lamedores*. Electuaries were more complex syrups that were very sugary to disguise the unpleasant taste of other ingredients, often purgatives. The most common were *diacatolicón* (a light purgative based on senna and rhubarb), *diafinicón* (a purgative based on dates and

39 Today *alambiques* are considered preferable because they enable the separation of the processes of vaporisation and condensation better and allow greater control of the process.

40 *Recopilación de las leyes, pragmáticas* (Muñoz) cap. XIV ordinance 5: 177.

41 For example, RGI 2: 7–8 Relación de la ciudad de Cuzco, Vasco de Contreras y Valverde 1 Jan. 1650.

42 This section is a modified version of Newson and Minchin, *From Capture to Sale*, 273–77.

43 Laval, *Botica*, 40, 138, 172, 202.

quince), *diamargaritón* (a compound including powder of ground pearls) and *girapliega* (based on a variety of ingredients that might include balsam wood, aloeswood, mastic and cinnamon), while *bol arménico* (Armenian bole) and powder of the myrobalan plant were also common components. These were all used as purgatives and laxatives.[44]

Other treatments for both internal and external afflictions took the form of plasters (poultices), which were placed on the outside of the body. Containing either "hot" or "cold" substances they were used to restore the balance in the humours.[45] Commonly used plasters were *diapalma*, which contained litharge and white and red vitriol and was employed to heal ulcers, wounds, dislocations and fractures, and *diaquilón*, which also contained litharge and was used as an emollient and to treat tumours.[46] Like electuaries, plasters contained a variety of ingredients.

Ointments were made of wax, animal fat, or resins with vegetable or mineral ingredients. They were primarily used to treat ailments of the skin, such as scabies (*sarna*), and less commonly for venereal disease and muscle strain.[47] These disorders commonly afflicted African slaves in Lima who were generally treated with yellow ointment, Zacarías ointment, and white ointment.[48] Yellow ointment was made of yellow wax, fat, oil, and resins and was used as a suppurative. Zacarías ointment was also made of yellow wax, but contained beef, pork and chicken fat, linseed oil and fenugreek and was used as an expectorant and to soften hard parts of the skin. White ointment served a variety of purposes, but was commonly used for the treatment of burns and skin infections.[49] The ointments used to treat non-African patients differed somewhat since their ailments were not dominated by skin infections. They included relatively simple ones based on rose, sandalwood, and in the case of *unguento basilicón*, the vegetable resin, *pez negra*. Other popular ointments were those composed of an assortment of ingredients, such as *unguento apostolorum* and

---

44 For electuaries see Oviedo, *Methodo*, 88–199 (for *diacatolicón* 122–29, *diamargaritón* 171–72, *diafenicón* 88–98); Laval, *Botica*, 105, 106, 141; Pilar Arrebola Nacle and Guillermina López Andújar, "Suministro de medicamentos a la gente de mar y guerra de la Real Armada (1636)," *Boletin de la sociedad española de historia de la farmacia* 27 (1986): 46.

45 Virginia Gutierrez de Pineda, *Medicina tradicional de Colombia: El triple llegado*, Vol. 1 (Bogotá: Universidad Nacional de Colombia, 1985), 94.

46 Oviedo, *Methodo*, 454–70; Laval, *Botica*, 108.

47 Anastasio Rojo Vega, *Enfermos y sanadores en la Castilla del siglo XVI* (Valladolid: Universidad De Valladolid, 1993), 75; Laval, *Botica*, 189–97.

48 Newson and Minchin, *From Capture to Sale*, 278–81.

49 Laval, *Botica*, 189–90, 195.

populeón,[50] as well as *unguentos opilativos* which were applied to remove obstructions of the liver, stomach or spleen,[51] as was *unguento agripa*, a recipe based on bryony and other roots supposedly invented by an unspecified king Agrippa.[52]

Using Purgatives and Emetics

Humoral medicine generally involved purging and vomiting to dispel malignant humours and restore bodily balance. The dominance of purgatives both among imports of *materia medica* and contained in pharmacies suggests that humoral medicine dominated in the sixteenth and early seventeenth centuries. It is significant that Robles Cornejo in his guide to the use of medicinal plants listed the purgatives first: rhubarb, rhapontic, agaric, turpeth and polypody in that order (See Figure 8).[53]

Of the 128 types of botanical simples, flowers and herbs, excluding roots and seeds, contained in the original *botica* of the Hospital of Santa Ana in 1551 and in three private *boticas*, only eight products were found in all four pharmacies: scammony, agaric, myrobalans (yellow and emblic), linaloe, colocynth, long pepper and red sandalwood, the first six of which were valued for their purgative qualities (Appendix B).[54] Among the seeds the only product found in all four pharmacies was cardamom, but its prevalence may relate to its culinary rather than medicinal use. The dominance of purgatives is also seen in the composition of cargoes of medicines sent to Lima in the late sixteenth and early seventeenth centuries (See Table 11), where senna and scammony alone accounted for about 40 percent of the value of simples dispatched to hospitals and only slightly less in the case of those destined for *boticarios*. The lower proportion imported by *boticarios* may be explained by the fact that they acquired more expensive purgatives in the form of complex electuaries.[55] Other important purgatives were myrobalans, aloe, and the fungus agaric. Aloe was said to be abundant in Peru, so the fact that it featured highly among imports

50 Oviedo, *Methodo*, 389, 410.

51 Ibid. 413.

52 Felix Palacios, *Palestra pharmaceutica chymico-galenica* (Madrid: Juan Garcia Infançon, 1706), 300; Laval, *Botica*, 191.

53 ARJBM División 1, leg. 17 fols. 1–21 Libro de examen de los simples medicinales Antonio de Robles Cornejo 1617.

54 All eight were also in Diego de Tineo's list of plant materials requested from Spain for his *botica* in Lima (JCB Mss codex Sp 136 Francisco Martínez y compañía obligación – Diego de Tineo y consortes 1555.

55 Electuaries accounted for 5.5 percent of the total value of imports in the case of private boticarios and 1.5 percent in the case of the Hospital of Santa Ana (see Table 8).

is perhaps surprising.[56] However, there were different varieties of aloe. The variety found abundant locally was probably the American aloe or agave,[57] while imported one was probably the true aloe which is native to Arabia and North Africa.[58]

Rhubarb was imported when the pharmacy was established in the Hospital of Santa Ana in 1551, but thereafter no cargoes destined directly for hospitals that are examined here contained rhubarb; small amounts were shipped to private *boticarios*, however. Even so, in the early years of its establishment the Hospital of Santa Ana was acquiring quite large quantities of rhubarb indirectly from merchants and from the *boticarios* Francisco de Alva and Juan de Bilbao.[59] Rhubarb was expensive. Cargo lists and purchases made by hospitals suggest that it was worth between 60–80 reals a pound, when other effective purgatives notably scammony could be purchased for 20–25 reals a pound and senna for only 4–6 reals. Rhubarb was highly esteemed as a mild but effective purgative so that when it was shipped fresh to Lima for the establishment of the pharmacy of the Hospital of Santa Ana, it was wrapped in an oil cloth and transported in a separate bag.[60] The special treatment afforded to rhubarb on this occasion suggests that it was probably the true rhubarb from China (*Reum rhabarbarum* L.). Writing in 1617 Antonio de Robles Cornejo, referring to Garcia de Orta, maintained that there was little rhubarb in Spain and that it was not known in the Indies. He claimed that in Peru it was substituted for by rhapontic (another type of rhubarb – *Reum rhaponticum* L.), which was native to the area around the Bosphorus,[61] or by myrobalans, or better still mechoacan or jalap (*Ipomoaea purga* Wender. Hayne).[62] However, it seems that rhubarb may have been better known in Spain than Robles Cornejo realised because as early as 1555 Diego de Tineo made a clear distinction between the two plants when he ordered eight pounds of rhubarb and two pounds of rhapontic from Seville for the establishment of his opulent pharmacy.[63] It is worth noting that

56 Lewin, *Descripción*, 77.

57 http://www.theplantlist.org/tpl/record/kew-293578 [Accessed 17 June 2016].

58 http://apps.kew.org/wcsp/namedetail.do?name_id=298116 [Accessed 17 June 2016].

59 ABPL 9080 fols.176–236, 254–57 Libro de cuentas del hospital de Santa Ana 1575 a 1585.

60 AGNP Protocolos Siglo XVI 63 Diego Gutiérrez fols. 167–88 Hospital de los españoles y naturales compañía con Francisco de Bilbao de la botica 20 Oct. 1551.

61 Clifford M. Foust, *Rhubarb: The Wondrous Drug* (Princeton: Princeton University Press, 1992), 12.

62 ARJBM División 1, leg. 17 fol. 5 Libro de examen de los simples medicinales Antonio de Robles Cornejo 1617.

63 JCB Mss codex Sp 136 fol. 864v. Francisco Martínez y compañía obligación – Diego de Tineo y consortes 1555.

TABLE 11 *Purgatives in pharmacies and in cargoes shipped to hospitals and apothecaries in Lima.*

	Cargoes for hospitals						
	Santa Ana[a]	San Andrés[b]	San Andrés[c]	San Andrés[d]	San Andrés[e]	Convento de la Caridad[f]	Santa Ana[g]
	1551	1596	1605	1612	1618	1618	1618
Senna	1.5	4.7	26.3	64.8	94.4	42.9	54.1
Scammony	19.9	34.5	14.5	0.0	0.0	0.0	0.0
Myrobalans	2.6	0.0	0.0	0.0	0.0	0.0	0.0
Aloe	1.8	5.9	0.0	9.7	0.0	6.9	0.0
Agaric	11.3	0.6	1.1	6.5	0.0	7.3	6.5
Rhubarb	2.9	0.0	0.0	0.0	0.0	0.0	0.0
Total percentage	39.9	45.6	41.8	81.0	94.4	57.1	60.6
Total value of simples (maravedís)	60,132	17,414	25,874	12,600	5,402	7,922	4,716
Number of types of simples (includes roots and seeds)	105	32	29	27	10	30	9

Sources

a AGNP Protocolos Siglo XVI 63 Diego Gutiérrez fols. 167–88 Hospital de los españoles y naturales compañía con Francisco de Bilbao de la botica 20 Oct. 1551. See also: Miguel Rabí Chara, "La primera botica de los hospitales de la ciudad de Lima en el siglo XVI," *Asclepio* 52 (2000): 276–77.

b AGI Contratación 1115 N4 676–77 Registro del navío La Magdalena and AGI Contratación 1116 N6 349–50 Registro del navío San Gregorio, Gaspar de Perales, 1596.

c AGI Contratación 1145A N2 341 Registro del navío San Vicente y San Juan Bautista and AGI Contratación 1145B N8 403–05 Registro del navío San Pedro, Rodrigo Arias de Buiça, 1605.

d AGI Contratación 1158 N1 R2 423–25 Registro del navío Nuestra Señora de Gracia, Ambrosio Villaza, 1612.

e AGI Contratación 1166 N4 277–80 Registro del navío San Francisco, Alonso González de la Canal, 1618.

f AGI Contratación 1166 N10 353–56 Registro del navío San Pedro, Adriano Legasso, 1618.

g AGI Contratación 1166 N4 297–98 Registro del navío San Francisco, Alonso González de la Canal, 1618.

Cargoes for apothecaries					Contents of *boticas*	
Juan de Bilbao[h]	Pedro de Bilbao[i]	Francisco Crespo[j]	Francisco Martín Reyna[k]	Gerónimo Pujadas[l]	Bartolomé Díaz Cabeza de Vaca[m]	Pedro de Bilbao[n]
1591–1601	1607–1621	1601	1615–1618	1618–1619	1608	1636
12.9	39.9	29.2	29.2	23.6	2.0	0.0
19.6	3.7	16.7	0.0	5.7	28.4	27.1
1.6	1.2	6.7	5.4	7.0	3.5	17.6
10.1	0.0	0.0	0.0	3.1	14.7	13.3
2.9	1.6	3.1	3.3	0.0	1.4	4.8
1.8	1.7	0.0	0.0	3.5	4.8	0.0
47.0	46.4	55.6	38.0	39.2	54.9	62.7
143,091	73,243	24,448	16,285	28,848	151,606	64,073
85	73	39	66	74	117	55

h AGI Contratación 1094: 332–35 Registro de San Pedro 1591; AGI Contratación 1107 N6: 197–99 Registro del navío Santa Catalina 1593–94; AGI Contratación 1138A 3: 113–16 Registro del navío San Nicolás 1601; AGI Contratación 1138B 6: 101–04 Registro del navío Nuestra Señora de la Concepción 1601.

i AGI Contratación 1151B N1: 683–84 Registro del navío La Magdalena 1607, AGI Contratación 1151B N6: 429–30 Registro del navío Santa Cruz 1607; AGI Contratación 1155B N8: 445–48 Registro del navío Nuestra Señora de la Concepción 1609; AGI Contratación 1163 N1 R2: 81–83 Registro del navío La Visitación de Nuestra Señora 1615; AGI Contratación 1164 N4: 295–96 Registro del navío San Francisco 1618; AGI Contratación 1172 N2 R7: 281–82 Registro del navío Nuestra Señora de la Candelaria 1622.

j AGI Contratación 1138B N9: 137–40 Registro del navío San Ignacio 1601.

k AGI Contratación 1162 N4: 43–46 Registro del navío San Pedro 1615; AGI Contratación 1166 N1: 403–06 Registro del navío El Espíritu Santo 1618.

l AGI Contratación 1166 N10: 425–30 Registro del navío San Pedro 1618; AGI Contratación 1168 N1: 191–96 Registro del navío San Salvador 1619.

m AAL Testamentos 5-1 fols. 44–49v Testamento de Bartolomé Díaz Cabeza de Vaca 1608.

n AGNP Protocolos Siglo XVII 1789 Sánchez Vadillo fols. 1372v–1374v Inventario de bienes de Pedro de Bilbao 25 Aug. 1636.

rhapontic was essentially an astringent rather than purgative.[64] The rhubarb found in *boticas* in Lima would not have been acquired locally being unsuited to the warm dry climate of the Peruvian coast; even in Europe early attempts to cultivate the plant were unsuccessful. While it could be grown in gardens and produce flowers and seeds, it did not yield roots of equivalent medicinal value to that produced in Asia.[65] The question remains open as to whether some of the rhubarb being used in Peru was in fact rhapontic. In any case it was less commonly used than other purgatives. Although mechoacan (Figure 10), a native of Mexico but known in Spain at an early date, was called the "rhubarb of the Indies" and on occasions was used as a substitute for rhubarb,[66] it was not employed as a purgative as extensively in Lima as other products such as senna and canafistula.

Using Native Plants

As noted in Chapter 4 the Spanish were interested in native plants and other natural resources from the perspective of finding profitable substitutes for the medicinal products they normally used, but which were either expensive or unavailable; at the same time there was a genuine curiosity about the natural world of the Americas that often came, though not exclusively, from members of the religious orders. While these commercial and intellectual considerations figured among those in the Americas, there practical considerations featured more prominently in the selection of *materia medica* employed. This was especially so in the early years of conquest, when pragmatism generally overrode any concern there may have been about the religious significance of particular native medicines, which often preoccupied missionaries and

64 ARJBM División 1, leg. 17 fol. 6v. Libro de examen de los simples medicinales Antonio de Robles Cornejo 1617; Oviedo, *Methodo*, 283.

65 Foust, *Rhubarb*, 179.

66 Nicólas Monardes, *Primera y segunda y tercera parte de la historia medicinal de las cosas que se traen de nuestras Indias Occidentales, que sirven en medicina* (Seville: Alonso Escriuano, 1574), 1 fols. 28v–37v; ARJBM División 1, leg. 17 fols. 204–11 Libro de examen de los simples medicinales Antonio de Robles Cornejo 1617. For Spanish adoption of mechoacan see: Barrera-Osorio, "Knowledge and Empiricism," 229–32. The *Protomédico general* Andrés Zamudio de Alfaro was recommending it to *boticarios* in Spain in the 1590s (Davis and López Terrada, "Protomedicato y farmacia en Castilla," 595) and it was found in pharmacies in Seville in the late sixteenth century (Fernández-Carrión and Valverde, *Farmacia y Sociedad*, 70, 81).

FIGURE 10 *Drawing of mechoacan in Antonio de Robles Cornejo's "Examen de los simples medicinales," 1617.*
COURTESY SPAIN. MINISTERIO DE CULTURA. ARCHIVO DEL REAL JARDÍN BOTÁNICO, MADRID. DIVISIÓN 1, LEG. 17 FOL. 204.

philosophers who were anxious about their role in indigenous rituals and witchcraft.[67]

Military personnel on expeditions and medical practitioners distant from sources of imports would have had little choice but to make use of local products. Hence, Captain Bernardo de Vargas Machuca, a seasoned soldier in the Indies, especially in present-day Colombia, wrote in his training manual for would-be conquistadores published in 1599:

---

67 Daniela Bleichmar, "Books, Bodies, and Fields: Sixteenth-Century Transatlantic Encounters with New World *Materia Medica*," in *Colonial Botany: Science, Commerce, and Politics in the Early Modern World*, eds. Londa Schiebinger and Claudia Swan (Philadelphia, PA: University of Pennsylvania Press, 2004), 98.

> In my experience I have made extraordinary cures and have tried out all these medicines, some learned from the Indians, great herbalists that they are. And I have acquired others through experience, as each one may do [in the field], discovering new medicines, being a new inventor of them through experiment as well as with good philosophy, for the health of his soldiers. Where there are no doctors, we must all make a choice. And even where there are this calls for careful study because of the simplicity of the medicines we apply, using no compounds.[68]

Vargas Machuca also provided a brief list of plants he recommended that soldiers should carry with them, which included a significant number of native products, including the resins *anime* and *caraña*, mechoacan, tobacco, bezoar stones, and *solimán crudo* (corrosive sublimate or mercury chloride).[69] In a non-military context where medicines from Spain were unavailable, Spaniards might similarly be forced to turn to native *materia medica*. Antonio de Robles Cornejo reported how Indians in the *yungas* had given a plant root to their *encomendero*, Don Diego Dávalos, a noble from Ecija in Spain, to treat his wife who was suffering from *flaquezas de cabeza* (weakness in the head) and was nearly blind.[70] Also, in 1610 when an Indian, Francisco García Julcapuma who had been working in the monastery of San Augustín in Trujillo, was found with herbs in his pocket and brought before the ecclesiastical court, he claimed that a priest had taught him to collect native herbs. It seems likely that the priest's interest in native plants was an attempt to find substitutes for familiar Old World ones that were unavailable. The plants included *tumbi* for dysentery, *pilpi* for abscesses, *tinti* for dropsy and *tulti* for use as anti-venom.[71]

68 Vargas Machuca, *Indian Militia*, 69. The Spanish version of the last two sentences reads, "adonde no hay médicos todos podemos tener voto y aún donde los hay, por ser simples los medicamentos que aplicamos, sin usar de compuestos, que es cosa que requiere particular estudio" (Vargas Machuca, *Milicia*, 140). This is better translated as "where there are no doctors everyone can have a say, because the medicines we apply are simples, not compounds, which is something that requires special study." On this work see also: Benjamín Flores Hernández, "Medicina de los conquistadores, en la *Milicia Indiana* de Bernardo de Vargas Machuca," *Boletín Mexicano de Historia y Filosofía de la Médicina* 6(1) (2003): 5–10.

69 Vargas Machuca, *Indian Militia*, 61.

70 ARJBM División 1, leg. 17 fol. 197 Libro de examen de los simples medicinales Antonio de Robles Cornejo 1617.

71 AAL Hechicerías 1: 17 fol. 1 Interrogatorio hecho a Francisco García Julcapuma 1610. It has not been possible to identify these native plants.

As constant complaints from hospitals attest, shortages of medicines were not unique to military campaigns or to rural areas, but also prevailed in major cities including Lima. As Chapter 4 has shown, the quantity of medicines imported failed to meet local demand by some margin such that apothecaries had to acquire *materia medica* from alternative sources. Resins and oils in particular were in short supply.

The presence of the cold Humboldt Current means that the coast of Peru receives little rain and is therefore dry and treeless. The demand for gums, resins and oils for the preparation of medicines, notably for ointments and plasters, was therefore particularly high in Lima, though shortages of these products were also common in other parts of the Americas.[72] In 1641 the surgeon of the Hospital of Santa Ana complained to the *mayordomo* that it was impossible to make yellow ointment (*unguento amarillo*), the most basic of ointments, since there was no "pine resin [pez] and virgin wax and resin [cera virgen y resina]."[73] As Table 8 shows, resins and oils figured significantly among imports from Spain. Some American resins and oils were also imported from regions further north. A schedule of import taxes on goods entering Callao from Mexico and Central America, included liquidambar, *copal, tacamahaca*[74] and *sangre de drago*, as well as mechoacan, *guaiacum* (lignum-vitae), tobacco, and pepper from Chiapas.[75] Liquidambar, *caraña, tacamahaca, anime, copal,* and *higuerilla*, were all known to Nicolás Monardes in Seville[76] and the first three were sufficiently well known in Spain to be officially recommended to pharmacists by the *Protomédico general* in the 1590s.[77] Even so, American resins and oils, indeed American *materia medica* in general, figured only rarely in *boticas* in Spain in the sixteenth century.[78]

---

72 Ronderos, *El dilema*, 104.

73 ABPL 9127 fol. 165 Don Pedro Garcia Capicha, cirujano, to Pedro Ramírez mayordomo of the hospital of Santa Ana 8 Jul. 1641.

74 De Vos, "Art of Pharmacy," 101–10. Tacamahaca was an American gum-resin extracted from a variety of trees whose name derived from the Aztec term, tecomahiyac (Francisco Hernández, *Quatro libros* (Mexico: Casa de la viuda de Diego Lopez Daualos, 1615), parte 1 lib. 2 cap. 16 fol. 16v.); Monardes, *Historia medicinal*, fols. 2–3.

75 See AGNP Cajas Reales: Lima H-3 Leg. 4 Libro 24-a fols. 34–39 Tasación de las mercaderías que vienen a esta ciudad de los reyes de las provincias de México (1617).

76 Monardes, *Historia medicinal*, fols. 2–7; Nicolás Monardes, *Joyfull Newes out of the New Found World* [1580], ed. John Frampton (London: William Norton, 1580), fols. 2–7v. See also Acosta, *Historia natural y moral*, lib. 4 cap. 29: 123–24.

77 Davis and López Terrada, "Protomedicato y farmacia," 594.

78 AMS Varios antiguos 370 Memoria de las medicinas yo Gaspar de Herrera…he dado para los presos probes [sic] de la cárcel..[Sevilla] 1562–1563; Estes, "European Reception," 13–14;

Apart from the American resins and oils, there is little evidence that licensed medical practitioners in Peru acquired or used native botanical materials. Evidence from hospital account books and from the medicines apothecaries used to treat patients shows that the only American plants being employed regularly were canafistula (*Cassia spp.*), which was used as a purgative, and sarsaparilla (*Smilax* spp.) which was used as a general tonic and to treat skin disorders, especially syphilis. There were two types of canafistula. One originated in India (*Cassia fistula* L.) and was known in Europe from the later Middle Ages; the other was native to the Americas (*Cassia grandis* L.f.) and was first encountered by the Spanish in the Caribbean. Already in the sixteenth century Santo Domingo and Puerto Rico were exporting large quantities of canafistula to Spain, where it was able to compete with that from Asia; the Asian variety was more delicate and often arrived in Europe in a deteriorated state.[79] By the beginning of the seventeenth century the American variety, which was thicker than the European, was being growing profitably in many parts of Spanish America, including around Lima and Paita on the coast of Peru.[80] In the 1590s the Hospital of Santa Ana was regularly acquiring canafistula in large quantities from sources as diverse as Indians, physicians, barbers, and the *boticario* Juan de Bilbao.[81] Bernabé Cobo suggested that the canafistula that was being used in pharmacies was generally imported rather than grown locally;[82] some of the canafistula acquired by the hospitals from *boticarios*, including from Juan de Bilbao, may have been the Asian variety. However, there is no evidence from the cargoes examined in this study that Juan de Bilbao or any other *boticario* in Lima was importing canafistula. On the other hand they

Risse, "Transcending Cultural Barriers," 32–33, 37–38. For some examples of Spanish pharmacy inventories see: Pastor Frechoso, *Boticas*, 114, 152, 164; Fernández-Carrión, *Farmacia y sociedad*, 80–96.

79 Monardes, *Historia medicinal*, fol. 20; Chaunu, *Seville et l'Atlantique*, vol. 6²: 1025–26; López Piñero, "Las 'Nuevas medicinas'," 28. As such the canafistula being used in Seville in the sixteenth century may have been the American variety. See: AMS Varios antiguos 370 Memoria de las medicinas yo Gaspar de Herrera…he dado para los presos probes [sic] de la cárcel..[Sevilla] 1562–1563; Fernández-Carrión, *Farmacia y sociedad*, 95, 97.

80 ARJBM División 1, leg. 17 fols. 416–17 Libro de examen de los simples medicinales 1617.

81 ABPL 9085 Libro de la razón…deste hospital de Santa Ana de los gastos que hace Rodrigo Arias de Buyca [Buiça] mayordomo…desde el primero del mes de agosto 1595 and fols. 59–64 1597; ABPL 9084 fols. 169–170 Libro de cuentas de gastos del hospital de Santa Ana año 1598; ABPL 9084 fols. 241–242v Libro de cuentas de gastos del hospital de Santa Ana año 1598 [1599–1600].

82 Cobo, *Obras*, 1: 252; Real Jardín Botánico, Madrid División 1, leg. 17 fols. 416–17 Libro de examen de los simples medicinales 1617.

were importing large quantities of senna (*Cassia senna* L.), another species of cassia. Probably most of the canafistula being used in Lima at that time was the American variety. Luis de Oviedo describes five ways in which canafistula could be employed either in syrups, often of fumitory (*Fumaria officinalis* L.), or in different types of electuary.[83]

Sarsaparilla was grown widely in the Americas, with different species being grown in Mexico, Honduras and Guayaquil.[84] Sarsaparilla (Figure 11) was particularly abundant in the Guayas Basin where according to Nicólas Monardes Indians employed it as a panacea and had taught Spaniards how to make it into a syrup, powder and water.[85] Robles Cornejo claimed that the sarsaparilla found around Guayaquil was not exported to Spain, but was employed in large quantities to treat syphilis (*bubas*) in Peru, where "because of the ease and freedom, vices are more serious."[86] In fact in the nineteenth century the sarsaparilla emanating from Ecuador was known as *Smilax syphilitica* Kunth., taking its name from the disease it was used to treat.[87] In the early colonial period the Hospital of Santa Ana in Lima was purchasing large quantities of sarsaparilla, very often from Indians.[88] Meanwhile the large cargoes of sarsaparilla that were exported to Spain in the sixteenth century did not come from Guayaquil, but rather from the Vice-Royalty of New Spain and especially from Honduras which was said to produce sarsaparilla of superior quality.[89] In Spain it was used as a substitute for the China root (*Smilax china* L.) which was similarly used to treat syphilis.[90]

Another item commonly used in medical treatments, foremost as anti-venom, was the bezoar stone (*piedra bezar*), which is a calcareous concretion developed in the stomach of ruminants (See Figure 12). The use of such calcareous concretions developed in Asia and they were being employed in

---

83 Oviedo, *Methodo*, 118.

84 López Piñero, "Las 'nuevas medicinas'," 33.

85 Monardes, *Historia medicinal*, fols. 14–17, 46, 72. See also Acosta, *Historia natural y moral*, lib. 4 cap. 29: 123; Pedro Cieza de León, *Obras completas* (Madrid: Consejo Superior de Investigaciones Científicas, 1984), 1: 78–79.

86 ARJBM División 1, leg. 17 fols. 194 Libro de examen de los simples medicinales Antonio de Robles Cornejo 1617.

87 López Piñero, "Las 'nuevas medicinas'," 33.

88 ABPL 9080 fols.176–236, 254–57 Libro de cuentas del Hospital de Santa Ana 1575 a 1585; ABPL 9083 fols. 122–23 Libro mayor de rentas del hospital de Santa Ana desde 1593 a 1629 [1609].

89 Hernández, *Quatro libros*, parte 2 lib. 3 cap. 41 fol. 165v; Monardes, *Historia medicinal*, fol. 72; Cobo, *Obras*, 1: 193; ARJBM División 1, leg. 17 fols. 194 Libro de examen de los simples medicinales Antonio Robles de Cornejo 1617; Chaunu, *Seville et l'Atlantique*, 6^2: 1022–25.

90 Winterbottom, "Of the China Root," 33–41.

FIGURE 11 *Drawing of sarsaparilla in Antonio de Robles Cornejo's "Examen de los simples medicinales," 1617.*
COURTESY SPAIN. MINISTERIO DE CULTURA. ARCHIVO DEL REAL JARDÍN BOTÁNICO, MADRID. DIVISIÓN 1, LEG. 17 FOL. 187.

FIGURE 12 *Spherical bezoar stone from unknown animal, 1551–1750.*
COURTESY SCIENCE MUSEUM, LONDON, WELLCOME IMAGES.

Europe in the sixteenth century, though they were expensive. Nicolás Monardes dedicated a whole section of his *Historia medicinal* to the bezoar stones that were produced in Andean alpacas, guanacos and vicuñas, amongst which he indicated the last were the most esteemed.[91] His knowledge was based on information and samples provided by one Pedro de Osma, a soldier and resident in Peru in 1568.[92] Those from the Andes had been used as anti-venom by the local population in pre-Spanish times[93] and local bezoar stones were rapidly adopted by Spaniards in Peru as a welcome substitute for those imported. In the late sixteenth century they were given to the Peruvian Viceroy, the Conde del Villar, whose chamber physician examined them favourably and found them to be quick acting and healthful.[94] Bernabé Cobo also relates the story of the Governor of Chucuito, the Conde de la Gomera, who in 1610 experimented with bezoar stones by feeding two chickens mercury (*solimán*) and then giving powdered bezoar stone to one and not the other. The result was that the former lived thereby demonstrating the product's beneficial anti-venom properties.[95] This account is interesting not only from the perspective of the use of bezoar stone, but also from the fact that Spaniards were employing some rudimentary scientific methods.

Apart from these few products, the sources examined in this study contain virtually no evidence for the adoption of other native plants especially herbs. One might speculate that this reflected their low value, but other low-priced items were included in the accounts. An examination of sixteen legal claims for payments for medicines supplied by *boticarios*, which include detailed lists of medicines prescribed, including the ingredients of *compuestos* and sometimes the individual *recetas* on which they were based, also suggest that the range of native plants employed was very restricted.[96] The sixteen claims list about 30,000 medicines, but native products only figured in about seven

---

91 ARJBM División 1, leg. 17 fols. 460–62 Libro de examen de los simples medicinales Antonio de Robles Cornejo 1617; Cobo, *Obras*, 1: 128; Acosta, *Historia natural y moral*, lib. 4 cap. 42: 137–39; RGI 2: 11 Relación de la ciudad de Cuzco, Vasco de Contreras y Valverde 1 Jan. 1650. For early Spanish observations on the bezoar stone see Marcia Stephenson, "From Marvelous Antidote to the Poison of Idolatry: The Transatlantic Role of Andean Bezoar Stones during the Late Sixteenth and Early Seventeenth Centuries," *Hispanic American Historical Review* 90 (1): 3–39.

92 Monardes, *Historia medicinal*, fols. 575–62v.

93 Acosta, *Historia natural y moral*, lib. 4 cap. 42: 137–38.

94 ARJBM División 1, leg. 17 fols. 641–42 Libro de examen de los simples medicinales 1617.

95 Cobo, *Obras*, 1: 130.

96 See AAL Causas Civiles leg. 17 exp. 18 (1616); leg. 20 exp. 5 (1618); leg. 32 exp. 14 (1627); leg. 33 exp. 19 (1628); leg. 39 exp. 16 (1631), leg. 41 exp. 15 (1632), leg. 42A exp. 2 (1633), leg. 48A

percent of them and similarly accounted for a small proportion of their total value.[97] Apart from the inclusion of canafistula, sarsaparilla, mechoacan, and the resins and oils, liquidambar, *higuerilla*, *tacamahaca* and *anime*, discussed above,[98] the only other noteworthy native botanical products found in the medicines were the Peruvian balsam quina quina (*Myroxylum periferum* L.), which was used to clean and heal wounds, and the oil of the mangrove root – *cativo de mangle* (*Prioria copaifera* Griesb.), which was used for respiratory and urinary infections.[99] *Cativo de mangle* was also found in the pharmacy belonging to Bartolomé Díaz Cabeza de Vaca.[100] Also represented was the calabash (*Cucurbita* sp.) which was frequently used to make ointments or thick syrups (*lamedores*) that were prescribed to treat worms and chest infections.[101] Although calabashes (squashes) are not exclusive to the New World, they were one of the main staples on the coast of Peru and therefore are likely to have been American varieties.[102]

The prescriptions described above were all for individuals associated with the Church who might be regarded as elites. However, an analysis of the prescriptions drawn up for claims against Manuel Bautista Pérez, which were primarily for the treatment of his African slaves, contain only slightly more evidence for the use of native botanical *materia medica*.[103] Of about 2,300 medicines prescribed only about 9 percent contained products of identifiable American origin and these accounted for only about 12 percent by value.[104] Like the medicines prescribed for elites, the native products used to treat slaves

exp. 11 (1639), 20 (1636), leg. 50 exp. 16 (1640), 24 (1640), 38 (1640), leg. 65 exp. 6 (1651); AAL Testamentos 6 exp. 1 (1594); Hospitales 2 exp. 3 (1622); AAL ASNT 30 CO 125–1091 (1656).

97 The total number of entries in the sixteen claims was 29,421 of which 2,238 were native products and the total value was 2,984 reals, of which 209 reals were for native products.

98 Canafistula figured in eleven claims, sarsaparilla in eight and mechoacan in seven. Among the oils and resins liquidambar was present in six, higuerilla in three, tacamahaca in two and anime in one.

99 Laval, *Botica*, 92.

100 AAL Testamentos 5-1 fol. 45v. Testamento de Bartolomé Díaz Cabeza de Vaca 1608.

101 Laval, *Botica*, 78.

102 Newson and Minchin, *From Capture to Sale*, 203, 215, 303.

103 The contracts that Manuel Bautista Pérez had with the apothecaries Pedro de Bilbao and Alonso de Carrión were for the treatment of his household as well as his slaves, so some persons of non-African origin were probably also treated with the listed medicines. See Newson and Minchin, *From Capture to Sale*, 259–60, 272, 305.

104 Medicines of American origin accounted for 211 entries out of 2,238 and cost 2,683 reals out of a total of 22,216. These figures are indicative rather than definitive.

were dominated by a very small number of plants. The purgatives – canafistula, cassia, and mechoacan – accounted for 36 percent of native products, and resins and oils – *higuerilla*, liquidambar, and turpentine – a further 44 percent. The only other items which appear more than five times were sarsaparilla and bezoar stones. Altogether, only eleven other items of American origin were recorded.[105]

These figures probably overestimate the importance of native *materia medica* marginally, since complex medicines, especially electuaries and syrups, contained several ingredients. Since it is impossible to disaggregate the value of different ingredients in these medicines, the medicine and its total value have been attributed to the item of American origin. In addition, in some cases it is not certain that the item was native to the Americas. This is the case with cassia and turpentine, which could have been imported from Europe.[106] Unlike hospital records, where only canafistula is mentioned, treatments for private patients indicate that at least some *boticarios* were distinguishing between canafistula, cassia, and senna (*Cassia senna* L.), all three species of *Cassia*. On what basis this distinction was made is not clear, particularly since there was no significant difference in price, even though the last, which was imported, would probably have been more expensive. Most of the claims associated with private patients date from the early seventeenth century and it is possible that the name cassia was later used to refer to the American variety of canafistula, though neither Antonio de Robles Cornejo nor Bernabé Cobo made this distinction.

A Few Experiments

Inasmuch as American botanical materials were employed by apothecaries in Lima, probably most were acquired in the market place. There is no evidence that *boticarios* possessed extensive private gardens where they experimented with native plants, though there is some evidence that some were conducting experiments. For example, the *protomédico* Doctor Iñigo de Homero was said to have experimented with papillas, a type of mechoacan, while Robles Cornejo, tried to make a sweeter version of a bitter native purgative, putallanco

105 Sarsaparilla appeared in the accounts eleven times and the bezoar stone in six. The other items were cera de Nicaragua, copal, resina de puna, caraña, quina quina, paico, suelda con suelda, ishpink, plantain, calabash, and palo santo (guaiacum).

106 In six cases cassia and mechoacan are listed together, but the medicine and its value have been counted only once under cassia.

(*Sicyos bryoniaefolius* Chod.).[107] Perhaps more interesting, due to the considerable scholarly attention it has attracted, is evidence for experiments with cinchona bark being undertaken in Lima at an early date. The Sevillian physician Gaspar Caldera de Heredia in his treatise *Acerca del polvo febrifugo de la india occidental* (1663) claimed that Jesuits from the province of Quito took the bark of the "quarango" to the apothecary Gabriel de España in Lima, who made it into a powder to treat fevers. Caldera de Heredia also noted that Gabriel de España, whose pharmacy was located next to the bridge over the river Rimac in Lima, was distributing it widely in the city for the treatment of fevers.[108] This information probably came from Spaniards returning to Seville and must refer to the early seventeenth century since Gabriel de España was described as a master apothecary in 1619.[109] At the same time the Jesuits were preparing the bark of the cinchona for distribution throughout the viceroyalty and eventually to Europe.[110] What interaction there might have been between the Jesuits and apothecaries in Lima is not known.

Native plants were no more common in hospital herb gardens. *Mayordomos* were exhorted to establish gardens to overcome shortages of imported medicines and to ensure that the *materia medica* available were fresh and cheap. Ordinance 45 of the Hospital of Santa Ana in 1609 required the *boticario* to make use of "drugs that are grown locally" (drogas que se dan en la tierra).[111] But this did not necessarily mean native plants for, as indicated in Chapter 4, inspectors instructed *mayordomos* to grow those with which they were familiar in Spain rather than those that were native to the region.

The gardens of Lima's monasteries may have contained a greater variety of native plants. The establishment of monastery gardens served both symbolic and practical functions.[112] Gardens symbolised the coming of civilisation and represented a Garden of Eden in a wilderness. Hence, even prior to the formal

107 RGI 2: 7–9 Relación de la ciudad de Cuzco, Vasco de Contreras y Valverde 1 Jan. 1650.

108 José María López Piñero and Francisco Calero (ed.s), *De pulvere febrifugo Occidentalis Indiae* (1663) de Gaspar Caldera de Heredia y la introducción de la quina en Europa (Valencia: Instituto de Estudios Documentales e Históricos sobre la Ciencia, Universidad de Valencia-C.S.I.C., 1992), 19,35–36: Rocco, *Miraculous Fever-Tree*, 60–61.

109 See Chapter 2, where he is called Graviel de España.

110 Rocco, *Miraculous Fever-Tree*, 76–79.

111 *Constituciones y ordenanzas*, 1609.

112 Meredith Beck Sayre, "Cultivating Soils and Souls: The Jesuit Garden in the Americas." MA diss., Simon Fraser University, 2005, 1–4; Andrew Cunningham, "The Culture of Gardens," in *Cultures of Natural History*, eds. Nicholas Jardine, James A. Secord and Emma C. Spary (Cambridge: Cambridge University Press, 2008), 38–47.

establishment of the College of San Pablo in 1568, the Jesuits had established a patio with orange trees and a kitchen garden some 400 feet long by 200 feet wide.[113] Later in 1605 when the Italian Jesuit and trained apothecary Agustino Salumbrino arrived at the College he set about establishing a pharmacy and herb garden.[114] However, with the exception of canafistula, relatively few native plants appear to have been grown the College's gardens. The inventory of its pharmacy when the Jesuits were expelled in 1767 suggests that native plants were still not being used extensively.[115] The College was, however, a major source of *materia medica* for the Viceroyalty supplying large quantities of canafistula to the Hospital of San Andrés,[116] and in the 1630s it was exporting bezoar stones and cinchona to Europe.[117]

The extent to which native plants appear to have been absent in professional medical practice is an unexpected finding, particularly given that visitors and residents in Peru, especially the Jesuit, Bernabé Cobo, observed and commented on the wide range of native plants to be found in the region and on their use by indigenous peoples for medicinal purposes.[118] Antonio de Robles Cornejo observed that *boticarios* showed little curiosity about native plants and were content to rely on imports from Spain, even though they were often of poor quality.[119] In the 1630s Antonio de la Calancha also observed that,

> This land produces innumerable beneficial medicinal herbs and roots, some known, which are the Indians' pharmacy and are used to treat Spaniards where there are no pharmacies, and many are used in cities where physicians use them, and hopefully they [the Spaniards] will try to

---

113 ARSI Peru Cartas anuas 12 fol. 1 Carta anua [1568] Diego de Bracamonte 21 Jan, 1569.

114 Martín, *Intellectual Conquest*, 99–100; Rocco, *Miraculous Fever-Tree*, 69–80.

115 AGNP Temporalidades. Inventarios leg. 2 cuad. 28 fols. 1–68 Inventario de la botica del Colegio de San Pablos, hecho a raíz de la expatriación de los PP. de la Compañía de Jesús 1768; Rocco, *Miraculous Fever-Tree*, 70–71.

116 AHIRA Maldonado A-III-310 fol. 31 Hospital Real de naturales de Santa Ana. Cuentas que dieron Francisco de Estrada y Fernando de Carvajal año 1645 y 1646.

117 AGNP Compañía de Jesús, Colegio San Pablo, Cuentas leg. 119 fols. 24, 107v–108v. Libro de viáticos y almacén 1628–1631; Martín, *Intellectual Conquest*, 99–102; Rocco, *Miraculous Fever-Tree*, 69–78; Anagnoustou, "Jesuits in Spanish America," 13–14.

118 Cobo, *Obras*, 1: 154–284; RGI 2: 9–11 Relación de la ciudad de Cuzco, Vasco de Contreras y Valverde 1 Jan. 1650.

119 ARJBM División 1, leg. 17 Prologo Libro de examen de los simples medicinales Antonio de Robles Cornejo 1617.

> understand those that have not been tried and will study to be herbalists, so that health might not be put in so much danger, nor simple medicines cause so much harm.[120]

The implication of this statement is that while some experimentation was taking place, native plants were not being widely employed a century after the Spanish had arrived in Peru.

Explaining the Failure to Adopt Native Botanical Materia Medica

The absence of native botanical *materia medica* in remedies prescribed by professional medical practitioners in early colonial Lima is not unique in Spanish America. Paula Ronderos in her study of a *botica* in Bogotá similarly found a lack of native products.[121] What can explain why professional medical practitioners appear to have shown so little interest in native products?

A limited explanation is that plants were sometimes misnamed. It was noted at the time that Spanish observers had the propensity to give native plants Spanish names if they possessed the same physical attributes as those found in Spain. This criticism has been levelled at Antonio de Robles Cornejo himself, who it was said followed Dioscorides and thus identified a number of plants as being of Old World in origin when in fact they were distinct indigenous ones. For example, he reported that in Peru there was an abundance of *cebollas albarranas* (maritime squill – Urginea maritima(L.) Stearn.), when it is argued that more likely it was the Caribbean spiderlily (*Hymenocallis caribaea* (*L.*) *Herb.*). Other plants might be labelled as indigenous, when in fact they were native to Africa or Asia, because they had been introduced and spread at an early date thereby giving the impression that they were native to the region.[122] More broadly, as noted above, it is not clear whether entries for canafistula referred to the American variety or to the true canafistula (*Cassia fistula L.*) of Asian origin, while some balsams, resins, and turpentine registered in pharmacy inventories and treatments were referred to generically making it impossible to know whether they derived from Old World or New World trees or plants. The acute observer Bernabé Cobo, who knew both Quechua and Aymara, recognised that many native plants had been misnamed on the basis of their similarities to European ones and was anxious to distinguish between

---

120 Calancha, *Crónica moralizada*, lib. 1 cap. 9: 59.

121 Ronderos, *El dilema*.

122 Colmeiro, "Noticias," 442–43. Colmeiro provides a list of other examples where Robles Cornejo may have mistaken the origin of plants. See also Anagnoustou, "Jesuits in Spanish America," 12.

them.[123] Hence in his *Historia del Nuevo Mundo* he set about describing the local plants in precise detail and devoted a whole section to explaining how he distinguished native species from foreign ones. His methods involved examining the origin of the names applied to plants and in interviewing old native informants as well as Spaniards who had been in Peru in the earliest years of colonial rule.[124]

While misnaming is likely to have contributed to an underestimation of the use of native plants, other obstacles to their adoption were probably more significant. Perhaps the most important impediment from the perspective of Spanish trained practitioners was that any new product had to be placed within the prevailing Galenic framework. Each new plant needed to be tested for smell, taste, and therapeutic function, for example, as a purgative, emollient, or sudorific medicine, as well as assessed for its efficacy.[125] Hence, the Spanish physician Matías de Porres claimed that in preparation for an uncompleted study entitled "Concordancias medicinales de entrambos mundos," he had studied native plants with "particular care and curiosity" obtaining information from old native informants who had learned by experience how to use them. Porres was educated at the University of Salamanca and became chamber physician to Francisco de Borja, Principe de Esquilache, when he was Viceroy of Peru. His arrival in Lima was facilitated by the *protomédico* at the time, Doctor Melchor de Amusco. Part of Porres's aim in writing his treatise was to demonstrate that it was unnecessary to import medicines from Spain since Peru possessed all the necessary things to support human life.[126]

Apothecaries who might be interested in using native products had few sources to turn to for guidance. Manuals imported from Europe sometimes included comments on the few American plants that had been studied in Spain, but often their humoral qualities were disputed. For example, sarsaparilla was described by Nicólas Monardes as hot and dry in the second degree

---

123 Prieto, *Missionary Scientists*, 189–94; Emmanuelle Moreau, "Le Père Bernabé Cobo et la nomenclature botanique," *Caravelle* 82 (2004): 195–204.

124 Cobo, *Obras*, 1: 154–58; Prieto, *Missionary Scientists*, 106.

125 Risse, "Transcending Cultural Barriers," 32, 34–35; Ronderos, *El dilema*, 131; Xavier Lozoya, "Natural History and Herbal Medicine in Sixteenth-century America, in *Science in Latin America: A History*, edited by Juan José Saldaña (Austin: University of Texas Press, 2006), 44–45. For native plants classified according to Galenic principles, such as wet or dry, hot or cold see: RGI 2: 7–9 Relación de la ciudad de Cuzco, Vasco de Contreras y Valverde 1 Jan. 1650, and ARJBM División 1, leg. 17 Libro de examen de los simples medicinales Antonio de Robles Cornejo 1617 *passim*.

126 Matías de Porres, *Breve advertencias para beber frío con nieve* (Lima, 1621), 32–32v; Rey Bueno, "Concordias," 351–52, 358.

and by Francisco Hernández as cold and dry.[127] Discrepancies in the qualities and application of different medicines were quite common at the time and they contributed to a general wariness in the use of American products for fear of the harm they might inflict.[128] As such Spaniards generally preferred to be treated with those medicines with which they were familiar.[129] Such uncertainties surrounding native plants also contributed to their delayed adoption in Europe, which was further discouraged by the low efficacy of imports that had to survive the Atlantic journey and by the existence of already effective medicines in Europe, as well as by pure prejudice against foreign products.[130]

The process of acquiring knowledge of the qualities and use of native plants might have been easier had Spaniards shown more interest in the way that indigenous peoples employed them. However, even the best ethnographic studies, such as that of Bernabé Cobo, focus on finding substitutes for products employed in Europe rather than on learning from indigenous people how they might be used. Reluctance to adopt native *materia medica*, particularly from the clergy, was based in part on the perception that native products were potentially harmful and probably more important were associated with indigenous ritual and heretical beliefs.[131] Hence, if native products were to be used in medical treatments they needed to be stripped of their cultural meaning.[132] The Jesuit, Pablo José de Arriaga, who was involved in the programme to suppress idolatry in the Andes in the early seventeenth century advised that priests should instruct popular healers, so that "removing what is superstitious and bad, advantage can be taken of what is good: as is the knowledge and use of some herbs, and other simples, which they use in their illnesses."[133] Nature thus became a "quarry" from which medicinal products could be extracted, placed within a Galenic framework, and fresh medicines made more widely

127 Monardes, *Historial medicinal*, f.22; Hernández, *Quatro libros*, parte 2 lib. 3 cap. 41 fol. 165v; Risse, "Transcending Cultural Barriers," 37–38.

128 McVaugh, "'Experience-Based Medicine'," 113–14.

129 Mauricio Sánchez-Menchero, "From Where They Are Now to Whence They Came From": News About Health and Disease in New Spain (1550–1615)," in *Medical Cultures of the Early Modern Spanish Empire*, eds. John Slater, María Luz López Terrada and José Pardo-Tomás (Farnham: Ashgate, 2014), 100–01.

130 Estes, "European Reception," 11–13; Slater, "Green Gold Fallacies," 106–112.

131 Brosseder, *Power of Huacas*, 193–98, 263–64; Prieto, *Missionary Scientists*, 6, 81–84.

132 Calancha, *Crónica moralizada* lib. 3 cap. 2 Section 12: 556; Bleichmar, "Books, Bodies, and Fields," 97–99; Crawford, *Andean Wonder Drug*, 34–39.

133 Pablo Joseph de Arriaga, *La extirpación de idolatrías del Perú* (Lima: Geronymo de Contreras, 1621), 74.

available at cheaper cost.[134] This approach was consistent with Christian world views of the time that saw nature as having been placed there by God to serve human needs.[135] Thus, as Paula De Vos shows for eighteenth century Mexico, even when indigenous medicines did penetrate pharmacies, apothecaries did not involve themselves in the healing practices that went with them.[136]

Even though there might be concern about the potential harm that native treatments might cause and their ritual significance, Spaniards often acknowledged that some Indians were great herbalists and were respected as "licenciados" within their communities. Hence bishops convened for the Second Council of Lima in 1567–1568 agreed that native healers might be given written licence to practise using herbs and roots as long as their healing practices had been examined by the bishop and any superstitious words and ceremonies removed.[137] Indigenous healing skills were also respected by Martín de Murúa who was equally concerned about the superstitious practices that accompanied them as well as their lack of knowledge of when and how to apply treatments for different ailments, circumstances that would encourage experimentation, sometimes with disastrous consequences.[138] Spanish recognition of the superior knowledge and experience of local plants possessed by indigenous people may have persuaded professional medical practitioners to leave the field to them.[139] This may have been especially true of *Conversos* who may have been more hesitant to employ native products since they were more vulnerable to charges of heresy and being brought before the Inquisition.[140] There

134 José Pardo-Tomás, "Two Glimpses of America from a Distance: Carolus Clusius and Nicolás Monardes," in Carolus Clusius: Towards a Cultural History of a Renaissance Naturalist, eds. Florike Egmond, Paul Hoftijzer and Robert Visser (Amsterdam: Edita, 2007), 179, 183, 192–93.

135 Robin Attfield, "Christian Attitudes to Nature," *Journal of the History of Ideas* 44 (3) (1983): 369–86; Per Binde, "Nature in Roman Catholic Tradition," *Anthropological Quarterly* 74(1) (2001): 16–18; Sayre, "Cultivating Soils and Souls," 3–4; Steven J. Harris, "Jesuit Scientific Activity in the Overseas Missions, 1540–1773," *Isis* 96 (1) (2005): 75.

136 De Vos, "From Herbs to Alchemy," 139.

137 Rubén Vargas Ugarte, *Concilios limenses (1551–1772)* (Lima: Imp. Juan Cardenal Guevara, 1951), 1: 255.

138 Murúa, *Historia general*, 2: 102–103. See also ARJBM División 1, leg. 17 Prologo. Libro de examen de los simples medicinales Antonio de Robles Cornejo 1617.

139 Ronderos, *El dilema*, 130.

140 For a detailed example of the influence of the Inquisition on the investigations conducted by a New Christian naturalist/medical practitioner see the case of the Portuguese physician Garcia de Orta in Hugh Cagle, "Cultures of Inquiry, Myths of Empire: Natural History

were also social status advantages for professional practitioners in distancing themselves from popular healers.[141] Timothy Walker has observed a similar process of separation between the professional and popular sectors during the Enlightenment in Portugal, when the former sought to associate themselves with ideas of rationalism and distanced themselves from popular healers who came under the increased scrutiny of the Inquisition.[142]

Not only were medical practitioners disinclined to adopt native plants, but indigenous healers were also reluctant to reveal their healing practices to them.[143] When Robles Cornejo was in La Paz he was anxious to ascertain the name and qualities of the root that some local people had given their *encomendero*, Don Diego Dávalos, to cure his wife, but was unable to extract more information from them despite offering gifts, money, and clothing.[144] For native healers their medical knowledge was often essential to their power and status within their community.[145] Such cultural resistance to the disclosure of methods of healing would have been reinforced by the experience of conquest, colonial rule, and fear of persecution. This was especially so in the late sixteenth and early seventeenth centuries Andes when the slow pace of Christian conversion led to large scale programmes to eradicate idolatrous practices.[146] While the unwillingness of native peoples to disclose knowledge of medicinal plants may have been a hindrance to their adoption by professional medical

---

in Colonial Goa," in *Medicine, Trade and Empire: Garcia de Orta's Colloquies on the Simples and Drugs of India (1563)*, ed. Palmira Fontes da Costa (Farnham: Ashgate, 2015), 107–128.

141 For this argument in the context of food see: Rebecca Earle, "'If You Eat Their Food': Diets and Bodies in Early Colonial Spanish America," *American Historical Review* 115 (3), 710–12.

142 Timothy D. Walker, *Doctors, Folk Medicine and the Inquisition: the Repression of Magical Healing in Portugal during the Enlightenment* (Leiden: Brill, 2005).

143 Schiebinger, *Plants and Empire*, 90–92, 104; Heidi V. Scott, *Contested Territory: Mapping Peru in the Sixteenth and Seventeenth Centuries* (Notre Dame, IN: University of Notre Dame Press, 2009), 56.

144 ARJBM División 1, leg. 17 fol. 198 Libro de examen de los simples medicinales Antonio de Robles Cornejo 1617.

145 Prieto, *Missionary Scientists*, 41–42. This refers to Jesuit work among the Mapuche but it is no less relevant to Peru.

146 For the extirpation of idolatries in the Andes see Pierre Duviols, *La lutte contre les religions autochtones dans le Pérou colonial: l'extirpation de l'idolâtrie, entre 1532 et 1660* (Lima: Institut Français d'Etudes Andines, 1971); Nicholas Griffiths, *The Cross and the Serpent: Religious Repression and Resurgence in Colonial Peru* (Norman: University of Oklahoma Press, 1995); Kenneth Mills, *Idolatry and Its Enemies: Colonial and Andean Religion and* Extirpation, 1640–1750 (Princeton: Princeton University Press, 1997); and Brosseder, *Power of Huacas*.

practitioners, in theory at least it could have encouraged them to experiment with native plants themselves.

Finally, the failure to adopt native plants may also be viewed in the context of the contemporary debate over the influence of astrology on the natural world and whether local or foreign remedies were the most appropriate to use.[147] The complexion or humoral balance of an individual was thought to be affected by the stars and climate and there was a belief that Spanish bodies should not be treated with exotic or strange plants that were raised under a different constellation. Paracelsians however argued for the universal application of remedies that were appropriate for the treatment of particular diseases rather than for different bodies. The fact that professional medical practitioners, whose business focused, though not exclusively, on Spanish and Creole elites, were continuing to use Old World *materia medica* rather than adopt native medicinal products suggests that at the beginning of the seventeenth century Paracelsian arguments for the universal applicability of remedies were not having a significant impact on medical practice in Lima. However, the procedures involved in testing medicinal products for their efficacy in treating particular ailments, where they did occur, had the effect of encouraging experimentation, a task that fell largely into the hands of apothecaries. Paradoxically, therefore, rather than bolster humoral medical practice that focused on individualised treatments based on a patient's particular constitution, these experimental processes fostered the search for universal remedies that ultimately served to undermine it.[148]

A Medical Marketplace?

These findings raise the question of the applicability of the concept of a "medical marketplace" to medical provision in early colonial Lima. Developed in the context of early modern England, the concept envisaged a unified market where a diversity of practitioners and products competed with each other for custom.[149] The concept emerged at a time when medical historians focussed on the professional and institutional sector, so by including a spectrum of

147 Andrew Wear, "Early Modern Debate," 149–51.

148 Harold J. Cook, "Markets and Cultures: Medical Specifics and the Reconfiguration of the Body in Early Modern Europe," *Transactions of the Royal Historical Society* 21 (2011): 123–45.

149 Roy Porter, "Patient's View," 175–98. For a critique of the concept of a medical market place see: Mark S.R. Jenner and Patrick Wallis, "*The Medical Marketplace," in Medicine and the Market in England and its Colonies, c.1450–1850*, eds. Mark S. R Jenner and Patrick Wallis (Basingstoke: Palgrave Macmillan, 2007), 1–23.

popular healers, it had the advantages both of integrating all forms of health provision under one umbrella and of contributing to knowledge from below. However, it has drawn criticism from a number of standpoints. First, and perhaps most pertinent for this discussion, it is argued that in early modern times the market was not such a competitive free-for-all as the concept implies. In the case of Lima, the market was constrained by the regulation of apothecary shops and the training and licensing of practitioners, even though this may not have been completely effective. In addition, the Inquisition and ecclesiastical courts attempted to control popular medical practices that were regarded as heretical.

Other critics of the concept have noted that the term "market" is less appropriate for the early modern period, when charitable provision and family care were prevalent and treatment of the sick was not driven so much by commercial considerations.[150] Neither were all options freely available to all. Elites with more substantial financial resources would have had a broader choice of treatments, while the care provided by charitable institutions was only available to the poor. Individuals' knowledge of the medical treatments available would have been filtered by their beliefs, tradition, experience, and the information they had acquired from family and friends. Residents of Lima were generally preoccupied with their health and in common with populations world-wide would have been risk-averse in their search for medical care.[151] Hence the sick would have been inclined to seek advice largely though not exclusively from the sector with which they were most familiar. In early modern Lima two broad medical sectors can be identified: a professional or corporate sector, which largely served the elite, and a diverse popular sector. While the two sectors appear essentially distinct and complementary in terms of their medical *practices*, their separation was slightly more blurred in the case of *clienteles*. Individuals, though seeking advice predominantly from medical practitioners in one sector, might also look for treatments in the other.

In the Americas, as in Europe, elites regularly consulted popular healers with specialist skills such as midwives, bonesetters, phlebotomists, tooth-pullers, and oculists, some of whom were partially trained and officially regulated.[152]

150 See Andrew Wear, *Knowledge and Practice,* 28–29.

151 Kenneth J. Arrow, "Uncertainty and the Welfare Economics of Medical Care," *The American Economic Review* 53 (5)(1963): 946.

152 Goodman, "Philip II's Patronage," 54–55; Lanning, *Royal Protomedicato,* 282–86, 290–97, 303; Solano Alonso, *Salud, cultura y sociedad,* 103–04; Brockliss and Jones, *Medical World,* 291–92.

As will be shown in Chapter 7, the medical profession, including *boticarios*, was not held in high regard and members of the elite were not therefore averse to consulting popular healers or herbalists when treatment by a licensed practitioner failed.[153] Within the popular sector several healers might be employed to treat a single condition. Hence, in one redhibitory case involving an African slave in 1608, his owner a carpenter, Juan López, consulted an Indian from El Cercado in the district of San Lázaro who knew how to cure dysentery, and an African slave, Beatriz Criolla, whose healing methods involved placing the beak of a bird on the slave, an African healing practice that aimed at transferring the illness from the person to the bird.[154]

Even though in reality the majority of Lima's population probably could not afford the services of professionally trained practitioners, especially physicians,[155] there is some evidence that members of the popular classes, including slaves, sought advice and treatment by licensed *boticarios*. One such person was Quiomar Biafara, whose name indicates she was of African heritage, who consulted the *boticario* Bernardo Gil in search of a cure for scabies.[156] Outside Lima, in the province of Cuzco, the *corregidor* of Chumbivilcas noted that Indians went to pharmacies in the towns, though they generally preferred to use herbs in their medicinal treatments.[157]

Overall, the structure of medical provision in Lima corresponds closely to that outlined for early modern France by Lawrence Brockliss and Colin Jones. They argue that the market in France was not unstructured, competitive, and open to all. Rather it comprised two sectors: a core comprising a professional and corporatist medical community of practice that was surrounded by a

---

153 Osorio, "'*Callejón de la Soledad*,'" 199–200, 217.

154 AAL Causas de Negros 1609 leg. 1 exp. 31 Juan López, carpintero, contra el Padre Diego de Ybarreta 13 Oct. 1608. For the use of birds and chickens in African healing rituals see Inés Sosadias, "El negro curandero en la Inquisición de Cartagena de Indias siglo XVII." MA in Anthropology, Universidad de los Andes, 1981, 174, 178; Luz Adriana Maya Restrepo "Botánica y medicina africanas en la Nueva Granada, siglo XVII," *Historia crítica* 19 (2000): 39–42; James H. Sweet *Domingos Álvares, African Healing, and the Intellectual History of the Atlantic World* (Chapel Hill: University of North Carolina Press, 2011), 114, 258 n. 32.

155 Juan Lastres (*Historia de la medicina* 2: 81) estimates that in sixteenth-century Lima a medical examination by a physician might cost six *ducados* or by a surgeon or *boticario* four *ducados*.

156 AAL Causas de Negros leg.4 doc. 7 Expediente de la demanda de redivitoria que sigue Hernando de Valdés 1620.

157 RGI 1: 319 Relación fecha por el corregidor de los Chumbibilcas, Don Francisco de Acuña 23 Feb. 1586.

"medical penumbra" of informal and unregulated healers that provided services for the majority of the population.[158]

Using Minerals and Chemicals

Those medical practitioners who advocated experimental techniques in the treatment of illness often employed chemical methods and incorporated minerals in some form. The use of such techniques would have been encouraged in Peru by the region's rich mineral deposits. Indeed, it is clear from Bernabé Cobo's writings that in the early colonial period indigenous people were employing some minerals for medicinal purposes and that some of their treatments dated back to pre-Spanish times.[159] However, Cobo states explicitly that the Indians lacked knowledge of certain minerals, such as copper, tin, magnetite (*piedra imán*) and lead carbonate (*albayalde*),[160] prior to Spanish arrival; in other cases, for example emeralds and pearls, he makes no comment on their indigenous use leaving the question open.

Among the minerals clearly being used by Indians for medicinal purposes was powder made from Armenian bole (*bol arménico*), which the Indians called *tacu*. This was being sold freely in the market and consumed in *chicha* to cure dysentery.[161] According to Robles Cornejo adulterated *bol arménico* was causing many deaths.[162] *Haquimasci*, which Cobo likened to *piedra judaica* (Jews' stone), was also being used to cure dysentery and prevent blood loss, especially from women who had just given birth; it was also employed to heal wounds and treat contagious diseases, in addition to being used as anti-venom and an anti-depressant.[163] Another mineral used by indigenous peoples for medicinal purposes was the copper-based *copaquira*, also known as *piedra lipes* (blue vitriol or copper sulphate), which was employed as an astringent and to treat toothache.[164] A copper-based green stone called *coravari* was also used

158 Brockliss and Jones, *Medical World*, 296. They prefer to use the term "core" and "penumbra" rather than elite and popular, due to the negative connotations of the latter (pp. 8–20, 228–29).

159 Cobo, *Obras*, 1: 108–53.

160 Ibid. 1: 153.

161 Ibid. 1: 116.

162 ARJBM División 1, leg. 17 fols. 619–20 Libro de examen de los simples medicinales Antonio de Robles Cornejo 1617.

163 Cobo, *Obras*, 1: 127.

164 Ibid. 1: 126; Calancha, *Corónic moralizada*, 54.

as an astringent and in the preparation of an eye-wash.[165] Other minerals were used for veterinary purposes. Sulphur was utilised to cure mange in llamas and sheep;[166] initially it was not used to treat humans, though it was later employed as a remedy for skin infections.[167] Some minerals were known to indigenous people but were not used for medicinal purposes. Hence, Cobo claimed that the Indians did not know how to extract *azogue* (mercury) prior to the arrival of the Spanish, but did use vermillion (mercury sulphate) with which it was generally found. However, the vermillion was not being used for medicinal purposes but rather employed as a red pigment in paints, ink, and dyes, while copper sulphate was used with the fruit of *tara* (*Tara spinosa* (Molina) Britton & Rose) to prepare a local ink called *colpa*.[168]

Cobo concluded that God had enriched Peru, not only with innumerable plants and herbs for the benefit of humankind, but also with a great diversity of minerals that could be used to cure illnesses, to manufacture weapons, tools and equipment, and to make ornaments to adorn themselves and their houses.[169] The interaction of native peoples with the mineral world may even have extended to geophagy or the consumption of earth. This is practised in parts of the Peruvian and Bolivian Andes where it is thought that the ingestion of mineral earths provides dietary supplements that counteract the phytotoxins found in local foods.[170]

In Europe, the use of minerals in medicines extended back to classical times. Dioscorides in his *De materia medica* devoted a whole section to the processing of minerals and their use in medicines.[171] Dioscorides was writing before the development of distillation methods, such that the processes he described were based on some form of heating. However, later commentators on Dioscorides sometimes included appendices outlining distillation methods.[172] Robles Cornejo judged that the use of *litargirio* (lead oxide) was so well known

165 Cobo, *Obras*, 1: 127.

166 Ibid. 1: 112.

167 Fresquet Febrer, "Uso de productos del reino mineral," 73.

168 Cobo, *Obras*, 1: 116.

169 Ibid. 1: 135–37.

170 David L. Browman, and James N. Gundersen, "Altiplano Comestible Earths: Prehistoric and Historic Geophagy of Highland Peru and Bolivia," *Geoarchaeology* 8 (5) (1993): 413–25.

171 Robert T. Gunther, *The Greek Herbal of Disocorides* (New York: Harner Publishing Co., 1959), 623–60.

172 Robert J Forbes, *A History of the Art of Distillation* (Leiden: E.J. Brill, 1970), 127, 133–34.

that he decided not to discuss it at length in his manual.[173] As noted in Chapter 5, those who developed chemical methods drew on the work of alchemists who were especially interested in mercury, sulphur and antimony, and in the preparation of mineral salts and spirits.[174] Antimony (*stibio* or *antimonio*) in particular became associated with the new experimental forms of medicine and the centre of debates between the Galenists and Paracelsians.[175] The extensive deposits of silver found in Peru and Bolivia were often accompanied by antimony, lead, sulphur, and arsenic; antimony was also found in small quantities on its own.[176] Antimony was one mineral that the Spanish *protomédico general*, Zamudio de Alfaro, indicated that *boticas* should possess,[177] even though it was not commonly used in Spain until the seventeenth century.[178]

A key difference between the Galenists and Paracelsians was that the former only used minerals in medicines that were applied externally, often in ointments and plasters that were used to treat skin infections and wounds, whereas the latter advocated their inclusion in medicines that were taken internally. Antimony was probably used in *unguento de plomo*,[179] but was also a strong purgative.[180] Robles Cornejo writing in Peru at the beginning of the seventeenth century observed that antimony was not only being used externally, but was being taken internally as a purgative, with perhaps 10 or 12, or even 15 grains being used in a single dose; it was also used to treat melancholy.[181] Other hints at the use of antimony by *boticarios* can be seen in Robles Cornejo's

---

173 ARJBM División 1, leg. 17 fol. 581 Libro de examen de los simples medicinales Antonio de Robles Cornejo 1617.

174 Debus, *Chemical Philosophy,* 1: 20–23; Multhauf, "Significance of Distillation," 330–33; Moran, *Distilling Knowledge*, 12–25. For the relationship between alchemy and chemistry see: Newman, "From Alchemy to 'Chymistry'," 497–517. For the dates at which the processes for separating particular metals and producing mineral acids were discovered see Georgius Agricola, *De re metalica*, trans. Herbert Clark Hoover and Lou Henry Hoover. (New York: Dover Publications, Inc., 1950), 354.

175 Laval, *Botica*, 52–53; Fernández-Carrión and Valverde, *Farmacia y sociedad*, 49.

176 Cobo, *Obras*, 1: 108–53; ARJBM División 1, leg. 17 fol. 588 Libro de examen de los simples medicinales Antonio de Robles Cornejo 1617; Georg G. Peterson, *Mining and Metallurgy in Ancient Perú. The Geological Society of America Special Paper 467* (Boulder, Col.: The Geological Society of America, Inc., 2010), 27–31.

177 Davis and López Terrada, "Protomedicato y farmacia," 593.

178 Fernández-Carrión and Valverde, *Farmacia y Sociedad*, 49, 80.

179 Davis and López Terrada, "Protomedicato y farmacia," 609; Laval, *Botica*, 195.

180 Fresquet Febrer, "Uso de productos del reino mineral," 68.

181 ARJBM División 1, leg. 17 fols. 588–62 Libro de examen de los simples medicinales Antonio de Robles Cornejo 1617.

observation that *boticarios* were not following recipes precisely, but adding ingredients such as *estibio*, scammony, and spurge (*lechetrezna*).[182] According to him, *estibio* was the crude mineral and *antimonio*, also referred to as alcohol, was the name given to the product when it was used in medical treatments.[183] The latter was probably antimony sulphide.[184] Small quantities of antimony were imported by the *boticarios* Juan and Pedro de Bilbao, and when the latter died in 1634 8.5 pounds of antimony and 2.75 pounds of *estibio* were inventoried in his pharmacy.[185] The *boticarios* Francisco Martín Reyna and Gerónimo Pujadas also imported about a dozen pounds of antimony with each fleet. Gerónimo Pujadas regularly imported minerals in larger quantities than other *boticarios*. Two cargoes consigned to him in 1618 and 1619 together contained 37 pounds of antimony, as well as sulphur and arsenic sulphide (*oropimente* and *rejalgar*).[186] Antimony was valued at 20 reals an *arroba* (25 pounds). However, there is no evidence of antimony being imported for hospitals or that they secured it from local suppliers.

Another toxic mineral acid promoted by Paracelsus was *solimán* (mercuric chloride). In Lima this was quite widely employed in the form of unctions to treat syphilis, which in the Hospital of Santa Ana were often applied by African slaves.[187] *Solimán* does not appear among imports for hospitals, but rather the Hospital of Santa Ana acquired it mainly from shops (*tiendas*), though it is not clear whether it was imported or obtained locally. Related to mercury but used externally were *polvos de juanes* or *vigo* containing red precipitate of mercury. These powders had been formulated as an astringent for the treatment of wounds by the Italian surgeon Giovanni da Vigo (ca. 1450–1525).[188] Despite

182 ARJBM División 1, leg. 17 Prohemiales 11 Libro de examen de los simples medicinales Antonio de Robles Cornejo 1617.

183 ARJBM División 1, leg. 17 fol. 588 Libro de examen de los simples medicinales Antonio de Robles Cornejo 1617.

184 Agricola, *De re metallica*, 110, 428.

185 AGNP Protocolos Siglo XVII 1789 Diego Sánchez Vadillo fols. 1372v.–1374v. Inventario de los bienes de Pedro de Bilbao difunto 8 Jul. 1636.

186 AGI Contratación, 1166 N10 425–29 Registro del navío San Pedro 1618; AGI Contratación 1168 N1 191–96 Registro del navío San Salvador 1619.

187 ABPL 9806 fols.104–05 Visita al Hospital de Santa Ana sin fecha [1588]; AAL Causas de Negros leg. 1 doc. 2 Expediente de los autos que sigue el Doctor Vásquez Fajardo contra Gaspar Guerrero 1593; Fresquet Febrer, "Uso de productos del reino mineral," 86.

188 Oviedo, *Methodo*, 441.

their high cost *polvos de juanes*, were acquired in quite large quantities when the *botica* of the Hospital of Santa Ana was established in 1551.[189]

There were more frequent imports of arsenic sulphide (*oropimente*) and red arsenic sulphide (*rejalgar*). These were very toxic minerals and they were generally used in medicines that were applied externally, including in an eyewash called agua de Lanfranc.[190] These mineral products and *solimán* were particularly common among the imports of three private *boticarios* – Francisco Martín Reyna, Gerónimo Pujadas, and Francisco Crespo – where they accounted for about 10 percent of the minerals they imported by value. The proportion is surprisingly high given that these minerals were widely available locally. In fact they were so readily available that in 1551 the *cabildo* of Lima found it necessary to prohibit the sale of "solimán or rejalgar or oropimente or any other product that might be fatal" to any slave and only to Spaniards over 15 years old, because some slaves had died from taking it.[191] Evidence from legal cases resulting from the sale slaves with undeclared illnesses, also reveal that slave owners were regularly using "agua de solimán" to treat their slaves who were suffering from syphilis[192] and employing "piedra lipes" (copper sulphate) to wash their lower parts.[193] In 1634 the *protomedico*, Doctor Juan de Vega, complained that everyone was selling *solimán*, and that the authorities should make it a monopoly and use the proceeds to support the university.[194]

Another mineral product employed in medicines that were taken internally was copper sulphate (*caparrosa*), which Robles Cornejo observed was abundant in the kingdom of Peru especially in Charcas in the valleys of Chaynata, Guantapita and Quilaquila. He suggested that a scruple taken in a drink safeguarded people in an epidemic and a dragma (three scruples) in fennel water

189 AGNP Protocolos Siglo XVI 63 Diego Gutiérrez fols. 167–88 Quenta de la botica que va en la nao 20 Oct. 1551. The high price of *polvos de juanes* is evident from the level of *almojarifazgo* charged on their arrival in Lima (AGNP Cajas Reales, Lima H-3 leg. 4 Libro 24-a fols. 1–30 Aranceles para cálculo del almojarifazgo (1617)).

190 Laval, *Botica*, 147–48; Davis and López Terrada, "Protomedicato y farmacia," 601.

191 *Libros de cabildos de Lima*, 3: 416 Ordenanza de solimán 24 Jul. 1551.

192 AAL Causas de Negros leg. 2 doc. 4 Bartolomé Verdugo pone demanda al Bachiller Juan Manuel Carrasco, albacea y tenedor de bienes del Padre Diego de Ybarreta, pues le vendió una negra, Isabel Bañol, con un hijo, 1610.

193 AAL Causas de Negros leg. 2 doc. 29 Diego de Vílchez, marido de Isabel de Landecho, madre de Mariana de Landecho, a su nombre, contra el Licenciado Don Diego Gatica, presbítero, acerca de la venta de Catalina Beafara.

194 AGI Lima 45 N 4 fols. 146–47 Conde de Chinchón 21 Apr. 1634.

helped against poisoning and killed fungus.[195] It is therefore clear that some experimentation with minerals and mineral acids was taking place and that they were being used in wide range of medicines that were taken internally.

Despite these observations on the widespread use of toxic minerals, which probably attracted attention due to their potentially harmful effects, in fact they comprised only a small proportion of the minerals that were either imported or used in recipes in Lima. The minerals that were most commonly utilised were *alumbre* (aluminium potassium sulphate), *cardenillo* (verdigris or basic copper acetate), *albayalde* (lead carbonate or white lead) and *almártaga* (lead oxide – *litargirio*). Hospitals regularly acquired these minerals from *tiendas* and occasionally from apothecaries, who had probably imported them. They were mainly employed in the preparation of ointments and plasters that were used primarily in the treatment of ailments of the skin, such as scabies, but less commonly for venereal disease and muscle strain;[196] those that contained minerals were used more specifically as astringents to dry, clean, and heal wounds and ulcers, or to harden and smooth the skin.[197]

Ointments containing minerals accounted for just over 20 percent of all ointments supplied to Manuel Bautista Pérez for his slaves and household; this was slightly higher than the 15 percent found in prescriptions for other patients, mainly priests.[198] These are very rough estimates that do not take account of the quantity of ointments prepared or of the unknown illnesses that were being treated that would have required different medicines and doses, but they do suggest that ointments containing minerals represented a relatively small proportion of the total. The main ointments containing minerals that were prescribed for slaves incorporated litharge or lead. They included white ointment (*blanco*) that contained lead carbonate. Other common ointments were *egipciaco* prepared from an oximel of copper, honey, sharp vinegar, and *cardenillo*[199] and the more complex Condesa ointment composed of oil and

195 ARJBM División 1, leg. 17 fols. 599–602 Libro de examen de los simples medicinales Antonio de Robles Cornejo 1617.

196 Rojo Vega, *Enfermos y sanadores*, 75; Laval, *Botica*, 189–97.

197 Laval, *Botica*, 45, 49–51, 84; Fresquet Febrer, "Uso de productos del reino mineral," 65–66, 77; Vargas Machuca, *Milicia*, 131; Santiago Díaz Piedrahita and Luis Carlos Mantilla, *La terapéutica en el Nuevo Reino de Granada: un recetario franciscano del siglo XVIII* (Bogotá: Academia Colombiana de Ciencias Exactas, Físicas y Naturales, 2002), 71, 116,131, 135,137, 148.

198 For the bills incurred by Manuel Bautista Pérez with Pedro de Bilbao and Alonso de Carrión see: AGNP SO CO Ca. 57 doc 431 1629, 1635–1640. For the sources of those dispensed to elites associated with the Church see note 96. The precise proportion was 22 percent.

199 Oviedo, *Methodo*, 409; Laval, *Botica*, 192.

berries of myrtle, yellow wax, galls, cypress nuts, pomegranate rind, seeds of senna and plantain, mastic and *alumbre*. The latter was used to treat slackness in the belly, thus for treating hernias and to prevent miscarriage.[200]

Plasters accounted for about five percent of all medicines contained in the prescriptions examined here and of those less than 10 percent contained a mineral. Among the latter the most common were *diaquilón* which contained litharge and was used as an emollient and for tumors, *emplasto confortativo* (containing litharge and *bol arménico*) employed to heal fractures and wounds, and *emplasto geminis* which contained *albayalde* and rose oil, which was also used to treat wounds.[201] These particular types of plaster were applied more commonly to treat Manuel Bautista Pérez's slaves.

Although there is evidence for the employment of minerals in a wide variety of medicines, in fact such treatments remained a minority. An analysis of the set of sixteen lists of prescriptions referred to above, which relate to elite persons mainly associated with the Church, indicates that medicines containing minerals numbered only about three percent of the total number of entries and two percent by value.[202] Those prescribed for internal consumption were generally small quantities of ground precious and semi-precious stones, which had not been produced by distillation but formed part of complex powders or electuaries, such as *confección jacint* and *diamargaritón*.[203] Minerals appear to have been employed more readily to treat Manuel Bautista Pérez's slaves and household. Of medicines supplied to him in 1629 and between 1635 and 1640, about 15 percent (of 2,328 individual medicines) constituted or contained a mineral or chemical.[204] Nearly half of the minerals were to be found in ointments and a further 10 percent in plasters.

Even though the majority of medicines containing minerals were therefore applied externally, they appear to have been employed more extensively in Lima than in Mexico where chemical medicines did not appear until the late seventeenth century at the earliest and the minerals used consisted primarily of precious and semi-precious stones.[205] Minerals also figured more prominently in pharmacies in Peru than in Seville at this time, where mercury was

200 Laval, *Botica*, 193.

201 Ibid. 107–08; Davis and López Terrada, "Protomedicato y farmacia," 610; Oviedo, *Methodo*, 508–10.

202 For the archival sources for the sixteen cases analysed see note 96.

203 Oviedo, *Methodo*, 163, 171.

204 For the bills incurred by Manuel Bautista Pérez with Pedro de Bilbao and Alonso de Carrión see: AGNP SO CO Ca. 57 doc 431 1629, 1635–1640.

205 De Vos, "From Herbs to Alchemy," 141, 144, 154.

not common.[206] In Spain the high cost or shortage of certain minerals may have been factors affecting the extent to which they were employed.[207]

Conclusion

An analysis of the medicines imported and used in Lima prior to the mid-seventeenth century suggests that they were prepared and applied within a humoral framework. Where possible apothecaries used imported *materia medica*; even when they employed locally-produced products, they were often of Old World origin. Their reluctance to adopt native plants was based on a combination of ideological, practical and professional considerations. It reflected the extent to which humoral medicine was entrenched in popular thought and the practical difficulties of working outside the Galenic framework, as well as the social status advantages that professional practitioners might derive from distancing themselves from indigenous healers whose knowledge of native plants was probably superior and whose practices were regarded with suspicion as being harmful and heretical. This is not to say that no experimentation or innovation occurred.

The books possessed by *boticarios* indicate that in preparing medicines they relied on practical manuals rather than scholarly medical texts. This reflected the growing professionalisation and specialisation of pharmacy within the medical profession. Even though apothecaries drew on traditional texts, such as Galen or Dioscorides, later authors who interpreted their works increasingly added their own commentaries based on personal observations and their experiences of the therapeutic value of remedies. They also began writing in the vernacular and simplifying remedies such that they could be applied by the population at large without the intervention of a physician.[208] While there is some evidence for experimentation with *materia medica* in Lima, generally the aim was to establish some consistency or standard of practice rather than

206 Fernández-Carrión and Valverde, *Farmacia y sociedad*, 85.

207 Some costs appear to have been higher in Spain. For example, alum cost 204 maravedís or 6 reals in Seville (Fernández-Carrión and Valverde, *Farmacia y sociedad*, 105) and 2 tomines or 2 reals in Lima (AGNP Cajas Reales, Lima H-3 leg. 4 Libro 24-a fols. 1–30 Aranceles para cálculo del almojarifazgo (1617). Although these lists were compiled to assess the tax to be paid on goods arriving from Spain, they reflected local values. Accounts for the Hospital of Santa Ana in the late sixteenth century suggest that it was paying 2.5 reals a pound for *alumbre* (ABPL 9080 Libro de cuentas del hospital de Santa Ana 1575 a 1585 fols. 254–57).

208 Solomon, *Fictions of Well-Being*, 6–10, 16–39.

prove or disprove propositions or hypotheses. While minerals appear to have been widely available, they were not used extensively in medicines and then generally in treatments that were applied externally. Paula De Vos has argued that the chemical methods that were employed in late seventeenth and eighteenth century Mexico drew on medieval alchemy and did not require the adoption of the new Paracelsian natural philosophy.[209] In fact they were being used alongside traditional methods of preparation. This study echoes her finding but suggests that experimentation with chemical medicines may have been more precocious in Peru than in Mexico reflecting the rich mineral deposits available in the Andes.

209 De Vos, "From Herbs to Alchemy." See also Palmer, "Pharmacy in the Republic of Venice," 117.

CHAPTER 7

The Social World of Apothecaries

> ...the good apothecary should be God-fearing, compassionate, pious, humane, affable, diligent, gentle, and generous to the poor, upright, and prompt to the needs and dangers of the sick; not a player [of cards], nor covetous, or a drinker, or distracted [by entertainment], because from these vices can only follow delays, mistakes, blunders and oversights; things which usually result in a decline in the reputation of doctors, to whom their malice and mistakes are attributed.[1]

∴

The social standing of apothecaries in Lima reflected both the general structure of society in Spain and Peru and the status of the medical profession as a whole within it. Lima's social structure differed in some respects from other colonial cities. As capital of the Viceroyalty of Peru, the city symbolised and promoted the authority of the Spanish state throughout South America and as such its social hierarchy reflected the traditional values that underpinned society in Spain.[2] It was not a mere replication of that in the Peninsula, however. Lima possessed an ethnically diverse population that needed to be incorporated within its social hierarchy, while the opportunities for social advancement were considerable, a process that was facilitated by the weaker exercise of royal authority resulting from Peru's distance from Spain.[3] As such the social hierarchy in Lima was not precisely delineated and individuals jostled to establish their positions in society.[4] The aim here is not to examine the social hierarchy as a means of state control, but rather provide a backdrop to understanding how the medical profession and apothecaries in particular were regarded by

1 Suárez de Figueroa, *Plaza universal*, 302.

2 For the way in which the authority of the king was promoted through ceremonies see Osorio, *Inventing Lima*.

3 Herzog, *Upholding Justice*, 5–8.

4 For studies of the development of social hierarchies and their limitations in Spanish America see Chapter 1 note 59.

 | DOI 10.1163/9789004351271_008

society as a whole and how they attempted to enhance their status, which had implications for the nature of pharmacy practice.

Social stratification in Spain was based on a system of estates, which defined the rights and obligations of different social groups and therefore their role and status.[5] There were three primary estates – the nobility, Church and commoners – and, with the exception of the clergy, the status of individuals was largely determined by birth. However, within each group the status, function, and wealth of individuals varied considerably and it was possible for individuals to ascend the social ladder through entering the Church or military or by providing other services for the Crown. Commoners aspired to become nobles or *hidalgos* and use the title of Don or Doña.[6] It was this hierarchical, estate-based corporative system that was introduced to the New World, but there the social structure was complicated by the presence of other ethnic groups – Native Americans, Africans and *castas*.

This estate system has been contrasted with a class-based form of social stratification, where the status of individuals is based on wealth and where there are no legal restrictions on social mobility. It has been argued that over time the basis of social stratification in the New World shifted from an estate-based system towards one constructed on class.[7] In colonial Lima several factors may have promoted an early shift towards such a class-based hierarchy. The city offered exceptional opportunities for individuals to improve their economic position due to the wealth that could be generated from mining and trade, opportunities that were not available in many other regions of Spanish America. Merchants and miners could become wealthier than many nobles, who became impoverished as their tribute income and access to native labour fell with the decline in the indigenous population.[8] Its distance from Spain and the lack of effective control in many spheres of government meant that social mobility was more easily effected in Peru. In 1637 Viceroy Chinchón advised his successor that because of the vastness of Peru (though possibly referring to the viceroyalty rather than the province), coupled with the relatively small number of Spanish settlers and the lack of an effective judicial system, social

5 See Lyle N. McAlister, "Social Structure and Social Change in New Spain," *Hispanic American Historical Review* 43 (3) (1963): 350–53.

6 Fred Bronner, "Elite Formation in Seventeenth-Century Peru," *Boletín de Estudios Latinoamericanos y del Caribe* 24 (1978): 10–12, 20-; Lockhart, *Spanish Peru*, 34–48.

7 McAlister, "Social Structure," 363; Mörner, *Race Mixture*, 7–8; John K. Chance and William B. Taylor, "Estate and Class in a Colonial City: Oaxaca in 1792," *Comparative Studies in Society and History* 19 (4) (1977): 454–55.

8 Bronner, "Elite Formation," 9–10, 17–19.

control was difficult and people moved freely up and down the social hierarchy.[9] It was this fluidity, which might also be interpreted as instability in the social structure, that it is argued encouraged the Crown to dispense titles to Lima's elites in order to maintain the social order.[10]

From the perspective of an individual raised in an environment where status was based on claims to nobility, the acquisition of wealth in itself was not regarded as sufficient to ensure high social standing. However, wealth could be transmuted into status through the payment of large dowries to secure strategic marriages or could be demonstrated through charitable donations, notably for the support for hospital or convents, or through the sumptuary accumulation and display of precious metals and jewellery. Charitable giving might not only enhance one's social standing, but be seen as a way of obtaining release from purgatory and securing a favourable position in the afterlife.[11] It will be demonstrated below how apothecaries used these mechanisms to boost their public profile and improve their social status.

The Status of the Medical Profession

The medical profession as a whole was not highly regarded. The roots of this lack of esteem lay in the domination of the profession in Spain by Jews and Muslims, even though in theory they were prevented from entering universities due to their inability to demonstrate purity of blood or *limpieza de sangre*.[12] With the expulsion of the Jews from Spain in 1492, some adopted Christianity and became New Christians or *Conversos*, as did some Muslims who converted from Islam to become *Moriscos*, only later to be expelled from Spain in 1609. While persons of Jewish or Muslim ancestry were not permitted to settle in the New World, many did so. They came under the increasing scrutiny of the Inquisition, such that Spaniards, not only those who were New

9 Ibid. 9–10.

10 Ibid. 10–11, 16; Paul Rizo-Patrón, "La nobleza de Lima en tiempos de los Borbones," *Bulletin de l'Institut Français d'Études Andines* 19(1)(1990): 135–36; Osorio, *Inventing Lima*, 24.

11 Magnus Mörner, "Economic Factors and Stratification in Colonial Spanish America with Special Regard to Elites," *Hispanic American Historical Review* 63 (2) (1978): 355–56; Bronner, "Elite Formation," 24–26; Suárez, *Desafíos transatlánticos*, 187–93.

12 Harry Friedenwald, *The Jews and Medicine*, vol. 2 (Baltimore: The John Hopkins Press, 1944), 620, 701–71; Goodman, *Power and Penury*, 219–21; Luis García Ballester, "The Inquisition and Minority Medical Practitioners in Counter Reformation Spain: Judaizing and Morisco Practitioners, 1560–1610," in *Medicine and the Reformation*, eds. Ole Peter Grell and Andrew Cunningham (London: Routledge, 1993), 156–66.

Christians, repeatedly stressed that they were of Old Christian heritage and claimed purity of blood for fear they might be discriminated against or worse brought before the Inquisition. The social value of such claims was enhanced by denying ethnic groups with indigenous, African or mixed blood access to official positions, guilds, religious orders, or entry to universities.[13]

As in Spain, elites in Lima encouraged their sons to enter the Church or the law rather than medicine on account of its low status as a profession that derived from its historic association with Jews and Muslims.[14] In theory the professions of physician, apothecary and surgeon were limited to those who could demonstrate legitimacy and *limpieza de sangre*,[15] but this stipulation was not adhered to consistently in Lima. While those *boticarios* in Lima who aspired to be *familiar* to the Inquisition, such as Francisco de Alva, Jerónimo Rodríguez, Juan de Bilbao, Pedro de Bilbao, and Mateo Pastor,[16] probably needed to demonstrate their *limpieza de sangre*, as the presence of African slaves practising as *boticarios* in Lima's hospitals and working as salaried employees indicates, this was often overlooked. In the early colonial period the shortage of medical

13 Lanning, *Royal Protomedicato*, 175–82; Cope, *Limits of Racial Domination*, 24–25; Elliott, *Empires of the Atlantic World*, 171.

14 Ramirez, *Provinical Patriarchs*, 176–77, 226; Zarate, *Inicios de la escuela de medicina*, 27–28, 30. See Chapter 1 for a discussion of the status of the medical profession and of professions involved in science in general in Spain. See also López Piñero, *Ciencia y técnica*, 47–88.

15 *Recopilación de las leyes, pragmáticas* (Muñoz) cap. 6 ordinance 1: 71; Poole, "Politics of *Limpieza de Sangre*," 363, 388; Lanning, "Legitimacy and *Limpieza de Sangre*," 43.

16 For Francisco de Alva and Juan de Bilbao see: *Libros de cabildos de Lima*, 7: 549–50 Título de familiar al bachiller Alba 21 Jan. 1574, *Libros de cabildos de Lima*, 8: 201–02 Familiar del santo oficio al Juan de Bilbao 15 and 24 Feb. 1576. For Pedro de Bilbao's services to the Santo Oficio see: AGNP SO CO 27–277 Pedro de Bilbao, ministro boticario del Santo Oficio contra Don Juan Arévalo de Espinosa, Caballero de Alcántara, Alguacil Mayor del Santo Oficio, (1629). Little is known of Jerónimo Rodríguez, but he made a donation of 30 pesos to the Crown in 1590 (AGI Lima 272 Servicio hecho al rey nuestro señor en la ciudad de los reyes 1590), following a request by Philip II the previous year for a *donativo de servicio* due to Spain's difficult financial position on account of wars with England, France, and Flanders (AGI IG 427, lib. 30, fols. 401r.–403v. real cédula 6 Mar 1589).

Mateo Pastor was *familiar* to the Inquisition in the mid-1630s when he purchased an African slave who was part of the estate of Tomé Cuaresma. The latter was a *Converso* surgeon who was tried by the Inquisition and put to death in 1635 (AGNP SO CO Ca 44 doc. 394 fols. 191–210 no date; Medina, *Historia del tribunal de Santo Oficio*, 2: 137). See also AGNP SO CO 126–1113 1658 El alférez Francisco Fajardo, defensor de los bienes de Mateo Pastor, familiar y boticario del Santo Oficio 14 Sep. 1654.

practitioners meant that the authorities in Lima were forced to be pragmatic and therefore did not press those seeking licenses to present their credentials. Even the respected *boticario*, Bernardo Gil, was not required to produce a certificate of *limpieza de sangre* when he was licensed to practise in 1609.[17]

The Middling Professional Status of the Apothecary

How then did *boticarios* fit into the prevailing social order, including within the medical profession? James Lockhart organises his social history of sixteenth-century Peru around the social categories used at the time, describing the characteristics, functions and status of different social groups.[18] In this study he regards physicians as professionals who fell between noblemen and artisans, whereas he classifies *boticarios* as artisans, though he notes that since they were literate and required specialist training and equipment, they were "near to the professional class."[19] Physicians were accorded higher status on account of their university education and their knowledge of natural philosophy which was deemed to be a higher form of learning than the practical knowledge and skills possessed by apothecaries and surgeons.[20] Although there was no requirement for *boticarios* to attend university, for licensed *boticarios* some knowledge of Latin was essential in order that they could understand physicians' prescriptions and read manuals on how to prepare medicines. Given the requirements for theoretical knowledge as well as technical skills, *boticarios* were regarded as superior to those who practised the more exclusively manual tasks of surgery or phlebotomy and as such they were regarded as having a "middling" status within the medical profession.[21]

Reflecting the availability of archival sources, more is known about those *boticarios* who owned their own pharmacies who came from bourgeois if not elite families. They often attended university and entered the profession to continue the family business. However, probably the majority of *boticarios* were

---

17 AHML Libro 4 de cédulas y provisiones. Título de Bernardo Gil, boticario 7 Nov. 1609. See Chapter 2 for the process of acquiring a title to practise.

18 Lockhart, *Spanish Peru*, 8–9.

19 Ibid. 103.

20 Lanning, *Royal Protomedicato*, 261–62; López Piñero, *Ciencia y técnica*, 50–51; Findlen, *Possessing Nature*, 261–72; Harold J. Cook, "Medicine," in *The Cambridge History of Science*, vol. 3, eds. Katherine Park and Loraine Daston (Cambridge: Cambridge University Press, 2006), 408–16.

21 De Vos, "Art of Pharmacy," 59, 62. See Chapter 2.

what Maher Memarzadeh has called wage pharmacists.[22] These apothecaries possessed the skills to prepare medicines and may even have been examined and possessed title, but lacked the financial resources to establish their own pharmacy. As Chapter 3 has shown the economic standing of *boticarios* would have varied considerably.

Some *boticarios* possessed substantial resources.[23] Francisco de Alva was said to be worth 50,000 *pesos ensayados*, and Mateo Pastor was able to provide goods valued at 2,000 *pesos ensayados* as security for the establishment of the private bank of Juan de la Cueva in 1636.[24] It is important to note that like many individuals at the time, *boticarios* did not earn their income exclusively from their pharmacy business or related trading activities. Juan de Bilbao derived a substantial income from renting out houses and shops. In 1592 he was renting a shop to two tailors for 800 pesos a year and made a contract with the merchant, Pablo de Collaguos, to rent a house for 300 pesos a year.[25] These were only two of the large number of properties he owned. Even though *boticarios* were discouraged from playing cards along with other games and drinking,[26] another *boticario*, Diego de Tineo joint-owned the monopoly for the sale of playing cards (*estanco de naipes*) with the merchant, Juan Ximénez de los Rios.[27] Prominent *boticario*s were often wealthy enough to employ several African slaves in their households. In early colonial Lima household slaves

22 Maher Memarzadeh, "Medical practitioners in Early Colonial Mexico," PhD diss., UCLA, 2005, 340–42, 453–54.

23 *Libros de cabildos de Lima*, 7: 549 Título de familiar al bachiller Alba 21 Jan. 1574, *Libros de cabildos de Lima*, 8: 201 Familiar del santo oficio a Juan de Bilbao 24 Feb. 1576 AGNP Real Audiencia. Causas Civiles leg. 16 cuad. 81 fol. 7v., 687 Pleito…contra las bienes y herederos del bachiller Francisco de Alva, boticario morador en la Ciudad de los Reyes. 1576.

24 AGNP Real Audiencia. Causas Civiles leg. 16 cuad. 81 fols. 119–26, 687–715 Pleito…contra las bienes y herederos del bachiller Francisco de Alva, boticario morador en la Ciudad de los Reyes. 1576; *Libros de cabildos de Lima*, 23: 398 Fianzas del banco 22 Dec. 1636. For the bank of Juan de la Cueva see: Suárez, *Comercio y fraude* and *Desafíos transatlánticos*.

25 For example, AGNP Protocolos Siglo XVI 52 Rodrigo Gómez Baeza fols. 2–2v. Arrienda de Juan de Bilbao 4 Jan. 1592; AGNP Protocolos Siglo XVI 52 Rodrigo Gómez Baeza fols. 17–17v. Arrienda de Juan de Bilbao 8 Jan. 1592. See also AGNP Protocolos Siglo XVI 47 Rodrigo Gómez Baeza fols. 731–731v. Arrienda de Juan de Bilbao 4 May 1588 where he rented a house to a shoemaker for 190 pesos a year. Similarly within ten years of arriving in Lima by 1552 the licenced physician, Álvaro de Torres, had acquired nine houses in central Lima which he sold to the city for 15,400 pesos (Lockhart, *Spanish Peru*, 66).

26 ARJBM División 1, leg. 17 Prohemiales 4 Libro de examen de los simples medicinales Antonio de Robles Cornejo 1617.

27 AGNP Protocolos Siglo XVI 51 Rodrigo Gómez Baeza fol. 285–285v. Compañía entre Diego de Tineo boticario y Juan Ximénez del Rio, mercador, sobre el estanco de naipes 8 Mar. 1591.

cost about 500 to 600 pesos or more, about half a year's salary of a minor royal official such as a *corregidor*.[28] When he died in 1576 Francisco de Alva had six female slaves and two male youths living in his household.[29] Clearly those who owned their own *boticas* made a comfortable living, but *oficiales* eked out a more modest existence on an income of several hundred pesos a year or slightly more in the case those who worked in hospitals.[30]

The social status and networks of these different types of *boticarios* were often distinct, but the standing of all was generally enhanced by their association with the medical profession; it also offered opportunities for social advancement for those of less privileged backgrounds, including migrants from Spain.[31] Those with greater financial resources were better placed to improve their social standing, but there were other mechanisms by which this could be achieved, which included the purchase of offices or titles or the payment of dowries to secure favourable marriages.

It was not uncommon for members of Lima's elite to become knights of military orders through government service or purchase, though they remained a minority of the city's *vecinos*.[32] Nevertheless, writing on early sixteenth century Peru, James Lockhart concluded that physicians generally had few pretentions to nobility and seldom sought public office. He suggests that they often came from families in which medicine was a tradition and they were content to dedicate themselves to their profession.[33]

Lockhart's observation would seem to apply equally to *boticarios*; certainly very few used the title Don. The title of Don or Doña was originally the prerogative of the high nobility and those who held high office in government or the Church. In 1560 there were only about 150 nobles in Peru,[34] but over time the title came to be used by a large number of other people, often relatives of true Dons, while some merely assumed the title feeling they had sufficient power

---

28 Newson and Minchin, *From Capture to Sale*, 225–31.

29 AGNP Real Audiencia. Causas Civiles leg. 16 cuad. 81 fols. 7v–8 Inventario de los bienes del bachiller Alva después de fallecido. 1576. The *boticarios* Juan Matías de Vera and Bernardo Gil both owned four slaves (AAL Testamentos 25-1 fol. 6v. Testamento de Juan Matías de Vera 18 Jun. 1640 (boticario); AAL Testamentos 52-21 fol. 12 Testamento de Bernardo Gil, boticario 1662).

30 See Chapter 3.

31 Memarzadeh, "Medical Practitioners," 347, 349.

32 Lockhart, *Spanish Peru*, 46–47; Bronner, "Elite Formation," 17, provides the figure of 149 *caballeros* (nobles and knights) out of 9–10,000 *vecinos* in mid-seventeenth century Lima.

33 Lockhart, *Spanish Peru*, 65–68.

34 Ibid. 39.

and prestige. In England apothecaries often sought public office to enhance their reputations, but this does not appear to have been the case in Lima.[35]

Even though *boticarios* may not have sought noble status or titles for themselves, they were sufficiently conscious of their social position to pay significant dowries for their daughters to secure husbands of high standing. The amount of dowry paid on marriage reflected the honour of the bride and the economic and social standing of the two families. Dowries could be extremely large leading women in Lima to complain that it was an obstacle to securing a suitable husband. In the seventeenth century dowries among elites were generally between about 10,000 and 40,000 pesos.[36] The *boticario* Pedro de Bilbao provided his daughter with a dowry of 32,000 pesos for her marriage to a lawyer and *relator* of the *Audiencia* Dr Andrés Téllez de Cabrera. Two years later Téllez de Cabrera became *corregidor* of the mining district of Conchucos before dying suddenly in 1634.[37] The post of *corregidor* of Conchucos was worth only 1,000 pesos a year, but it brought status and provided considerable opportunities for enhancing one's income.[38] Daughters of medical practitioners in general, like those of merchants, were commonly married to lawyers or lesser government officials. The dowry paid by Pedro de Bilbao is suggestive of his high economic status, particularly when it is compared to that of 6,000 *pesos de a nueve* paid by the physician Pedro Ramón for the marriage of his daughter Isabel Guzmán to one Antonio del Valle in 1604.[39] However, the dowries paid by even the

---

35 Penelope J. Corfield, "From Poison Peddlers to Civic Worthies: The Reputation of the Apothecaries in Georgian England," *Social History of Medicine* 22(1)(2009): 7–9.

36 Van Deusen, *Between the Scared and Worldly*, 105. For dowries in Peru see: Martín, *Daughters*, 117–27; María Emma Mannarelli, *Private Passions and Public Sins: Men and Women in Seventeenth Century Lima*, trans. Sidney Evans and Meredith D Dodge (Albuquerque: University of New Mexico Press, 2007), 103–05; Jane E. Mangan, *Transatlantic Obligations: Creating the Bonds of Family in Conquest-Era Peru and Spain* (New York, Oxford University Press, 2015), 122–33.

37 *Libros de cabildos de Lima*, 22: 343 Cabildo de Lima receives cédula of appointment of Andrés Téllez de Cabrera as *corregidor* (20 Aug. 1633) on 13 Sep. 1633; Suardo, *Diario de Lima*, 1: 198, 199, 284; 2: 41. He died in 1634 leaving two children as his heirs.

38 Antonio Vázquez de Espinosa, *Compendio y descripción de las Indias occidentales*. Biblioteca de autores españolés 231 (Madrid: Ediciones Atlas, 1969), 507. Salaries for minor officials ranged from 800 to 2,000 pesos (pp. 506–08).

39 AAL Testamentos leg. 7 exp. 18 Testamento de Pedro Ramón, médico 1 Apr.1618. A *peso de a nueve* was worth 306 maravedís compared to the *peso de a ocho* which was 272 maravedís (Bowser, *African Slave*, xiii). It was therefore equivalent to 6,750 pesos. For the marriage see "Perú, matrimonios, 1600–1940," database, *FamilySearch* (https://familysearch.org/ark:/61903/1:1:FNR6-H1M: accessed 21 December 2015), Antonio Del Valle and Ysabel De Gusman Y Sotomayor, 17 Nov 1604; Lima, Peru; FHL microfilm 1,110,202. One Antonio del

wealthiest apothecaries were less than those paid by members of other professional groups, particularly merchants whose daughters might receive a dowry of 60,000 or 80,000 pesos.[40] In the 1630s the largest dowry provided in Peru was by Diego de la Cueva who provided a dowry of 120,000 on the marriage of his daughter to a merchant and hat mill owner, don Antonio Mioño, a Spaniard and knight of the military Order of Santiago.[41]

Criticisms of the Medical Profession

Licensed medical practitioners as a group had a higher status than many other professionals due to their university education, formal training and because their activities were officially regulated. Nevertheless, they were still not held in high esteem. This is exemplified by a letter from Jorge López de Paz, a resident of Arequipa, to a business associate in Lima in which he complained that the medical practitioners attending to him, who had administered syrups, purgatives and other remedies, were "matasanos" (killers of the healthy) who knew no more medicine than an ass.[42] Such feelings towards medical practitioners were widespread and were encapsulated in the late seventeenth century satirical poetry of Juan del Valle y Caviedes. In some fifty poems he condemned professional doctors referring to them as "verdugo en latín" (executioner in Latin), "doctor de la sepultura" (doctor of the grave) and more generally "médicos matantes" (killer doctors).[43] He also levelled charges of corruption against the university for awarding degrees on payment of a fee, suggesting that one doctor Don Melchor Vázquez had paid 3,000 pesos for the title

---

Valle was given the title of silkworker (*sedero*) in 1608, though it is not certain that this was the same person (*Libros de cabildos de Lima*, 15: 708 Cartas de examen de gorreros y sederos 28 Nov. 1608).

40 For dowries provided by merchants see: Suardo, *Diario de Lima*, 1: 45, 162, 164, 2: 43, 99.

41 Ibid. 1: 164; Manuel de Mendiburu, *Diccionario histórico-biográfico del Perú* (Lima: Imprenta de J. Francisco Solís, 1876) 2: 426.

42 AGNP Tribunal de la Inquisición. Contencioso – Siglo XVII leg. 21 Jorge López de Paz a Manuel Bautista Pérez, Arequipa, 11 May 1635.

43 Daniel R. Reedy, *The Poetic Art of Juan Valle Caviedes* (Chapel Hill: University of North Carolina, 1964), 60–79; Uriel García Cáceres, *Juan del Valle y Caviedes: Cronista de la medicina* (Lima: Banco Central de Reserva del Perú and Universidad Peruana Cayetano Heredia, 1999), 55–119 *passim*. Carlos F. Cabanillas Cárdenas, ed., *Juan del Valle y Caviedes: guerras físicas, proezas medicales, hazañas de la ignorancia* (Madrid: Iberoamericana Editorial Vervuert, 2013), 72–84, identifies the physicians and surgeons in the poems with those practising in Lima at the time.

of *médico*.[44] Although fewer criticisms were directed specifically at *boticarios*, they are generally depicted as accomplices of physicians since they prepared the medicines that physicians prescribed. As such Valle y Caviedes refers to the apothecary as an artilleryman – *boticario artillero*.[45] Such public criticism of medical practice was not exceptional for the time, indeed it was commonplace in Europe, where it was similarly satirised by writers such as Rabelais, Molière and Shakespeare.[46]

These universal complaints probably had as much to do with the limitations of humoral medicine as the shortcomings of physicians or apothecaries, though clearly there were many cases where *boticarios* prepared harmful medicines. In fact Antonio de Robles Cornejo considered this to derive from *boticarios*' ignorance of Latin which led them to rely on their own experience rather than published texts, and thereby produce harmful or ineffective remedies that resulted in further treatments and rising costs. This he argued contributed to the poor reputation of the medical profession as a whole.[47] Even when administered according to traditional manuals, medical treatments had unpredictable outcomes. Medical care was therefore approached from a position of suspicion rather than confidence.

It was not easy for a patient to select a medical practitioner or supplier given their differing abilities and the wide range of *materia medica* available of varying cost and quality.[48] The reputation of particular *boticarios* did not rely wholly on their own skills, for they were supposed to follow the prescriptions issued by physicians. However, they also dispensed medicines on their own account to people who sought their advice and treatment without having previously consulted a physician. In both cases, the actual preparation of medicines required knowledge of the particular qualities and quantities of *materia medica* that were to be employed and combined, and necessitated skill and care in composing them.[49]

The "laws" (*leyes y debidas ordenanzas*) specified by Antonio de Robles Cornejo as being essential for a good *boticario* were that he should purchase

44 Cabanillas Cárdenas, *Juan del Valle y Caviedes*, 344–45, 412.

45 Ibid. 227.

46 García Cáceres, *Juan del Valle y Caviedes* 34–35; Tanya Pollard, *Drugs and Theater in Early Modern England* (Oxford University Press: Oxford, 2005), 23–54.

47 ARJBM División 1, leg. 17 Prologo. Libro de examen de los simples medicinales Antonio de Robles Cornejo 1617.

48 Arrow, "Uncertainty," 951 notes that uncertainty about the quality of medicines is more intense than for any other commodity.

49 Michael R. McVaugh, *Medicine before the Plague: Practitioners and their Patients in the Crown of Aragon 1285–1345*. (Cambridge, Cambridge University Press, 1993), 117–18.

good quality simples, prepare compound medicines faithfully according to recipes, and store them correctly so that they did not lose their potency. Other "laws" stressed that they should not overcharge for medicines, noting that sometimes they demanded ten times their value.[50] As the confessional for physicians and surgeons written by Martín de Azpilcueta acknowledged, their sins did not stop at overcharging for medicines but extended to prolonging an illness in order to increase their income or, in collusion with *boticarios*, purchasing more medicines than were needed.[51] Medicine as a practice was a Christian calling and it was expected that medical practitioners' prime interest was in healing patients rather than making profits.[52] In England advice to apothecaries was that they should charge sufficiently high prices for treatments so as to be able them to provide medicines free for the poor.[53] Whether or not this was a consideration for *boticarios* in Lima is not clear, but the popular view was certainly that the prices they charged were excessive.

In an environment where there was a general lack of confidence in the medical profession and where the knowledge and skills of particular physicians or apothecaries was imperfect, the choice of who to turn to for medical care was a hazardous task. Those with reputations for killing or harming patients, overcharging, or whose *boticas* contained old, adulterated products might be avoided.[54] Those aside, the *boticario* to whom an individual might resort would depend on influences other than demonstrated medical expertise or cost; these might include family tradition, knowledge or recommendations from relatives or friends, or merely social standing.[55] Patients might believe that they would receive quality treatment if practitioners possessed official qualifications or were treating people of high status. Bartolomé Díaz Cabeza de Vaca supplied medicines to Diego Núñez de Figueroa, a rich Limeño with an

50 ARJBM División 1, leg. 17 Prohemiales 5, 8, 9, 11,12, 13 Libro de examen de los simples medicinales Antonio de Robles Cornejo 1617. Valle y Caviedes in twelve of his poems made a similar charge against physicians, in one referring to "sentencia de muerte injusta por culpas de mí dinero"(Cabanillas Cárdenas, *Juan del Valle y Caviedes*, 217).

51 Martín de Azpilcueta, *Compendio y summario de confessores y penitentes* (Alcalá: Casa de Iuan Iníguez de Lequerica, 1580).

52 Arrow, "Uncertainty," 949–51.

53 Robert Ralley, "Medical Economies," 29–31.

54 Patrick Wallis, "Consumption, Retailing, and Medicine in Early-Modern London," *Economic History Review* 61(1) (2008): 26, 42–43. For Mexico see De Vos, "Art of Pharmacy," 159–60.

55 See Chapter 6. For England see: Jenner and Wallis, "Medical Marketplace," 12; Ralley, "Medical Economies," 36, 41. For Italy see David Gentilcore's, *Healers and Healing*, 2.

estate worth 200,000 pesos who was appointed governor of Huaylas,[56] as well as a number of royal officials and members of the *cabildo*.[57] Similarly, on his death Pedro de Bilbao was owed some 8,211 pesos by more than 150 individuals, of which a large number held official positions and another 30 possessed the title of Don or Doña.[58] Elites probably continued to seek the advice of professional practitioners despite the debatable efficacy of their humoral treatments and that herbal remedies prepared by popular healers probably inflicted less harm.[59] Indeed, as has been noted in Chapter 3, elite families often made annual contracts with particular physicians and surgeons to treat members of their households.

The Christian Calling of an Apothecary

The status of the medical profession as a whole was enhanced by the fact that it was regarded as a calling by God, within which *boticarios* were seen to have particular responsibilities. According to Ecclesiasticus 38: 4–8, God had created medicinal herbs and given certain people knowledge of how to use them for healing and the relief of pain, which would ensure human health throughout the world and serve to praise God.[60] As such the first of the fourteen "laws" listed by Antonio de Robles Cornejo as being essential for a good *boticario* was that he was to fear and love God with all his heart and comply not only with human laws but also divine ones.[61] The Christian calling of the apothecary was also emphasised in other "laws" which exhorted them to be humane, willing to please, happy, affable, and charitable.[62] According to Robles Cornejo they should not be a player of cards or dice or someone who ate or drank

56 Cook, *Demographic Collapse*, 182–83.

57 AAL Testamentos leg 5 exp. 1 fols. 28–35v. Testamento de Bartolomé Díaz Cabeza de Vaca boticario y autos sobre su cumplimiento con el albacea Juan Manuel Carasco 1608–1626.

58 AGNP Protocolos Siglo XVII 1789 Sánchez Vadillo fols. 2068–2072 Inventario de bienes de Pedro de Bilbao 25 Aug. 1636.

59 Michael Stolberg, *Experiencing Illness and the Sick Body in Early Modern Europe* (Basingstoke: Palgrave Macmillan, 2011), 62–63.

60 Suárez de Figueroa, *Plaza universal*, 300. Different versions of the bible give different translations, but the essence of the verse is the same.

61 ARJBM División 1, leg. 17 Prohemiales 1 Libro de examen de los simples medicinales Antonio de Robles Cornejo 1617.

62 ARJBM División 1, leg. 17 Prohemiales 3 Libro de examen de los simples medicinales Antonio de Robles Cornejo 1617. See also Suárez de Figueroa, *Plaza universal*, 302; Aguilera, *Exposición*, fols. 17v., 21.

excessively.[63] Such behaviour it was argued would lead to "delays, mistakes, blunders, and oversights."[64] *Boticarios* were also exhorted to marry a woman of good blood (*limpia de buena sangre*) or at least an Old Christian of good conduct and reputation.[65]

Given these publically recognised qualities of a good *boticario*, an apothecary might enhance his status by outward expressions of piety and charity, which were seen as a fundamental Christian responsibility.[66] Religion permeated the lives of people in the early modern period, especially in Lima which was capital of the Viceroyalty and the seat of one of the only three archdioceses in Spanish America. Apart from the secular Church, the city housed a tribunal of the Inquisition and possessed monasteries and convents of many religious orders; more than ten percent of the population was associated with the Church.[67] *Cofradías* abounded, religious buildings dominated public space, and religious ceremonies were central to public life in the city. Meanwhile, private houses possessed religious objects, paintings and books.[68]

Piety was publically displayed through membership of *cofradías*, participation in religious ceremonies, and in charitable works. The concept of purgatory featured strongly in religious thought at the time encouraging people to prepare for a "good death." This included provisions in their testaments for the saying of masses for the soul of the deceased and for those in purgatory, or burial in the habit of a particular religious order and the provision of bread, wine and candles.[69] They might also specify donations that were to be made to the church, hospitals, *cofradías*, and other charitable institutions. The desire to demonstrate religiosity and piety was not unique to *boticarios* or the medical profession, but it prevailed among the population at large. However, the particular responsibility of *boticarios* before God for the healing of the sick probably

63 ARJBM División 1, leg. 17 Prohemiales 4 Libro de examen de los simples medicinales Antonio de Robles Cornejo 1617. See also, Aguilera, *Exposición*, fols. 23v–24.

64 Suárez de Figueroa, *Plaza universal*, 302.

65 ARJBM División 1, leg. 17 Prohemiales 6 Libro de examen de los simples medicinales Antonio de Robles Cornejo 1617.

66 Siraisi, *Medieval and Early Renaissance Medicine*, 7–9; Lindemann, *Medicine and Society*, 123–29; Risse, *Mending Bodies*, 73–109; Sowell, *Tale of Healer*, 14–15.

67 Nancy E. Van Deusen, *The Souls of Purgatory: The Spiritual Diary of a Seventeenth-Century Afro-Peruvian Mystic, Ursula de Jesús* (Albuquerque: University of New Mexico Press, 2004), 7.

68 Ibid. 1–49 for insight into the religious life of Lima and its inhabitants.

69 Van Deusen, *Souls of Purgatory*, 36–37; Gabriela Ramos, *Death and Conversion in the Andes: Lima and Cuzco, 1532–1670* (Notre Dame, Indiana: University of Indiana Press, 2010),148–49: Cussen, *Black Saint*, 86–91.

raised public expectations of their actions. It was said that Limeños expended more money on white wax for convents, *cofradías*, masses in one month than the largest cities in Europe did in eight. Antonio de la Calancha described it as the most charitable city in all Christendom.[70]

Boticarios demonstrated their religiosity and social standing through the membership of *cofradías* and participation in public processions. A large number of *cofradías* were founded in Lima, often for separate ethnic and professional groups. In 1619 the city possessed forty-six *cofradías* and there were few people who were not associated in some way with one of them.[71] Certain *cofradías* were considered to be more prestigious than others. Their status depended on their wealth, the Christian devotion of their members publically demonstrated in processions, and on their acts of charity. Individual members might enhance their piety, honour, and status through occupying posts in the *cofradía*, for example as *mayordomo* or *hermanos venticuatro*.[72] At the outset the *hermanos veinticuatro* were the original founders of a *cofradía*. They were responsible for organising its activities, including processions, and they enjoyed certain privileges. Following the death of the original founders they were replaced by new *hermanos venticuatro*, but the privileges they enjoyed were more limited.[73] These posts were always filled by members of the elite.

In the 1570s the *boticario* Francisco de Alva was a *hermano venticuatro* of an unnamed *cofradía*.[74] This was before the foundation in 1603 of the *cofradía* of Nuestra Señora de la Soledad which was particularly associated with medical

70 Calancha, *Crónica moralizada*, lib. 1 cap. 38: 245.

71 Olinda Celestino and Albert Meyers, *Las cofradías en el Perú: región central* (Frankfurt: Editorial Klaus d. Vervuert, 1981), 119–22 and cuadro 6; Walter Vega Jácome, "Cofradías limeñas," in *Lima en el siglo XIV*, ed. Laura Gutiérrez Arbulú (Lima: Pontificia Universidad Católica del Perú, Instituto Riva-Agüero, 2005), 706–07, 731–33. For Indian *cofradías* see: Paul Charney, "A Sense of Belonging: Colonial Indian Cofradías and Ethnicity in the Valley of Lima, Peru," *The Americas* 54(3) (1998): 379–407.

72 Judith Mansilla Justo, "Poder y prestigio social en las cofradías españolas, siglo XVII y XVIII," in *Corporaciones religiosas y evangelización en Iberoamérica, siglos XVI–XVII*, eds. Diego Lévano Medina and Kelly Montoya Estrada (Lima: Universidad Nacional Mayor San Marcos, 2010), 229–32. See Beatriz Garland Ponce, "Las cofradías en Lima durante la colonia: una primera aproximación," in *La venida del reino. Religión, evangelización y cultura en América, Siglos XVI–XX*, ed. Gabriela Ramos (Cuzco: Centro de Estudios Regionales Andinos "Bartolomé de las Casas," 1994), 212–26 for the organisational structure of *cofradías* in Lima.

73 Garland Ponce, "Las cofradías," 217–18.

74 AGNP Real Audiencia leg.16 cuad. 81 fol. 8v. Inventario de los bienes del bachiller Alva después de fallecido 9 Feb. 1576. Among his clothes were a cape and tunic of a *venticuatro* signifying that he was a member of a *cofradía*.

practitioners. The *cofradías's* patron saints were Cosmas and Damián, who were the saints of physicians, pharmacists and surgeons. The *cofradía* de Nuestra Señora de la Soledad was attached to the convent of San Francisco (See Figure 13); its chapel had five altars and was embellished with expensive sculptures. It became one of the most prestigious *cofradías* in the city, playing a leading role in religious celebrations especially during Holy Week, when its members displayed their religious fervour by re-enacting the crucifixion of Christ and flagellating themselves throughout the *cofradía's* procession (See Figure 15).[75] Members of the *cofradía* had to be honourable and of good reputation, while

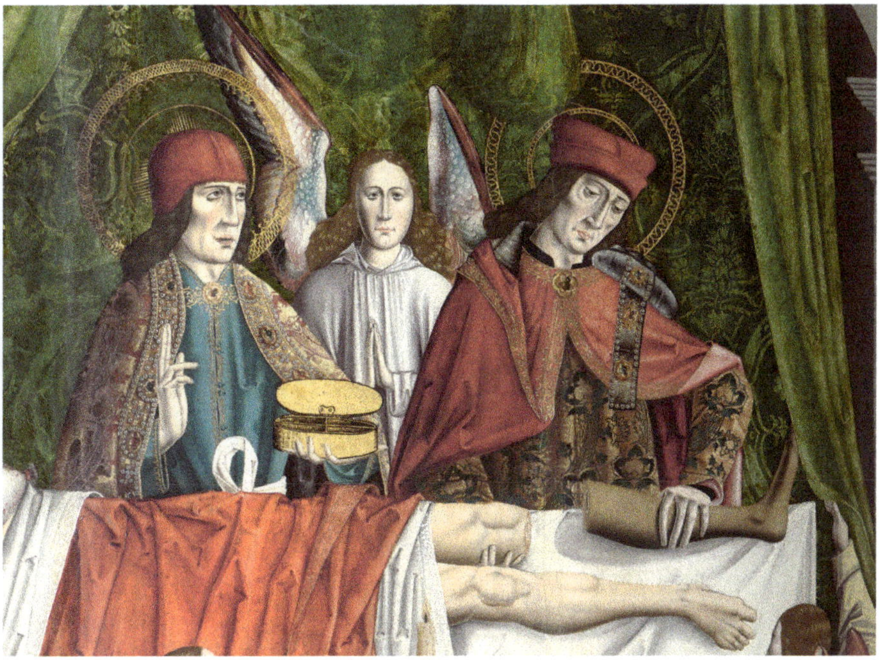

FIGURE 13 *A verger's dream: Saints Cosmas and Damian performing a miraculous cure by transplantation of a leg. Oil painting attributed to the Master of Los Balbases, ca. 1495.*
COURTESY WELLCOME LIBRARY, LONDON.

75 Diego de Córdoba y Salinas, *Crónica franciscana de las provincias del Perú*, ed. Lino Gómez Canedo (Washington DC: American Academy of Franciscan History, 1957), 536–57; Suardo, *Diario de Lima*, 2: 166; Rubén Vargas Ugarte, *Historia de la iglesia en el Perú*. Tomo II (1570–1640) (Burgos: Imprenta de Aldecoa, 1959), 496–97; Kelly Montoya Estrada, "Una procesión de Viernes Santo en Lima XVII," in *Corporaciones religiosas y evangelización en Iberoamérica, siglos XVI–XVII*, eds. Diego Lévano Medina and Kelly Montoya Estrada (Lima: Universidad Nacional Mayor San Marcos, 2010), 145–46.

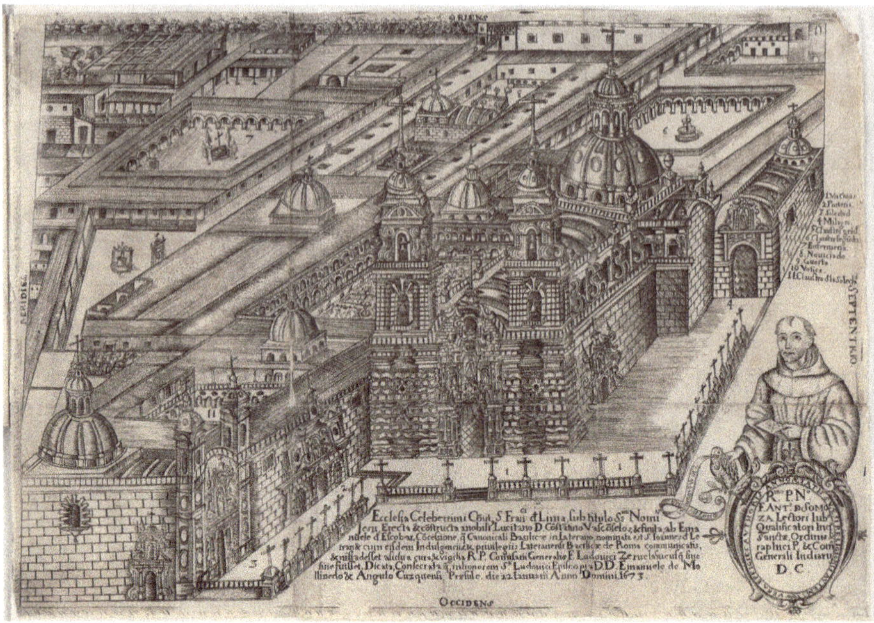

FIGURE 14 *Chapel of the* Cofradía *of Nuestra Señora de la Soledad* (bottom left) *in 1675. Miguel Suárez de Figueroa,* Templo de N. grande patriarca San Francisco de la provincia de los doze apostoles de le Peru en la Ciudad de los Reyes. *Lima, Juan de Benavides, 1675.*
COURTESY OF THE JOHN CARTER BROWN LIBRARY AT BROWN UNIVERSITY.

Blacks, Mulattoes, Mestizos, and Moriscos were, at least in theory, excluded.[76] The obligations of its members were contained in their "carta de esclavitud," which specified the payment of alms, contributions to the *cofradía*, as well as attendance at church services and processions. It also indicated the form in which their religiosity and commitment to the *cofradía* were to be expressed.[77]

Among the founding members of the *cofradía* de Nuestra Señora de la Soledad were a number of known physicians and *boticarios*. They included Francisco Martín Reyna who in 1604 was one of its two *mayordomos*; he was probably

76 Montoya Estrada, "Procesión de Viernes Santo," 147. Montoya Estrada suggests that for the eighteenth century at least, persons from other ethnic groups were buried in the *cofradía's* church.

77 See for example: AAL Cofradías 16 no. 15 Autos seguidos por Juan de Gutiérrez de la Sal contra Pedro Fernandez Valdés 1684. This includes the *carta de esclavitud* of the wife of Gutiérrez de la Sal, Doña Juana Picón.

FIGURE 15 *Holy week procession of the* Cofradía *of Nuestra Señora de la Soledad in the late seventeenth century.*
The procession leaving the chapel of the Cofradía with the re-enactment of the crucifixion on the right.
PHOTO: KELLY MONTOYA. COURTESY COFRADÍA DE NUESTRA SEÑORA DE LA SOLEDAD, LIMA.

one of its original founders.[78] By 1637 some of the original *hermanos venticuatro* had died, including Francisco Martín Reyna, but among those who remained were known *boticarios* including Bernardo Gil, Francisco de Sandoval, Alonso de Carrión, and Juan Matías de Vera; one of its two *mayordomos* was no less than the *protomedico*, Doctor Juan de la Vega.[79] At that time the *mayordomos* and *hermanos venticuatro* were complaining that the *cofradía* was suffering from an influx of "surgeons, barbers and apothecaries," many of whom were not paying the specified contributions. Not only that, but these new members, who outnumbered the original founders, were electing *mayordomos* from among the guild of surgeons and barbers to detriment of the original founders, who it claimed had committed significant funds to the establishment of the *cofradia*.[80] Apothecaries did not belong to this guild and the documentation suggests that the complaint was essentially against surgeons who were advocating that one of the posts of *mayordomo* should be reserved for them. The association of *cofradías* with particular professions or trades was commonplace in Lima; it served to guarantee support at a time when the number of *cofradías* was expanding rapidly and competition between them was increasing.[81]

Apothecaries might also enhance their social standing through charitable activities. Christian charity given free of charge would signify that an apothecary was a caring person concerned with the relief of suffering and not driven by monetary considerations. Apart from regular alms giving, *boticarios* might make substantial donations to the church or charitable institutions, particularly on their deathbeds. The *boticario* Bernardo Gil expressed his desire to be buried in the chapel of the *cofradía* of Nuestra Señora de la Soledad to which he paid alms.[82] Whether or not they belonged to the *cofradía* of Nuestra

78 AGNP Protocolos Siglo XVII 230 Ramiro Bote fols. 1473–1473v. Escritura de Francisco Martín de Reyna 23 Mar. 1604; *Libros de cabildos de Lima*, 14: 713–14 Piden los mayordomos de la cofradía de la Soledad ...vayan los regidores a la regir 12 Apr. 1604. He is mentioned as a *mayordomo* in AAL Testamentos leg 5 exp. 1 fols. 137–45 Testamento de Bartolomé Díaz Cabeza de Vaca 28 Apr. 1608. In 1604 the other *mayordomo* was Hernán Sánchez.

79 He participated in the *cofradía's* Holy Week procession in 1637 (Suardo, *Diario de Lima*, 2: 166). He died in 1640 (AAL Testamentos 25 no.1 Testamento de Juan Matías de Vera, 1641–1642).

80 AAL Cofradías 16 no. 3 Autos seguidos por los hermanos de venticuatro de la cofradía de Nuestra Señora de la Soledad 1637; AAL Cofradías 16 no. 5 fols. 5–6v., 13–13v. Autos de elección de nuevos mayordomos de la cofradía de Nuestra Señora de la Soledad 1643.

81 Diego Lévano Medina, "La administración de los bienes temporales en las cofradías limeñas del siglo XVII," in *Corporaciones religiosas y evangelización en Iberoamérica, siglos XVI–XVII*, eds. Diego Lévano Medina and Kelly Montoya Estrada (Lima: Universidad Nacional Mayor San Marcos, 2010), 120–22.

82 AAL Testamentos 52–21 fol. 6 Testamento de Bernardo Gil, boticario 1662.

Señora de la Soledad, members of the medical profession often affiliated with the church of San Francisco. The *boticario* Bartolomé Díaz de Cabeza de Vaca, indicated his desire to be buried in a Franciscan habit in the convent de San Francisco where a requiem mass was to be said; he also requested one hundred masses to be said for his soul and another fifty for his ancestors and those who had helped him during his life. Any residual estate was to be used to found a *capellanía* for the saying of masses in the convent of San Francisco.[83]

Capellanías or chaplaincies were capital endowments to a church, generally entrusted to a particular priest, for the saying of masses for the donor's soul or to carry out charitable works. Those founded by elites in Lima reached several thousand pesos.[84] The *boticario* Juan de Urrutia, founded two *capellanías* of unknown value, one to support "the chapel for souls in purgatory" in the Cathedral and the other for the chapel of the Hospital de Nuestra Señora de Atocha for orphan children.[85] More often death-bed donations were much smaller and were pledged for the saying or masses or given to charitable institutions without the establishment of a *capellanía*. In 1618 the physician, Pedro Ramón, who was *familiar* to the Inquisition left 20 pesos to the *cofradía* del Santísimo Sacramento attached to the Cathedral, 100 pesos to the Hospital of San Andrés, 50 pesos to the Hospital of Santa Ana, 50 pesos to the Hospital of La Caridad for the poor, 100 pesos to the female Convent of Las Descalças (San José), and 8 pesos for orphan children.[86] Such donations were not confined to elites and could be quite small. Examples include an Indian tailor from Chachapoyas who in 1578 donated two pesos to the Hospital of Santa Ana and another two pesos to the *cofradía* de Nuestra Señora del Rosario for Indians, while in 1597 a free Black carpenter left 4 pesos to each of the city's four hospitals.[87]

---

83 AAL Testamentos leg. 5 exp. 1 fols. 12–12v., 20 Testamento de Bartolomé Díaz Cabeza de Vaca 15 Apr. 1608. The surgeon Martín Sánchez, similarly specified that he wished to be buried in a Franciscan habit in the convent of San Francisco and that masses be said in the cathedral for his soul and those of his deceased ancestors (AGNP Protocolos Siglo XVI 154 Pedro de Salinas fols. 258–258v., 721v–722 no date [1547/1548]).

84 Josephe and Francisco Mugaburu, *Chronicle of Colonial Lima: The Diary of Josephe and Francisco Mugaburu, 1640–1697*, trans. and ed. Robert R Miller (Norman: University of Oklahoma Press, 1975), 8, 91, 235.

85 AAL Monasterio de la Concepción leg. 4 exp. 8 Causa seguida por Alonso de Castro en nombre de doña Maria Magdalena, viuda de Gerónimo Pujadas, contra el monasterio de la Concepción 1628.

86 AAL Testamentos leg. 7 exp. 18 Testamento de Pedro Ramón médico 1 Apr. 1618.

87 Quiroz, *Artesanos y manfactureros*, 116–17. For small donations by Indians see: Paul Charney, "'For My Necessities': The Wills of Andean Commoners and Nobles in the Valley of Lima, 1597–1607," *Ethnohistory* 59 (2)(2012): 344.

Despite the large number of people making donations, the financial contribution they made collectively to the running of hospitals was small; most income came from *censos* and properties. Those who lacked the ready cash to make a donation might impose a *censo* on a property. This pledged to the institution an annual income usually set at 5 percent of its capital value. When hospitals were first founded they often received tribute income from *encomiendas*, but as the native population declined they relied increasingly on *censos* and on the rental income from properties that had been donated or acquired; donations made a relatively small contribution.[88] Whereas in 1619 donations to the Hospitals of Santa Ana and San Andrés were 500 pesos and 2,500 pesos respectively, their corresponding income from rents and *censos* were 12,179 pesos and 8,500 pesos.[89] The *boticario* Juan de Bilbao and his wife had a *censo* on four pairs of houses (*pares de casas*) worth 1,558 *pesos ensayados* which generated an annual income of 111 pesos 3 reals for the Hospital of Santa Ana.[90] Over time, *censos* and rents from properties, some of which were donated, came to constitute the main source of income not only for hospitals but also *cofradías*.[91]

Apart from death-bed donations, there were other ways in which charity might be dispensed. As Chapter 3 has demonstrated, *boticarios* often allowed debts to build up, especially those incurred by hospitals and other charitable institutions. In the case of Mexico City, the hesitance of apothecaries to

88 David Cahill, "Financing Health Care in the Viceroyalty of Peru: the Hospitals of Lima in the Late Colonial Period." *The Americas* 52 (2) (1995): 136–54. In fact the growth in the native population towards the end of the eighteenth century meant that tribute income was increasing, but it still represented a minor source of hospital income. Even in the sixteenth century *censos* contributed about a quarter of the income of the hospital of San Andrés (AGI Lima 126 El hospital real de los españoles de la advocación de San Andrés, fols. 9–11 Posesiones, censos y situaciones que tiene el dicho hospital 4 Apr.1582). This may not have included all income, since by 1630 the income of the hospital was recorded as 18,000 ducados. This compared with 24,000 pesos for the hospital of Santa Ana and 6,000 for La Caridad. The most substantial income was of 28,000 for the convents of La Concepción and 16,000 for La Encarnación (Salinas y Córdova, *Memorial*, 196–97). Details of all the income from censos, renting properties and alms from 1575 to the mid seventeenth century can be found in the account books of the Hospital of Santa Ana in the ABPL legajos 9080, 9083–87, 9095–96, 9099–9100.

89 AGI Lima 301 Relación de las ciudades, villas y lugares…20 Apr. 1619. For the sources of income of each hospital in Lima see Guerra, *El hospital*, 429–75 *passim*. The *mayordomo* of the hospital was responsible for setting the value of rents of properties belonging to the hospital and collecting the income (Domingo Angulo, "Las ordenanzas del hospital de Santa Ana." *Revista del Archivo Nacional del Perú* 11 (1938): 144).

90 ABPL 9080 Libro de cuentas del hospital de Santa Ana 1575 a 1585 fols. 45, 123 1583.

91 Lévano Medina, "Administración de los bienes temporales," 128–33.

call in debts has been interpreted as a strategy for enhancing their charitable reputations.[92] The evidence for Lima is more equivocal. While some *boticarios* may have acted in a similar way, there is more evidence for their pursuit of debtors through the courts and for specifying in their wills that monies that they were owed should be collected by their executors.

Apothecaries might also demonstrate their religiosity through objects contained in their pharmacies and homes. Calvaries were commonly displayed in *boticas*,[93] though such displays were not restricted to apothecaries. Francisco de Alva possessed eight paintings, six depicting the history of Samson, one Descent from the Cross and one of Mary Magdalene, as well as eight old canvases depicting the history of Abraham.[94] Even surgeons and barbers of more modest means possessed one or two paintings, almost always depicting the saints or religious scenes.[95] One surgeon, Bartolomé Ortega, had ten canvases, which included paintings of Christ, San Juan Evangelista, Santa Lugarda, and a large painting of San Pedro which was auctioned for 41 pesos.[96] Such religious items may have been displayed more prominently by apothecaries of Portuguese descent and *Converso* status who were at greater risk of being investigated by the Inquisition and charged with Judaizing. It may be significant that when goods belonging to Andres Gonzales, a prominent Portuguese trader in medicines turned *boticario* based in Cartagena, were seized by the Inquisition in 1620, he had four paintings of saints in his possession.[97] However, religious paintings were not exclusive to medical practitioners, but were prominent in elite households in general and were highly valued.[98]

---

92 De Vos, "Art of Pharmacy," 171–81.

93 Bartolomé Díaz Cabeza de Vaca's accounts indicate that he commissioned a calvary for the pharmacy (AAL Testamentos leg. 5 exp. 1 fol. 14 Testamento de Bartolomé Díaz Cabeza de Vaca boticario y autos sobre su cumplimiento con el albacea Juan Manuel Carasco 1608–1626).

94 AGNP Real Audiencia leg. 16 cuad. 81 Inventario de los bienes del bachiller Alva después de fallecido 9 Feb. 1576. The presence of images of Old Testament figures suggests that he might have been a *Converso*, but there is no other evidence that this was the case.

95 AAL Testamentos leg. 15 exp. 13 Testamento de Domingo Pérez Pereira cirujano 16 May 1630; María Dolores Crespo Rodríguez, *Arquitectura doméstica de la Ciudad de los Reyes (1535–1750)* (Seville: Consejo Superior de Investigaciones Científicas, 2006), 353 Testament of Juan Ramírez, barber-surgeon 14 Aug. 1629; AAL Testamentos 46-5 fols. 36–37, 101v.–102 Testamento de Bartolomé de Ortega 1658.

96 AAL Testamentos 46-5 fols. 36–37, 101v.–102 Testamento de Bartolomé de Ortega 1658.

97 AGI Escribanía de Cámara 589B Causa contra Andrés Gonzales, boticario, portugués 1620.

98 Crespo Rodríguez, *Arquitectura doméstica,* 329–37.

Projecting Professionalism

A *boticario* could project his professionalism and social standing to the public through his personal attire. The Spanish Crown and residents of Lima alike were aware that clothing made important statements about status and identity.[99] From the fifteenth century, the Spanish Crown introduced sumptuary laws that restricted the wearing of luxury clothing, bearing of arms, and riding of horses to particular groups. Such laws reflected not only the desire to distinguish social groups, especially the elite from the popular classes, but at the same time embodied a distaste for luxury expressed by moralists and theologians.[100] In Spain, Philip II issued sumptuary laws eight times between 1563 and 1594, and Philip III and Philip IV issued them four times each.[101]

In the Americas, sumptuary laws acquired new dimensions. These laws were guided by concerns to maintain the superior status of Spaniards as conquerors and "civilised" persons while incorporating into the social hierarchy new ethnic groups who were at the same time subjects of the Crown, but were associated with illegitimacy and slavery.[102] In 1571 a law was passed for the Indies, that on pain of confiscation, no Black, whether slave or free, or Mulatto could wear gold, pearls or silk, except if they were married to a Spaniard in which case they could wear some gold earrings with pearls and a necklace, and a dress with velvet trimmings, but not a mantle of crêpe or another cloth, only a small mantilla.[103] Neither were Blacks, Mulattoes or *Zambos* allowed to carry arms, though Mestizos could do so with a licence.[104] As with much colonial legislation, laws followed practice. By then the municipality in Lima had already banned African slaves from wearing a cape and sword twice in 1539

99 Karen B. Graubert, *With Our Labor and Our Sweat: Indigenous Women and the Formation of Colonial Society in Peru, 1550–1700* (Stanford: Stanford University Press, 2007), 121–57.

100 Saúl Martínez Bermejo, "Beyond Luxury: Sumptuary Legislation in 17th-Century Castile," in *Making, Using and Resisting the Law in European History*, eds. Günther Lottes, Eero Medijainen, and Jon Vidar Sigurdsson (Pisa: Plus-Pisa University Press, 2008), 93–108.

101 Martínez Bermejo, "Beyond Luxury," 97. For Spanish legislation relating to clothing issued by Philip II see: *Recopilación de las leyes destos reynos, hecha por mandado de la Magestad Catholica del Rey don Philippe Segundo nuestro Señor* (Alcalá de Henares: Casa de Iuan Iñiguez de Lequerica, 1598), lib. 7 tit.12 ley 1: 117–24. For details of dress in Spain in the early sixteenth century see Ruth Matilda Anderson, *Hispanic Costume 1480–1530* (New York: Hispanic Society of America, 1979).

102 McAlister, "Social Structure," 350–55; Mörner, *Race Mixture*, 41–45, 51–68; Cope, *Limits of Racial Domination*, 16.

103 *Recopilación de las leyes de las indias*, lib. 7 tit. 5 ley 28: 290v. 12 Feb. 1571.

104 Ibid. lib. 7 tit. 5 leyes 14–18: 287–88 12 Feb. 1571.

FIGURE 16
Elite formal dress in Lima in the seventeenth century. Extract from painting of the marriage of captain Martín de Loyola to Beatriz Ñusta, ca. 1675–1690. Elite male formal dress was similar to that of Don Juan Henríquez de Borja, who is dressed in black with a ruff and wearing trousers typical of the sixteenth century. He also wears a cape with the emblem of the order of Santiago.
COURTESY MUSEO PEDRO DE OSMA, LIMA.

and 1549.[105] Control over the wearing of particular clothes in Lima was also exercised indirectly through the guilds. Tailors were required to use and sell only silk and woollen cloth, while Blacks, whether slave or free, could be employed in workshops, but could not be masters.[106] The city's markets for imported and locally produced cloth and clothing remained geographically separate. The former operated on the Calle de Mercaderes, where there were some twenty stores (*almacenes*) and two hundred shops, while locally-produced cloth was sold on the Calle de Mantas.[107]

Influences over clothing in Lima were diverse. Legislation emanating from the Counter-Reformation and the Council in Lima in the late sixteenth and early seventeenth centuries generally discouraged the wearing of luxury clothing.[108] (Figure 16).

---

105 *Libros de cabildos de Lima*, 1: 287 8 Jan. 1539 and Libros de cabildos de Lima, 3: 56 21 Jan 1549.

106 *Libros de cabildos de Lima*, 4: 587–94 Ordenanzas de sastres confirmadas 5 Mar. 1557; Ricardo Cantuarias Acosta, "Las modas limeñas," in *Lima en el siglo XIV*, ed. Laura Gutiérrez Arbulú (Lima: Pontificia Universidad Católica del Perú, Instituto Riva-Agüero, 2005), 296–307 *passim*.

107 Salinas y Córdova, *Memorial*, 255–56.

108 Vargas Ugarte, *Concilios limenses*, 1: 131, 332, 243, 251.

This along with luxury furnishings and the use of silverware was viewed by Licenciado Martel Santoyo, lawyer to the *cabildo* in Lima, as provoking covetousness and discouraging Christianisation.[109] This objection to luxury clothing was reinforced by the personal preference of the Spanish monarchs for sobriety and the desire to protect Spain's textile industry from competition from exotic foreign imports.[110] In Lima an additional deterrent was the high cost of luxury clothing. This arose in part from high transport costs and *almojarifazgo* taxes on imports. The best quality brocade (*brocados de tres altos*) was taxed at 20 pesos a yard, velvet breeches at 40 pesos, and an outer skirt of taffeta at 15 pesos.[111] The scarcity of imports also served to maintain the high demand and both the monetary and non-monetary value of luxury clothing from Europe.[112]

Since imported cloth and clothing were expensive in Lima, from an early date a brisk trade in secondhand clothes emerged.[113] When the *boticario* Bartolomé Díaz Cabeza de Vaca died, his clothing, to be detailed below, most of which was worn, was auctioned and netted some 200 pesos.[114] The items of clothing which commanded the highest prices were those fashioned from luxury cloths imported from Spain, such as a cape of black cloth from Segovia which sold for 21 pesos and two pairs of black velvet collars from Castile, one worn and the other new, together valued at 25 pesos. It is significant that many of these items of used clothing were purchased by his business partner, the *boticario*, Francisco Martín Reyna, suggesting that the second-hand clothes market, particularly for high-priced European goods, was not limited to the popular classes. In fact the market in secondhand clothing was so brisk that it threatened the local textile industry and persuaded the *cabildo* in 1551 to try

109 AGI Patronato 185 no. 31 Relación que hace a su Magestad el Licenciado Martel Santoyo sobre lo que debe proveer y remedian en los reynos del Perú y en otras partes 1542 transcribed and published by Emilio Lissón Chaves, in *La iglesia de España en el Perú: colección de documentos para la historia de la iglesia en el Perú* (Sevilla: Editorial Católica Española, 1943), vol. 1, Part 3, 99–120.

110 Aileen Ribeiro, "Dress in the Early Modern Period, c.1500–1780," in *The Cambridge History of Western Textiles*, ed. David Jenkins (Cambridge: Cambridge University Press, 2003), 665, 672–73; Scarlett O'Phelan, "El vestido como identidad étnica e indicador social de una cultura material," in *El barroco peruano* II, ed. Ramón Mujica Pinilla (Lima: Banco de Crédito del Perú, 2003), 103.

111 AGNP Caja reales: Lima H-3 leg. 4 lib. 24-a fols. 1–30 Aranceles para cálculo del almojarifazgo (1617).

112 Velvet and taffeta were particularly esteemed (Enrique Otte, *Cartas privadas de emigrantes a Indias, 1540–1616* (Mexico: Fondo de Cultura Económica, 1993), 388, 413).

113 *Libros de cabildos de Lima*, 4: 354 Ropa vieja, corredor de tabla 29 Nov. 1555.

114 AAL Testamentos leg. 5 exp. 1 fols. 38–41 Testamento de Bartolomé Díaz Cabeza de Vaca 28 Apr. 1608.

and ban it completely. However, this proved ineffective and two years later it permitted two licensed secondhand dealers to operate.[115]

After 1570 when the Spanish established a permanent presence in the Philippines, cheap silk, satin and damask from China via the Philippines and Mexico made luxury clothing more widely available in Peru. Attempts to restrict this trade proved fruitless.[116] Hence, in 1602 the Mexican Viceroy, the Conde de Monterrey, with an interest in promoting trade between Mexico and Peru reported to the Crown that there [in Peru],

> All these people live very luxuriously. All wear silk, and of the most fine and costly quality. The gala dresses and clothes of the women are so many and so excessive, that in no other kingdom of the world are found such; so that if four merchant-vessels went to Peru annually, all the cloth goods would be sold…[117]

In the early seventeenth century the trade in cloth and clothing with Mexico and China was valued at more than one million pesos, while that from Castile was estimated to be only 5,000 to 6,000 pesos.[118] It was suggested that a *regidor* be commissioned to destroy any Chinese cloth, but this was strongly opposed by Lima's citizens, who added that much of it was used to adorn churches and convents. Chinese fabrics were cheaper than those imported from Europe and were therefore worn mainly by the popular classes, while Spaniards wishing to demonstrate their superior economic and social position preferred clothing from Europe.[119] An anonymous observer of Creole women on the coast of Peru at the beginning of the seventeenth century wrote:

> They are elegantly and expensively dressed; they all generally wear silk and very exquisite cloth and velvet with fine gold and silver. They have thick gold necklaces, strings of pearls, rings, necklaces and headbands of diamonds, rubies, emeralds and amethyst and other valuable and esteemed stones.

115 *Libros de cabildos de Lima*, 3: 460 [Ropavejeros] 16 Oct. 1551, *Libros de cabildos de Lima*, 4: 25 Ropavejeros que haya dos no más 19 May 1553; Cantuarias Acosta, "Modas limeñas," 292.

116 *Libros de cabildos de Lima*, 13: 267–68 Ropa de China 25 May 1599; William Lyle Schurz, "Mexico, Peru and the Manila Galleon," *Hispanic American Historical Review* 1 (1918): 394–402; Borah, *Early Colonial Trade*, 116–27.

117 Emma H. Blair and James A. Robertson, *The Philippine Islands, 1493–1803* (Cleveland, OH: A.H. Clark, 1911), 12: 63 Conde de Monterrey 15 May 1602; Schurz, "Mexico, Peru and the Manila Galleon," 396–97.

118 Salinas y Córdova *Memorial*, 256.

119 Lewin, *Descripción*, 115; *Libros de cabildos de Lima*, 13: 267–71 Ropa de China 25 May 1599.

> While the women are beautiful and graceful, the men are elegant and gallant. They generally all wear good quality clothes of silk and fine cloths from Segovia and collars with costly lace from Flanders. They all wear stockings of silk, are discrete, affable and well bred…they all go through the city on horseback.[120]

The wardrobes belonging to two *boticarios*, Francisco de Alva and Bartolomé Díaz Cabeza de Vaca, exemplify many of these observations, in particular their preference for sober European clothing.[121] While their capes, tunics, doublets, and breeches were generally black, they were often made of satin or taffeta and lined with damask, trimmed with velvet or embroidered with silver thread. The latter also wore black velvet collars. It is also noteworthy that their wardrobes contained little Chinese cloth or clothing and when their Spanish clothes were auctioned following their deaths these commanded high prices. Francisco de Alva also possessed two harquebuses, as well as a sword with a velvet sheath and a dagger with a golden handle. Alva was *familiar* to the Inquisition which gave him licence to wear both offensive and defensive weapons in public and private both day and night and throughout its jurisdiction.[122] His higher status was also evident in his possession of a cape and tunic that signified his position as a *hermano veinticuatro* of a *cofradia*. Not all *boticarios* would have possessed such fine clothing. Wage pharmacists and those who worked in hospitals, who were probably the majority, were of more modest means and in most cases did not write wills.[123] What can be said is that the clothing of surgeons was significantly poorer than that owners of *boticas*. The wardrobe of the surgeon, Domingo Pérez de Pereyra, included a ruff which had become fashionable at the time of his testament in 1637, but few of his clothes were made of luxury fabrics; in fact one set consisted of a suit (*vestido*), waistcoat and long trousers made of "Indian cloth."[124]

---

120 Lewin, *Descripción*, 39. For similar contemporary observations see Fray Diego de Ocaña, *Memoria viva de una tierra de olvido: Relación del viaje al Nuevo Mundo de 1599 a 1607* (Barcelona: CECAL, 2013), 290–91.

121 AGNP Real Audiencia. Causas Civiles leg. 16 cuad. 81 fols. 8v–10 Pleito de ejecución…contra las bienes y herederos del bachiller Francisco de Alva, boticario morador en la Ciudad de los Reyes, 13 Feb. 1576; AAL Testamentos leg. 5 exp. 1 fols. 25–26 Testamento de Bartolomé Díaz Cabeza de Vaca 17 Apr.1608.

122 *Libros de cabildos de Lima*, 8: 202 Familiar del santo oficio a Juan de Bilbao 15 Feb. 1576. This refers to the appointment of Juan de Bilbao to replace Francisco de Alva as familiar to the Inquisition.

123 De Vos, "Art of Pharmacy," 72.

124 AAL Testamentos leg. 15 exp. 13 Inventario de Domingo Pérez de Pereyra 16 May 1637; Ribeiro, "Dress in the Early Modern Period," 662.

Social distances between different ethnic groups in Lima may not have been that discernible from their dress for there were constant complaints that Blacks and those of mixed ethnic origins were wearing luxury clothing and carrying swords and daggers, which resulted in proclamations forbidding these practices to be issued with increasing frequency in the seventeenth century.[125] In 1629 Black and Mulatto women were said to be wearing fine linen blouses with lace, each worth 70 pesos, expensive petticoats, underskirts (*manteo*) worth 200 pesos, and mantles (*manto*) of silk with excessive lace which might reach 100 pesos.[126] At that time the Fiscal reported that ordinances prohibiting Black and Mulatto women from wearing luxury clothing were never observed.[127] This was somewhat exaggerated for in the same year two *Mulatas* were fined for wearing blue silk that was highly decorated with gold trimmings, which was confiscated.[128] Even if there were some prosecutions, the fact that the ordinances were reiterated so frequently suggests they did not act as a deterrent to what was clearly common practice. In 1631 *Mulatas* in Lima, whether slave or free, were not allowed to wear clothes of any kind made of silk or Castilian cloth nor wear slippers with silver decoration.[129] Again in 1667 orders were given that no *Mulata, Negra*, or *Zamba*, free or slave, should without exception wear a dress of silk, nor trimming of gold or silver, nor black trim of silk or linen. The penalty for first offence was confiscation of the clothes, for the second a fine of fifty pesos, and for the third two hundred lashes and eight years exile.[130] Even so, slave owners often attempted to display their elevated status by dressing their slaves in fine clothing.[131] In fact although slaves were forbidden from wearing capes or arms, they could do so when accompanying their owners.[132]

In the early seventeenth century, there were some 30,000 Blacks and Mulattoes in Lima and its hinterland,[133] and legislation against Blacks or Mulattoes bearing arms derived essentially from the fear they would revolt.[134] Exceptions

---

125 Mugaburu, *Chronicle of Colonial Lima*, p. 32 (1653), p. 59 (1661), pp. 82–83 (1663), pp. 91–92 (1665), p. 124 (1667).

126 AGI Lima 160 Fiscal Licenciado Varahona Encinillas 30 May 1629.

127 Ibid.

128 Suardo, *Diario de Lima* 1: 17.

129 Ibid. 1: 155.

130 Mugaburu, *Chronicle of Colonial Lima*, 124.

131 Tamara J. Walker, "'He Outfitted his Family in Notable Decency': Slavery, Honour and Dress in Eighteenth-Century Lima," *Slavery and Abolition* 30 (3) (2009): 390–91; O'Phelan, "El vestido," 107–08.

132 *Libros de cabildos de Lima*, 1: 287 Ordenanza sobre los negros 8 Jan. 1539; *Libros de cabildos de Lima*, 2: 56 Negros de cofradía 21 Jan. 1549.

133 AGI Lima 160 Fiscal Licenciado Varahona Encinillas 30 May 1629.

134 Lewin, *Descripción*, 39.

were made for Blacks who were servants of officials who were permitted to wear swords and for Blacks who needed them for their occupation, such as wood cutters and muleteers, who were allowed to carry knives.[135] In terms of style the population at large tried to emulate the upper classes, which in turn followed the fashion in Spain, though the particular fabric and trimmings might be different.[136] Rebecca Earle argues that in the early colonial period, despite sumptuary legislation, there was very little difference in the dress of elites and the popular classes who dressed extravagantly, but that distinctions grew in the Enlightenment period.[137]

Other possessions indicative of a person's social status were jewellery, furniture and furnishings. Perhaps surprisingly, cloth and furnishings figured more commonly than jewellery among Limeños' possessions and they commanded higher prices.[138] Most elite households possessed some pieces of silverware, such as plates, tankards and cutlery, no doubt reflecting the general availability of silver. Gold items were generally rarer and consisted of only small pieces of jewellery, trinkets, or religious items.[139] The apothecary Díaz Cabeza de Vaca possessed over 100 pesos worth of silver, mainly in the form of plates, but only an amethyst ring, gold toothpick, and a purple stain bag embroidered with pearls containing a small gold cross and ornamental chain of steel, which together sold at auction for only 15 pesos 4 reals. Linen sheets on the other hand were each worth 2 to 4 pesos and bedspreads could command over 20 pesos.[140]

The fluidity of early colonial society in Lima meant that individuals employed forms of public display to indicate and enhance their status. This was evident for all social groups, such that Antonio de la Calancha observed that on feast days the artisan would appear as a *regidor*, wearing a cape and velvet, while labourers from Spain pretended to be nobles, and commoners studied

135 Ibid. 40.

136 O'Phelan, "El vestido," 99–100.

137 Rebecca Earle, "Two Pairs of Pink Satin Shoes!!' Race, Clothing and Identity in the Americas (17th–19th centuries)," *History Workshop Journal* 52 (2001): 175–95.

138 For an excellent introduction to the furniture, furnishings and decorations to be found in colonial houses in Lima see Crespo Rodríguez, *Arquitectura doméstica*, 297–340.

139 Ibid. 307–11.

140 AAL Testamentos leg. 5 exp. 1 fols. 38–41 Testamento de Bartolomé Díaz Cabeza de Vaca 28 April 1608. For the belongings of other medical practitioners see: AAL Testamentos 25–1 fols. 4–5v. Testamento de Juan Matías de Vera 18 Jun. 1640 (boticario); AAL Testamentos leg. 15 exp. 13 Testamento de Domingo Perez Pereira cirujano 16 May 1630 (surgeon); AAL Testamentos 28A-1 fols. 79v–81v. Testamento de Juan Gerónimo Navarro 1648 (physician); Crespo Rodríguez, *Arquitectura doméstica,* 353 Testament of Juan Ramírez, barber surgeon 14 Aug. 1629.

how to look like *hidalgos*.[141] This meant that in public places it was difficult to distinguish, the gentleman from the artisan or commoner. This Calancha explained was not policy, but was encouraged by the opportunities of the land.[142] In effect elite ostentation competed with plebeian mimicry.[143] Dress as a symbol of status was generally more important in the New World than the Old,[144] and especially in Lima. The preference of *boticarios* for black and sober clothing thus reflected a desire to emulate the Spanish monarchs and adhere to Church teachings that disdained the wearing of luxury clothing, while their preference for expensive European textiles over cheaper imported Asian cloths both reflected and reinforced their superior social standing.

Conclusion

In many respects, the social standing, religious beliefs and moral compass of *boticarios* in Lima, including their involvement in the Church, demonstrated significant similarities to the life and status of their counterparts in Spain and indeed to other professional groups. However, the opportunities for wealth creation and display that Lima offered as well as the presence of sizeable African, indigenous and *casta* populations, produced a different and more fluid social environment. Apothecaries in Lima sought to reinforce their position among the elite through seldom wearing luxury clothing and maintaining an air of sobriety. They also actively participated in *cofradías*, displayed their piety, and made charitable donations according to public expectations of their profession. While apothecaries in Mexico City were involved in similar religious and charitable activities, those in Lima seem to have been more conscious of the economic gains their businesses could generate.[145] As noted in Chapter 6, apothecaries in Lima also sought to enhance their status by distancing themselves from popular healers by adhering to humoral practices and *materia medica* of Old World origin. Despite such efforts, apothecaries in Lima were not always held in high regard and as in Spain continued to command only a middling status within the medical profession beneath that of university-educated physicians.

141 Calancha, *Crónica moralizada*, lib. 1 cap. 10: 67.

142 Ibid. lib. 1 cap. 38: 245.

143 Osorio, *Inventing Lima*, 149.

144 Solange Alberro, *De gachupín al criollo: o de cómo los españoles de México dejaron* de *serlo* (México: El Colegio de México, 1992), 177–84.

145 De Vos, "Art of Pharmacy," 156–90.

CHAPTER 8

Persistent Practices

> Medical practitioners in New Spain, New France and New England may have borrowed at times from indigenous healers and incorporated some native remedies into their pharmacopoeias, but their theoretical framework and their actual practices remained virtually unchanged. Even when they prescribed novel medicines, they did so for the same reasons that had governed Western medical practice for centuries.[1]

∴

Apothecaries in early colonial Lima embodied both tradition and innovation. Inadequacies in the mechanisms used to control and regulate medical practice in Peru and the opportunities afforded by a whole new world of *materia medica*, should have encouraged scientific enquiry and experimentation. Yet, this study suggests that in the early colonial period humoral medicine predominated. Professional *boticarios* in Lima worked within a humoral framework employing purgatives, laxatives, and other treatments aimed at restoring a balance in the humours. In addition, whether they were treating private clients, slaves or patients in hospitals, they relied largely on Old World plants that were either imported from Europe or were grown locally. They adopted a few native plants as substitutes for Old World imports, such as canafistula and sarsaparilla, and relied on American resins and gums imported from Spanish American regions further north to meet the shortages of these essential products on the Peruvian coast. However, contrary to what might be expected, *boticarios* made relatively little use of local botanical *materia medica*. On the other hand, some *boticarios* began to experiment with minerals to the extent that their use may have been more prevalent in Lima than in Spain at this time. Distillation methods were widely employed and minerals and chemicals were used in medical preparations. However, very few medicines of this kind were employed and they were rarely applied internally.

A complex range of material, social and cultural factors contributed to the absence of native *materia medica* in medicines prepared by apothecaries in

1 Numbers, "Conclusion," 155–56.

 | DOI 10.1163/9789004351271_009

Lima. Of considerable significance was the delay and controversy involved in fitting American plants into the prevailing Galenic framework; extensive testing was required both to identify their appropriate therapeutic use and to allay fears that they might be dangerous or harmful. At the same time, professional medical practitioners, even though they often respected the herbal knowledge of indigenous healers, saw social status advantages in distancing themselves from them and from other popular healers by adhering to Spanish humoral practices and employing *materia medica* of Old World origin. Social status considerations may have been particularly significant in Lima where the social hierarchy was still in the process of consolidation, where royal authority was limited, and where mining and trade provided exceptional opportunities for social advancement. In addition to distancing themselves from popular healers though their professional activities, they also adopted publically visible modes of dress and expressions of piety that emulated the aristocracy.[2]

Despite the predominance of humoral medical practice, this study suggests that some apothecaries were employing empirical and experimental methods, even if this did not signal the adoption of a new natural philosophy. There was a curiosity about Peru's natural world even if it did not involve the extensive adoption of New World *materia medica*. Questions emerged around the efficacy of botanical imports and of Old World plants grown under a different constellation that required reconsideration of the most appropriate products and dosages for the treatment of particular diseases, whose prevalence in Lima also differed from that in Europe, even if they were not unfamiliar. There were some pioneers such as Antonio de Robles Cornejo who was intent on bringing pharmaceutical methods to non-university educated practitioners and whose manuscript, organised by product rather than ailment signified growing specialisation of pharmacy within the medical profession as it diverged from medicine as a practical science. The format of his work also reflected the shift from the diagnosis of illness and personalised treatment following humoral principles to the study of the effectiveness of particular *materia medica*. This process was encouraged by a lack of knowledge and agreement between physicians on the efficacy of humoral treatments and by patients lacking classical training who could grasp the reality of illness and healing and who might treat themselves without the intervention of a physician. Despite the experimental tendencies evident in Robles Cornejo's work, it was written in the Galenic tradition.

2 Lockhart, *Spanish Peru*, 226–27; Bronner, "Elite Formation," 25–26. For Spain see Pike, *Aristocrats and Traders*, 81–93.

From this discussion it will be evident that in the period prior to 1650 apothecaries in Peru saw no contradiction in the employment of experimental methods and humoral beliefs. At this time, their experimental methods focussed on the composition and efficacy of medicines based on Old World plants and their substitutes; they were not, for the most part, directed at testing native products, which with a few exceptions, for example, cinchona, remained largely the preserve of indigenous and popular healers. The focus of apothecaries on Old World *materia medica* was reinforced by their professional training, by the dominance of medical manuals based on Old World products, and by pharmacy inspections. This emphasis effectively separated professional medical practice from other forms of healing, although the distinction was never sharp or absolute.

This finding has two conceptual implications. First, it suggests that medical practitioners did not compete for clients in no single "marketplace," but rather in terms of provision the market was segmented, with a regulated professional sector being largely distinct from the popular sector. Second, recent research on medical practice in early colonial Spanish America has stressed its hybrid nature and even the fusion of healing traditions of European, African and indigenous groups, through a process of medical *mestizaje*.[3] Such studies of the exchange and the interweaving of medical ideas, practices and products, which have generally been grounded in specific social, historical and geographical contexts, have added greatly to knowledge of colonial medicine. However, there is sometimes a tendency to assume that such exchanges were free and simple, when as this study suggests, the process was commonly slowed and constrained by cultural or institutional obstacles. Furthermore, it underlines the necessity not to lose sight of the unequal distribution of power that characterised the social context in which the exchange of ideas and practices took place.

Accounting for the Prevalence of Humoral Medicine

What can account for the prevalence of humoral medicine in Lima? Many interrelated factors were involved. These included attempts by the state working with the Church and Inquisition to bolster humoral beliefs through regulation

3 Martha Few, "Medical *Mestizaje* and the Politics *of Pregnancy in Colonial Guatemala," 1660–1730,* in *Science in the Spanish and Portuguese Empires, 1500–1800*, eds. Daniela Bleichmar, Paula De Vos, Kristine Hufine and Kevin Sheehan (Stanford CA: Stanford University Press, 2009), 133.

and censorship. Perhaps more significant, however, was the fact that humoralism was already embedded in popular Spanish culture.

The state and Church were reluctant to promote new medical practices that implied the adoption of a new natural philosophy, especially one that might challenge the basis of the Catholic faith.[4] It therefore sought to uphold humoral medical practice by establishing a medical infrastructure that was similar to that being developed in Castile. However, as this study has shown, even in Lima, which was capital of the Viceroyalty of Peru, this was only partially achieved. Instead the state relied on the clergy and particularly the religious orders to deliver healthcare.[5] Nevertheless, the fact that the Crown actively promoted the foundation of hospitals served to foster a paternalistic view of Spanish rule and provide legitimisation for its colonial mission.[6]

In the task of maintaining medical orthodoxy, the state was aided by the Inquisition and, in the case of the indigenous population, the ecclesiastical courts. These aimed at suppressing healing practices deemed to be heretical and at controlling the publication and dissemination of literature contrary to the interests of the state and Church. The Inquisition was most active in Lima in the early seventeenth century, but its impact on medical practices in Lima is difficult to evaluate. Only two physicians and two surgeons were brought before the Inquisition and this was for Judaizing or promoting heretical beliefs rather than unorthodox medical practices. Some *curanderos* were also tried by the Inquisition, but due to their healing skills and the general shortage of doctors, they appear to have been allowed to practise as long as they did not employ superstitious or idolatrous methods.[7]

The role of the Inquisition in the realm of censorship is equally difficult to gauge. The medical texts and manuals most widely used by apothecaries were written in the Galenic tradition and there is little evidence for the ownership of medical texts that promoted experimental methods and the use of chemical medicines. It cannot be concluded, however, that the absence of Paracelsian or "modern" texts was due to the effectiveness of the Inquisition's scrutiny.

---

4 Shapin, *Scientific Revolution*, 136.

5 See Cahill, "Financing Health Care" for the role of state finance in medical care in eighteenth-century Lima. For the role of the religious orders in the dissemination of humoral medicine see: Foster, "Origin of Humoral Medicine," 363–64 and *Hippocrates' Latin American Legacy*, 153–55.

6 Risse, "Medicine in New Spain," 42; Warren, *Medicine and Politics*, 20; Ramos, "Indian Hospitals," 204–05.

7 Griffiths, "Andean *Curanderos*," 185–97. For cases against *hechiceras* see, AAL Hechicerías leg. 1 exp. 7 fol. 2 Causa seguida contra Pedro Sayo, acusado de curar enfermos con hierbas 1621; Mannarelli, *Hechicheras*, 38; Sánchez, *Amancebados*, xxxv.

Indeed in Europe as well such texts were not widely available. More likely, as Richard Palmer has noted for Venice, the slow progress of Paracelsianism reflected the extent to which humoralism was entrenched both among medical practitioners and the public at large.

Palmer claims that in Italy Paracelsianism only began to emerge in the late sixteenth century with chemical drugs only becoming widely available in the early seventeenth century. Even then, he argues, they added to rather than replaced traditional remedies.[8] Adding to the delay in the adoption of a new natural philosophy in Spain was existence of a strong alchemic tradition which enabled the development of chemical methods without engagement with Paracelsian philosophy. The existence of copies of the works of the physician and alchemist Arnau de Vilanova in pharmacies in Lima suggests that there too apothecaries were using chemical methods that had been developed by alchemists and were applying chemical medicines within a Galenic framework. In fact the philosophical distinction between Paracelsianism and Galenism may not have been that obvious or seemed so novel at this time.[9] Indeed, Paracelsus himself declared that he had been influenced by the alchemic texts of Vilanova, Llull and Rupescissa. In any case, apothecaries in Lima appear to have been less concerned with challenging the philosophy underlying their methods and more focused on the practical task of preparing medicines that were appropriate for the treatment of particular illnesses.

The significant role played by the clergy in providing medical care had a number of implications not only for its delivery, but also the type of medicine practised. Even though papal decrees in the early thirteenth century discouraged priests from formally studying medicine because it distracted them from their spiritual goals,[10] their Christian responsibility for the relief suffering meant they were actively involved caring for the sick. The knowledge and skills they used would have been acquired through the private study of classical texts such as those of Galen and Hippocrates.[11] In addition to healing the sick themselves,[12] they instructed Indians how to let blood and prepare

8 Palmer, "Pharmacy in the Republic of Venice," 110–17.

9 López Pérez, "Spanish Paracelsus Revisited," 353–57.

10 Siraisi, *Medieval and Early Renaissance Medicine*, 26, 43–44.

11 John Lloyd Mecham, *Church and State in Latin America: A History of Politico-Ecclesiastical Relations* (Chapel Hill: University of North Carolina Press, 1934), 36.

12 Real Academia de Historia, Madrid, Jesuitas 129 fol. 383v. Letras anuas de la provincia del Perú de 1618 años.

medicines,[13] and compiled *recetarios* for use by the population at large.[14] As such, several authors have argued that priests played a more significant role in transmitting humoral medicine to the New World than trained medical practitioners.[15]

Despite the Church's general support for humoral medicine and for the suppression of healing practices deemed heretical, it is clear that some priests, particularly the Jesuits, were using empirical and experimental methods. In fact they saw no contradiction between the methods of scientific enquiry and their religious beliefs, which promised to reveal God's work in nature and support efforts to heal the sick, especially in remote regions where professional medical practitioners were lacking. Their observational and experimental approaches were therefore largely practical in intent and, since they were not seeking to find the *causes* of phenomena, did not challenge the Catholic belief in God's creation.[16]

Professional medical practitioners, the clergy, Inquisition, *protomedicato*, and municipal authorities, were not the only agents promoting humoral medicine. In Europe, medicine was just developing as a specialist occupation, prior to which everyone took responsibility for their own health and acquired a basic knowledge of treatments. Within a community certain individuals, such as a barber or grocer, might develop special skills or knowledge, or might become renowned for the production of simple medicines, such as waters or syrups with effective healing properties. Literate citizens might also collect well-tried recipes in hand-written *recetarios* for domestic use.[17] Unfortunately only a few such *recetarios* exist for Peru and then only from the eighteenth century.[18] Meanwhile some secular elites holding positions of authority, for example, as administrators of hospitals or members of *cofradías*, would have been able to influence the nature medical practice beyond the home. In remote regions they might even teach rudimentary healing skills themselves. In 1586 the *corregidor*

---

13 AAL Hechicerías 4 27 fol. 3v Cabeza de proceso de oficio por hechicero contra Pablo Ato 3 Dec.1660; Ramos, "Indian Hospitals," 198.

14 See Chapter 6.

15 Guerra, "Role of Religion," 180–81; Foster, *Hippocrates' Latin American Legacy*, 150–51.

16 William B. Ashworth, "Catholicism and Early Modern Science," in *God and Nature: Historical Essays on the Encounter between Christianity and Religion*, eds. David C. Lindberg and Ronald L. Numbers (Berkeley: University of California Press, 1986), 153–58.

17 Elaine Leong, "Making Medicines in the Early Modern Household," *Bulletin of the History of Medicine* 82 (2008): 157–63.

18 See Valdizán and Maldonado, *Medicina popular*, 3: 108–316, 421–87 for various examples of *recetarios* from the late eighteenth or nineteenth century.

of Yauyos, Diego Dávila Briceño, reported that he had been teaching the local residents in his jurisdiction how to let blood.[19] Humoral beliefs were not, therefore, the preserve of a particular occupational group, but came to be widely diffused and understood by the population at large.[20] This extended to the indigenous population, where the failure of native remedies to deal with the major epidemics in the Andes in the late sixteenth and early seventeenth centuries, probably undermined the authority of native healers within their communities, even though the humoral treatments the Spanish introduced were no more effective.[21] This was especially so since the epidemics that afflicted native peoples left Europeans largely unscathed. A final factor that might have facilitated the spread of humoral medicine was that Spanish and indigenous healing traditions were based on similar principles of bodily balance and the humours, even if the qualities attributed to particular substances used in treatments were not always the same. Among Andean peoples the body is viewed as a hydraulic system in which the fluids of air, blood and fat need to be kept in circulation and are affected by whether they are hot or cold.[22]

Accounting for the Slow Adoption of Experimental Methods

State regulation, the Inquisition and entrenched cultural traditions may partially explain why humoral medicine predominated into the early seventeenth century and why experimental approaches appear to have developed less

19 *Relaciones geográficas de Indias* 1: 159 Descripción y relación de la provincia de los Yauyos, Diego Dávila Briceño 15 Jan. 1586.

20 Henry, "Doctors and Healers," 199–202. For a study of the enthusiasm for medical knowledge in Spanish Valencia prior to the professionalization of medicine see McVaugh, *Medicine before the Plague*.

21 Linda A. Newson, "A Historical-Ecological Perspective on Epidemic Disease," in *Advances in Historical Ecology*, ed. William Balée (New York: Columbia University Press, 1998), 56; Daniel T. Reff, *Disease, Depopulation, and Culture Change in Northwestern New Spain, 1518–1764* (Salt Lake City: University of Utah Press, 1991), 260–64; Luz Maria Hernández Sáenz and George Foster, "Curers and Their Cures in Colonial New Spain and Guatemala: The Spanish Component," in *Mesoamerican Healers* eds. Brad R. Huber and Alan R. Sandstrom. (Austin: University of Texas Press, 2001), 42; Cueto, *Medicine and Public Health*, 16–18. For Andean epidemics see: Cook, *Demographic Collapse* and *Born to Die;* Suzanne A. Alchon, *Native Society and Disease in Colonial Ecuador* (Cambridge: Cambridge University Press, 1991); and Linda A. Newson, *Life and Death in Early Colonial Ecuador* (Norman: University of Oklahoma Press, 1995).

22 Joseph W. Bastien, *Healers of the Andes: Kallawaya herbalists and their medicinal plants* (Salt Lake City: University of Utah Press: 1987), 45–52.

rapidly than might be expected. Studies in the history of medicine have suggested that in order for a significant change in scientific practice to occur it is necessary for a community of scientists to emerge, who can exchange and discuss ideas and methods. Hence George Basalla's model for the spread of western science argues that for colonial science to develop beyond the stage of exploration there needs to be an expansion in the number of scientists and for their ideas to be accepted and validated by a dominant external culture.[23] The number of professional medical practitioners in Lima at this time was small. They would almost certainly have known each other professionally, for example, through trading *materia medica*, lending each other medical texts, inspecting each other's premises, or valuing their medicines for legal cases; they would also have known each other socially even to the extent of sometimes intermarrying. There is also evidence for interactions between medical practitioners and Jesuits at the College of San Pablo. Medical practitioners could borrow books from the College's extensive library which contained both traditional and modern medical texts, many brought to Peru by Jesuits arriving from Europe.[24] On the other hand, the Jesuits might approach apothecaries for the testing of new medicinal products, as was the case with cinchona.

While there appears to have been a growing number of scientists, both secular and ecclesiastical, in Lima in the early colonial period and evidence for at least some exchange in ideas and practices between them, the opportunities for wider dissemination would have been limited by inadequacies in local printing and by scrutiny by the Inquisition. As the seventeenth century progressed, the Inquisition became less active, knowledge of the natural history of Peru accumulated, and the numbers involved in scientific activity, many of whom were Creoles, expanded. This period is outside that studied here, but there are indications from later pharmacy inventories and from the libraries of physicians that experimental methods, including the use of minerals and mineral salts, were becoming more widespread in Lima from the mid-seventeenth.[25] As time progressed local scientists accumulated greater knowledge of the region's geography, flora and mineral resources, and were therefore better placed

23 Basalla, "Spread of Western Science," 611–17.

24 Luis Martín, "La biblioteca del Colegio de San Pablo," 24–29; Guibovich, "Bibliotecas de médicos," 302. See also Chapter 5.

25 See Biblioteca Nacional Perú 1560 Cuenta que da el Cap. Nicólas Navarro mayordomo del hospital de San Andrés 1674–1676; Biblioteca Nacional Peru 1562 Cuenta del mayordomo de San Andrés, Cap. Don José Bernardo de León 1676. See also the inventory of the pharmacy of the Colegio San Pablo when the Jesuits were expelled: AGNP Santo Oficio. Temporalidades. Inventarios leg. 2 cuad. 28 fols. 1–68 1768. See also Chapter 5.

to challenge scientific discourses emanating from the metropolis, making science more relevant to local needs and with scientists spurred on by patriotic sentiments.[26] Such knowledge was founded on empirical methods developed locally rather than in Europe, though the seeds had been sown by the Columbian encounter which forced a revaluation of many ideas and practices and the adaptation of Old World policies, frameworks and institutions to colonial circumstances.[27] This chapter in the history of pharmacy in colonial Lima has yet to be written.

The late seventeenth century is probably the least researched period in colonial Latin American history, yet this study suggests that it may be critical in understanding the development of science, medicine and pharmacy in Peru, and indeed Latin America more generally. It would enable further exploration of the role of different institutions, groups and individuals in scientific practice countering the European narrative that dominates current historiography. The putative role of apothecaries as pioneers in the history of medicine needs further study. An examination of the late seventeenth century when Habsburg power was at its nadir would also throw light on the role of Atlantic connections in understanding colonial medical practice, since in theory weakened state control and reduced intellectual exchange should have encouraged greater local innovation. As such it would throw light on the relationship between empire and medicine more generally.

---

26 Cañizares-Esguerra, "New World, New Stars"; Lafuente, "Enlightenment Connections."

27 Saldaña, "Science and Public Happiness," 56–59. The Bourbon reforms of the eighteenth century saw the activities of scientists finding validation through alliances with non-medical elites and in the emergence of learned societies and institutions dedicated to it.

Appendices

APPENDIX A

Books Shipped from Spain by the Apothecary Juan Sánchez in 1591

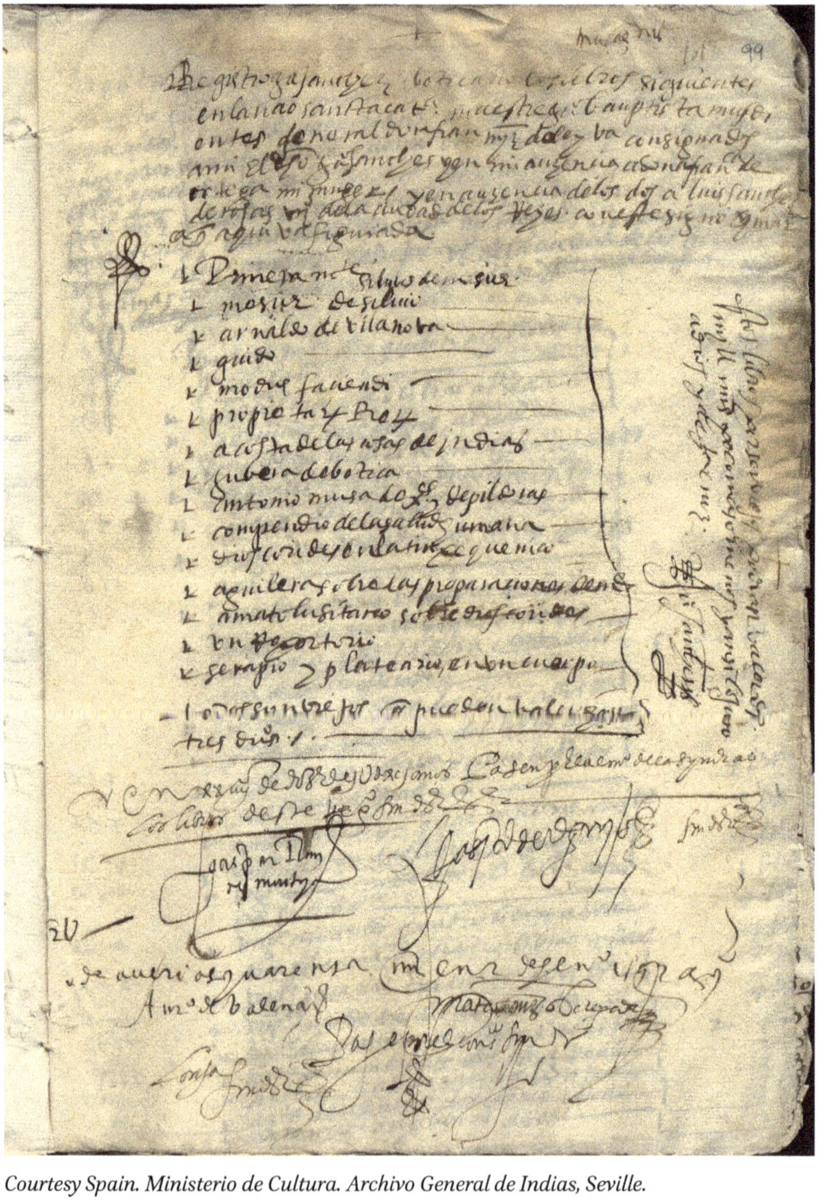

Courtesy Spain. Ministerio de Cultura. Archivo General de Indias, Seville.
SOURCE: AGI CONTRATACI

 | DOI 10.1163/9789004351271_010

Juan Sánchez, apothecary, registers the following books in the ship Santa Catalina, shipmaster Juan Bautista Musdientes and general [of the fleet] Don Francisco Martínez de Leyva consigned to me, the said Juan Sánchez and in my absence to my wife Francisca de Ortega and in the absence of both to Luis Sánchez de Rojas, citizen of the city of kings with this sign and mark entered here.

Primeramente Silvio de Mesue [*Joannis Mesuae Damaceni* ... Jacobo Sylvio [Jacques Dubois]], (Paris, 1544)

Mesue de Silvio [ibid.]

Arnaldo de Vilanova [*Antidotarium* (Valencia, 1495)]

Guido [Guy de Chauliac (Guido de Cauliaco) possibly *Chirugía* (Venice, 1480)].

Modus faciendi [Bernardino de Laredo (Seville, 1527)]

Propietarum rerum [Bartolomaeus Anglicus (Bartolomé Anglicus), *De propietatibus rerum* (Cologne, 1472)]

Acosta de las cosas de las indias [José de Acosta, *Historia natural y moral de las indias* (Seville, 1590)]

Jubera de botica [Alfonso Jubera, *Dechado y reformación de todas las medicinas compuestas usuales* (Valladolid, 1578)]

Antonio Musa de jarabes y piladoras [Antonio Musa Brasavola, *Examen omnium syruporum* (Lyon, 1540)]

Compendio de la salud humana [Johann von Ketham (Zaragoza, 1494)]

Dioscórides en latín pequeña. Unspecified small version of Dioscorides in Latin.

Aguilera sobre las preparaciones de Mesue [Antonio de Aguilera, *Exposición sobre las preparaciones de Mesue* (Alcalá de Henares, 1569)]

Amato Lusitano sobre Dioscórides [Amatus Lusitanus, *Index Dioscoridis* (Antwerp, 1536)]

Un reportorio [Jerónimo de Chaves, *Chronographia, o, reportorio de los tiempos* (Seville, 1548)]

Serapio y plateario en un cuerpo [possibly *Practica Io. Serapionis dicta breviarium. Liber Serapionis de simplii medicina. Liber de simplii medicina, dictus circa instans. Practica platearii* (Venice, 1503)]

They are all old and are worth about three ducats [about four pesos]

Source: AGI Contratación 1097 N 5 Image 197 24 Dec. 1591.

APPENDIX B

List of *Materia Medica* Found in Pharmacies in Spain and Lima

	Andrés Zamudio de Alfaro[a] 1590s	Hospital of Santa Ana[b] 1551	Diego de Tineo *botica*[c] 1555	Francisco de Alva *botica*[d] 1576	Bartolomé Díaz Cabeza de Vaca *botica*[e] 1608
Organic Simples					
Abarraz [Albarraz]					X
Abrótano		X	X		
Acacia		X			X
Aceitunas gordas		X			
Acíbar de levante					
Acíbar [de India]	X	X	X	X	X
Acíbar [h]epático			X		
Acíbar secotrino		X	X		
Agallas			X		
Agárico	X	X	X	X	X
Ajalidis					
Alarges [Alarguez?]			X	X	
Alcarabea		X	X		X
Alchechengi					X
Alheña			X	X	
Alholvas		X			X
Alhucema			X		X
Almendras			X		
Almendras amargas			X		X
Almoradux			X		
Ámbar	X		X		

 | DOI 10.1163/9789004351271_011

Pedro de Bilbao *botica*[f] 1636	Hospital of Santa Ana[g] 1606	Fernando de Cartagena Cuzco[h] 1618	Common name	Latin name
	X		Stavesacre, larkspur	Delphinium staphisagria L.
			Southern wood	Artemisia abrotanum L.
			Acacia	Acacia sp.
			Large olives	Olea europaea L.
		X	Aloe from the Levant	Aloe vera (L.) Burm.f.
		X	Aloe	Aloe vera (L.) Burm.f.
			Aloe for the liver	Aloe vera (L.) Burm.f.
			Aloe	Aloe from the island of Socotora
X			Galls	
X	X	X	Agaric, mushroom	Polyporus sp.
		X	Unidentified	
		X	Dog-rose	Rosa canina L.
X			Caraway	Carum carvi L.
			Bladder cherry, winter cherry, Chinese lantern	Physalis alkekengi L.
X			Privet	Ligustrum vulgare L.
		X	Fenugreek	Trigonella foenum graecum L.
			Lavender	Lavandula spica L.
			Almonds	Amygdalus spp.
			Bitter almonds	Amygdalus amara Hayne
			Marjoram	Origanum majorana L.
			Amber	

	Andrés Zamudio de Alfaro[a] 1590s	Hospital of Santa Ana[b] 1551	Diego de Tineo *botica*[c] 1555	Francisco de Alva *botica*[d] 1576	Bartolomé Díaz Cabeza de Vaca *botica*[e] 1608
Ameos		X	X		
Anacardos		X			
Anoscastos [Agnus-castus]					
Apio				X	
Arcollo			X		
Aristoloquias [luengo and redondo]		X	X		X
Arrayán		X			X
Aspálato			X		
Avellanas		X	X		
Azafrán		X	X		X
Azafrán romi					X
Azahar					X
Azufeifas [Jujubas]		X	X		
Beleño		X			X
Ben blanco		X	X		
Ben rubio		X	X		X
Bryonia			X		
Cabezuelas de rosas			X		
Cálamo aromático		X	X		
Camepitios					X
Canafistula	X				X
Canela		X	X	X	
Carlina			X		
Cassia [avellanos de cassia]					X

Pedro de Bilbao *botica*[f] 1636	Hospital of Santa Ana[g] 1606	Fernando de Cartagena Cuzco[h] 1618	Common name	Latin name
			Bishop's weed	Ammi majus L
			Cashew	Anacardium sp.
X		X	Chaste tree, Chaste berry, Monk's pepper	Vitex agnus-castus L.
		X	Celery	Apium graveolens L.
			Unidentified	
X		X	Birthwort, Aristolochia	Aristolochia sp.
			Myrtle	Myrtus communis L.
			Rosewood	Dalbergia latifolia Roxb. Simple
			Hazelnut	Corylus avellana L.
			Saffron	Crocus sativus L.
			Bastard saffron, rumin	Carthamus tinctorius L.
			Orange blossom,	Citrus sinensis L.
			Jujube, Chinese or Red Date	Ziziphus zizyphus (L.) H. Karst.
X			Henbane	Hyoscyamus niger L.
			White ben, behen	Moringa arabica Pers.
			Red ben, behen	Moringa arabica Pers.
			Bryony	Bryonia sp.
			Rose buds	Rosa sp.
		X	Calamus, Sweet flag	Acorus calamus L.
			Chamaepitys, cam-epitys, ground pine	Teucrium chamaepitys L
	X	X	Cassia, golden shower tree, drum-stick tree	Cassia fisula L.
		X	Cinnamon	Cinnamomum verum J. Presl
			Carline thistle	Carline sp.
			Cassia	Cinnamomum or Senna sp.

	Andrés Zamudio de Alfaro[a] 1590s	Hospital of Santa Ana[b] 1551	Diego de Tineo *botica*[c] 1555	Francisco de Alva *botica*[d] 1576	Bartolomé Díaz Cabeza de Vaca *botica*[e] 1608
Cebollas albarranas			X		
Cedoaria			X		
Cíperos			X		X
Ciruelas pasas		X	X		
Clavos and madres de clavos		X	X		
Coca de levante		X			
Coloquíntidas	X	X	X	X	X
Cominos rústicos			X	X	
Contra yerba					X
Coriandro			X		
Coronilla del rey		X			
Costo amargo		X	X		
Costos					X
Cubebas		X	X	X	X
Culantro preparado					X
Dauco					X
Dialtea				X	X
Díctamo crético	X		X	X	X
Díctamo real		X	X		X
Doronicus			X	X	
Eléboro blanco		X	X		X
Eléboro negro		X	X		X
Escamonea	X	X	X	X	X
Esclarimente		X			
Espica céltica	X	X	X		
Espicanardo	X	X	X	X	X

Pedro de Bilbao *botica*[f] 1636	Hospital of Santa Ana[g] 1606	Fernando de Cartagena Cuzco[h] 1618	Common name	Latin name
			Maritime squill	Urginea maritima (L.) Baker
			Zedoary	Curcuma zedoaria (Christm.) Roscoe
		X	Cipero sedge	Cyperus spp.
			Prunes	Prunus domestica L.
			Clove	Syzygium aromaticum L.
			Indian berry, levant nut	Anamirta cocculus (L.) Wight & Arn
X		X	Colocynth, bitter apple	Citrullus colocynthis (L.) Schrad
		X	Cumin	Cuminium cyminum L.
			Contrahierba, snakewort	Dorstenia contrajerva L.
			Coriander	Coriandrum sativum L.
			Crown vetch?	Coronilla minima L.
			Bitter costus	Saussurea lappa Decne
			Costus	Costus arabicus L.
		X	Cubeb, Javanese tailed pepper	Piper cubeba L.
	X		Cilantro, coriander	Coriandrum sativum L.
			Carrot	Daucus carota L.
	X		Dialthea, marshmallow	Althaea officinalis L.
		X	Dittany from Crete	Origanum dictamnus L.
			White dittany	Dictamnus albus L.
			Leopard's bane	Doronicum L.
X			White hellebore	Helleborus album L
X			Black hellebore	Helleborus niger L
X	X	X	Scammony	Convulvulus scammonia L.
X			Yellow amber	
X		X	Yellow or Roman Valerian	Valeriana celtica L.
	X	X	Spikenard	Nardostachys grandiflora DC, Nardostachys jatamansi DC

	Andrés Zamudio de Alfaro[a] 1590s	Hospital of Santa Ana[b] 1551	Diego de Tineo *botica*[c] 1555	Francisco de Alva *botica*[d] 1576	Bartolomé Díaz Cabeza de Vaca *botica*[e] 1608
Esquenanto [paja de meca]	X		X	X	X
Estafisagra					X
Ésula			X	X	X
Filipéndula			X		X
Fumaria					
Garbanzos negros			X		X
Guayacán					
Gengibre		X	X	X	X
Habas de mar			X		
Lengua avil [lenguavis]					
Lináloe	X	X	X	X	X
Linaza		X			X
Lingua cervina			X		
Lupinus		X	X		
Macis	X	X		X	X
Malagueta				X	
Maná	X		X	X	X
Maná (Magna) de calabria		X			
Matías			X		
Milium solis		X		X	
Mirabolanos-negros				X	
Mirobolanos-chébulos	X	X	X	X	
Mirobolanos					
Mirobolanos-beléricos	X	X	X	X	
Mirobolanos-citrinos	X	X	X	X	X
Mirobolanos-émblicos	X	X	X	X	X
Mirobolanos-indios	X	X	X	X	

Pedro de Bilbao *botica*[f] 1636	Hospital of Santa Ana[g] 1606	Fernando de Cartagena Cuzco[h] 1618	Common name	Latin name
X	X		Camel grass	Cymbopogon schoenanthus (L.) Spreng.
X			Stavesacre, larkspur	Dephinium staphisagria L.
X			Leafy spurge	Euphorbia esula L.
			Dropwort	Spiraea filipendula L.
X			Fumitory	Fumaria officinalis L.
X			Black chickpeas	Cicer arietinum L.
		X	Lignum vitae	Guaiacum officinale L
		X	Ginger	Zingiber officinale L.
			Sea beans, samphire	Salicornia sp.
X			Unidentified	Bird's tongue?
X	X	X	Linaloe	Bursera aloexylon (Shiede ex Shlecht)
		X	Linseed	Linum usitatissimum L.
			Hart's tongue	Asplenium scolopendrium L.
			Lupin	Lupinus sp
		X	Mace	Holarrhena antidysenterica (Roxb. ex Fleming) Wall. ex A. DC
			Malagueta pepper	Capsicum frutescens L.
X		X	Manna	Fraxinus ornus L.
			Manna from Calabria	Fraxinus ornus L.
			Unidentified	
	X		Gromwell	Lithospermum officinale L.
			Black myrobalan	Terminalia chebula L.
X		X	Chebulic myrobalan	Terminalia chebula L.
	X		Myrobalan unspecified	Terminalia sp.
X		X	Beleric myrobalan	Terminalia bellirica (Gaertn.) Roxb.
X		X	Yellow myrobalan	Terminalia chebula L.
X		X	Emblic myrobalan	Emblica officinalis Gaertn.
X		X	Indian myrobalan	Terminalia chebula L.

	Andrés Zamudio de Alfaro[a] 1590s	Hospital of Santa Ana[b] 1551	Diego de Tineo *botica*[c] 1555	Francisco de Alva *botica*[d] 1576	Bartolomé Díaz Cabeza de Vaca *botica*[e] 1608
Neguillas				X	
Nenúfar			X		
Nísperos			X		
Nuez moscada	X	X	X	X	
Orosus [Regaliz]		X	X	X	
Oruga		X			
Panis pores			X		
Pasas de almunécar			X		
Pelitre					X
Pentafilon			X		X
Peonía		X	X		
Pimienta			X		X
Pimienta blanca	X	X	X	X	
Pimienta larga	X	X	X	X	X
Pimienta negra	X				
Piñas secas			X		
Piñones			X		
Polipodio		X	X	X	X
Rapontico [reupontico]			X		
Ruibarbo	X	X	X	X	X
Sándalos blancos	X	X	X	X	
Sándalos cetrinos		X	X		
Sándalos colorados	X	X	X	X	X
Santónica					
Sasifragia					X
Scordeon			X		
Scrophularia [scrofularia]			X		
Sebesteno			X		

Pedro de Bilbao *botica*[f] 1636	Hospital of Santa Ana[g] 1606	Fernando de Cartagena Cuzco[h] 1618	Common name	Latin name
		X	Nigella, fennel flower	Nigella sativa L.
			Water lily	Nymphaea spp.
			Medlar	Eriobotrya japonica (Thunb.) Lindl.
		X	Nutmeg	Myristica fragrans Houtt.
X		X	Liquorice	Glycyrrhiza glabra L
			Rocket	Eruca sativa (L.) Mill.
			Common stinkhorn?	Phallus impudicus L.
			Raisins from Almuñécar	
	X		Pellitory	Anacyclus officinarum Hayne
			Cinquefoil	Potentilla reptans L.
X			Peony	Paeonia sp.
			Pepper	Unspecified
		X	White pepper	Piper nigrum L.
X		X	Long pepper	Piper longum L.
			Black pepper	Piper nigrum L.
			Dired pine cones	Pinus sp.
			Pinenuts	Pinus sp.
			Polypody	Polypodium vulgare L.
			Rhapontic	Rheum rhaponticum L.
	X	X	Rhubarb	Reum rhabarbarum L.
	X	X	White sandalwood	Santalum album L.
		X	Yellow sandalwood	Santalum citrimum J.B.
X	X	X	Red sandalwood	Santalum rubrum L.
		X	Santonica, wormseed	Artemisia cina Berg & C.F. Schmidt ex Poljakov
			Saxifrage	Saxifraga sp.
			Water germander	Teucrium scordium L.
			Figwort	Scrophularia sp.
			Sebesten plum	Cordia myxa L.

		Andrés Zamudio de Alfaro[a] 1590s	Hospital of Santa Ana[b] 1551	Diego de Tineo *botica*[c] 1555	Francisco de Alva *botica*[d] 1576	Bartolomé Díaz Cabeza de Vaca *botica*[e] 1608
	Selamente				X	
	Sen – oja	X				X
	Sen de luna			X		
	Sen- folículos	X	X		X	
	Sentalis				X	
	Seselios		X			X
	Sínfito			X		
	Stecados					X
	Suelda			X	X	X
	Tamarindo	X	X			
	Taragontía		X			
	Tormentila			X		X
	Turbi[t]		X	X	X	
	Turbit de la tierra					
	Turbit de levante			X		
	Vistorta			X		
	Yeros					
	Yervata			X		
Total	144	28	62	92	42	58
Seeds						
	Acederas			X	X	X
	Adormideras					
	Adormideras blancas	X	X			
	Adormideras montesinas		X			
	Adormideras negras	X	X		X	X
	Agnocasto [anus castus]	X		X		X

Pedro de Bilbao *botica*[f] 1636	Hospital of Santa Ana[g] 1606	Fernando de Cartagena Cuzco[h] 1618	Common name	Latin name
			Unidentified	
	X	X	Senna – leaves	Cassia senna L., Cassia acutifolia Delile
			Senna picked under the moon?	
			Senna – folicules	Senna sp.
			Unidentified	
			Hartwort	Tordylium offcinale L.
			Comfrey	Symphytum sp.
			French lavender	Lavandula stoechas L.
		X	Suelda con suelda, comfrey	Symphytum officinale L.
			Tamarind	Tamrindus indica L.
			Wild arum, lords and ladies	Arum maculatum L
			Tormentil	Potentilla erecta Uspenski ex. Ledeb.
		X	Turpeth	Operculina turpethum L.
	X		Local turpeth	Operculina turpethum (L.) Silva Manso
	X		Turpeth from the Levant	Operculina turpethum (L.) Silva Manso
			Bistort	Persicaria bistorta (L.) Samp.
	X		Bitter vetch	Vicia ervilia L. Willd.
			Unidentified	
30	19	44		
	X	X	Sorrel	Rumex acetosa L.
	X		Poppy unspecified	Papaver sp.
			White poppy	Papaver somniferum L.
			Mountain poppy	Papaver sp.
		X	Black poppy	Papaver somniferum L.
		X	Chaste tree, Chaste berry, Monk's pepper	Vitex agnus castus L.

	Andrés Zamudio de Alfaro[a] 1590s	Hospital of Santa Ana[b] 1551	Diego de Tineo *botica*[c] 1555	Francisco de Alva *botica*[d] 1576	Bartolomé Díaz Cabeza de Vaca *botica*[e] 1608
Albahaca	X	X			X
Alcaparras					
Alegría [sesame]	X		X		
Alexandria			X		X
Alholva	X				
Alquequenje	X				
Anís	X		X	X	
Apio		X			
Arrayán	X		X		X
Bayas del laurel	X		X		X
Beleño	X				
Berbero	X		X		
Brusco	X		X		X
Calabaza	X			X	X
Canamones			X		X
Cardamomo [mayor]	X	X	X	X	X
Cártamo	X		X		
Cebollino				X	X
Cidra	X				X
Cípres [nuez de]	X		X		
Cohombro	X				
Coles		X			
Cominos rústicos					
Cubeba	X				
Culantro	X				
Cuscuta	X		X	X	X
Dauco	X		X		
Endivia	X				
Eneldo	X	X			
Epítimo					

Pedro de Bilbao *botica*[f] 1636	Hospital of Santa Ana[g] 1606	Fernando de Cartagena Cuzco[h] 1618	Common name	Latin name
		X	Sweet basil	Ocimum basilicum L.
X			Caper	Capparis spinosa L.
			Sesame	Sesamum indicum L.
X		X	Alexandrian senna	Senna alexandrina Mill.
			Fenugreek	Trigonella foenum graecum L.
X			Bladder cherry, winter cherry, Chinese lantern	Physalis alkekengi L.
			Aniseed	Pimpinella anisum L.
		X	Celery	Apium graveolens L.
			Myrtle	Myrtus communis L.
	X	X	Bay laurel	Laurus nobilis L.
	X		Henbane	Hyoscyamus niger L.
	X	X	Barberry	Berberis vulgaris L.
		X	Butcher's broom,	Ruscus asculeatus L.
			Calabash	Cucurbita or Lagenaria sp.
			Hemp seed	Cannabis sativa L.
X	X	X	Cardamom	Elettaria cardomomum (L.) Maton
X		X	Safflower	Carthamus tinctorius L.
			Chives	Allium schoenoprasum L.
		X	Citron	Citrus medica L.
		X	Cypress tree	Cupressus sempervirens L.
			Squirting cucumber	Ecballium elaterium (L.) A. Rich.
			Cabbage	Brassica oleracea L.
		X	Cumin	Cuminium cyminum L.
		X	Cubeb	Cubeba officinalis Miquel
			Cilantro, coriander	Coriandrum sativum L.
X		X	Dodder	Cuscuta L.
X			Carrot	Daucus carota L.
			Endive, chicory	Cichorium endivia L.
			Dill	Enethum graveolens L.
		X	Clover dodder	Cuscuta epithymum Murray

	Andrés Zamudio de Alfaro[a] 1590s	Hospital of Santa Ana[b] 1551	Diego de Tineo *botica*[c] 1555	Francisco de Alva *botica*[d] 1576	Bartolomé Díaz Cabez de Vaca *botica*[e] 1608
Escarola	X				
Escarola [fría menor]	X				
Espárrago	X		X		X
Fresno	X				
Fumaria					
Granada	X				
Harmel			X		
Hinojo	X	X		X	X
Jujuba	X				
Junípero [[e]nebro]	X		X		X
Juzquyamo		X	X		
Lechuga	X				X
Lentium			X		
Linaza [lino]	X		X		X
Llantén			X		X
Malvas	X			X	X
Mastuerzo		X			
Melón	X				
Membrillo [cithoniore]	X		X		X
Milium solis [mijo del sol]	X		X	X	X
Nabos		X			
Nasturci[o]			X		
Neguilla [Nigele]			X		X
Orobo			X		
Ortiga	X	X	X		
Peonía	X	X	X		X
Pepino	X		X		X
Perejil	X	X		X	
Petrosilini macedonice			X		

Pedro de Bilbao *botica*[f] 1636	Hospital of Santa Ana[g] 1606	Fernando de Cartagena Cuzco[h] 1618	Common name	Latin name
			Curly endive	Cichorium endivia L.
			Curly endive	Cichorium endivia L.
X		X	Asparagus	Asparagus acutifolius L.
			Ash tree	Fraxinus spp.
X		X	Fumitory	Fumaria officinalis L.
			Pomegranate	Prunus granatum L.
			Harmel	Peganum harmala L.
		X	Sweet fennel	Foeniculum vulgare Mill.
			Jujube	Ziziphus jujuba Mill.
			Juniper	Juniperus communis L.
			Jusquiame, henbane?	Hyoscyamus niger L.
			Lettuce	Lactuca sativa L.
			Mastic?	Pistacia lentiscus L.
		X	Linseed	Linum usitatissimum L.
			Plantain	Plantago major L.
			Mallow	Malva spp.
		X	Garden cress	Lepidium sativum L.
			Melon	Cucumis melo L.
			Quince	Cydonia maliformis Mill.
		X	Gromwell	Lithospermum officinale L.
			Field mustard	Brassica rapa L.
			Watercress	Nasturtium officinale W.T. Aiton
			Nigella, black cumin, fennel flower	Nigella sativa L.
			Spring vetch	Orobus vernus L.
			Nettle	Urtica spp.
		X	Peony	Paeonia sp.
			Cucumber [cocumeris]	Cucumis sativa L.
X		X	Parsley	Petroselinum hortense Hoff.
			Macedonian parsley seed	Petroselinum crispum (Mill.) Nyman

		Andrés Zamudio de Alfaro[a] 1590s	Hospital of Santa Ana[b] 1551	Diego de Tineo *botica*[c] 1555	Francisco de Alva *botica*[d] 1576	Bartolomé Díaz Cabeza de Vaca *botica*[e] 1608
	Psilio [zargatona]	X	X	X	X	
	Puerros					X
	Rábanos		X			X
	Ruda			X		X
	Sauco					
	Verdolaga	X	X	X		
	Verdolaga fría	X				
	Zanahorias					X
Total	73	46	18	34	12	30
Roots						
	Alcaparras	X		X		X
	Aristoloquia larga	X				X
	Aristoloquia redonda	X				X
	Ásaro	X	X	X	X	X
	Boloños secas		X			
	Biznaga [viznaga]	X				
	Cálamo aromatico	X			X	
	Cautibo [cautivo] de mangle					X
	Cipero	X				
	Cohombrillo amargo secas		X			
	Cortezas de cidras	X				
	Díctamo blanco [real]	X			X	
	Ébolo	X				
	Eléboro blanco	X			X	
	Eléboro negro	X			X	

Pedro de Bilbao *botica*[f] 1636	Hospital of Santa Ana[g] 1606	Fernando de Cartagena Cuzco[h] 1618	Common name	Latin name
		X	Psyllium	Plantago pysllium L.
			Leek	Alium porrum L.
			Radish	Raphanus L.
			Rue	Ruta graveleons L.
		X	Elderberry	Sambucus negra L.
X		X	Purslane	Portulaca oleracea L.
			Purslane	Portulaca oleracea L.
			Carrot	Daucus carota L.
11	6	28		
			Caper bush root	Capparis spinosa L.
		X	Long birthwort, aristolochia root	Aristolochia longa L.
		X	Round birthwort, aristolochia root	Aristolochia rotunda L.
	X	X	Asarabacca, European wild ginger, hazelwort root	Asarum europaeum L.
			Dried root of henbane	Hyoscyamus niger L.
			Viznaga, khella root	Ammi visnaga Lam.
X		X	Calamus, sweet flag root	Acorus calamus L.
			Mangrove tree root	Prioria copaifera Griesb.
		X	Cyperus sedge root	Cyperus spp.
			Dried root of bitter cucumber	Ecballium elaterium (L.) A. Rich.
	X		Rind of citron	Citrus medica L.
		X	Dittany root	Dictamnus albus L.
			Red elderberry root	Sambucus racemosa L.
		X	Whitw hellebore root	Helleborus album L
		X	Black hellebore root	Helleborus niger L

		Andrés Zamudio de Alfaro[a] 1590s	Hospital of Santa Ana[b] 1551	Diego de Tineo *botica*[c] 1555	Francisco de Alva *botica*[d] 1576	Bartolomé Díaz Cabeza de Vaca *botica*[e] 1608
	Elemí [helenio, enula]	X				
	Esparraguera		X			
	Espicanardo				X	
	Esquila	X				
	Filipéndula					
	Galanga	X	X	X		X
	Genciana	X		X	X	
	Hermodactil	X	X	X		
	Lignum crucis	X				
	Lirio	X	X	X		X
	Malvavisco			X		
	Mechoacan	X				
	Palo [santo]	X				
	Pelitre	X		X		X
	Peonía	X				X
	Pes columbinus	X				
	Peucédano [perisidano]					
	Pie de palomylla		X			
	Polipodio	X				
	Racimillos de zumaque		X	X		
	Radix mandragore			X	X	
	Rubia [emtores]		X			X
	Rubia tinctorum			X		
	Saxifragia	X		X		
	Sello de santa maría	X				
	Texa blascas [rasuras de]		X			
	Turbit	X			X	X
	Valeriana	X		X	X	
	Zarza	X				
Total	44	30	11	13	10	11

Pedro de Bilbao *botica*[f] 1636	Hospital of Santa Ana[g] 1606	Fernando de Cartagena Cuzco[h] 1618	Common name	Latin name
		X	Elecampane root	Inula helenium L.
			Asparagus root	Asparagus acutifolius L.
X	X		Spikenard root	Nardostachys grandiflora DC, Nardostachys jatamansi DC
			Squill root	Scilla maritima L.
		X	Dropwort	Spiraea filipendula L.
X	X	X	Galangal root	Alpinia galanga (L.) Willd.
		X	Gentian root	Genciana spp.
	X		Hermodactyl, autumn crocus root	Colchicum autumnale L.
		X	Mistletoe root	Viscum album L.
			Lily root	Lilium candidum L.
			Marshmallow	Althaea officinalis L.
			Mechoacan root	Ipomoea purga (Wender.) Hayne
			Lignum vitae root	Guaiacum officinale L
		X	Pellitory root	Anacyclus officinarum Hayne
			Peony root	Paeonia sp.
			Crane's bill root	Geranium columbinum L
		X	Hog's fennel	Peucedanum officinale L.
			Fumitory root	Fumaria officinalis L.
	X		Polypody root	Polypodium vulgare L.
		X	Sumac root	Rhus spp.
			Mandrake root	Mandragora officinarum L.
X		X	Madder root	Rubia spp.
X		X	Madder root	Rubia tinctorum L.
			Saxifrage root	Saxifraga sp.
			Solomon's seal root	Polygonatum odoratum (Mill). Druce
			Yew shavings?	Taxus baccata L.
			Turpeth root	Operculina turpethum L.
			Valerian root	Valeriana offcinalis L.
			Sarsaparilla root	Smilax spp.
5	6	18		

		Andrés Zamudio de Alfaro[a] 1590s	Hospital of Santa Ana[b] 1551	Diego de Tineo *botica*[c] 1555	Francisco de Alva *botica*[d] 1576	Bartolomé Díaz Cabeza de Vaca *botica*[e] 1608
Herbs						
	Agrimonia	X		X		
	Ajenjo	X				X
	Artemisa	X				
	Betónica	X				
	Calamento	X		X		
	Camedrio	X				
	Cardo bendito	X				
	Centaurea					
	Centaurea menor	X				X
	Doradilla	X		X		
	Escordia	X				
	Espliego	X				
	Eufrasia	X				
	Hierba artética	X		X		
	Hipérico	X				
	Hisopo	X		X		X
	Hoja de laurel	X	X			
	Lengua de ciervo	X				
	Mastrantes		X			
	Mejorana [amoradux]	X	X	X		
	Meliloto			X		
	Menta	X				
	Orégano de la tierra			X		X
	Orminio [ormino]					
	Poleo	X	X	X		
	Romero		X	X		
	Ruda	X	X	X		

Pedro de Bilbao *botica*[f] 1636	Hospital of Santa Ana[g] 1606	Fernando de Cartagena Cuzco[h] 1618	Common name	Latin name
			Agrimony	Agrimonia eupatoria L.
X		X	Absinthe wormwood	Artemisia absinthium L.
			Wormwood	Artemisia vulgare L.
			Hedge nettle, betony	Stachyis officinalis Trev.
		X	Calamint	Calamintha officinalis Moench.
			Wall germander	Teucrium chamaedrys L.
			Blessed thistle	Cnicus benedictus L.
	X		Centaury	Centaurea centaurium L.
			Cornflower	Centaurea cyanus L.
			Rustyback fern	Asplenium ceterach L.
			Water germander	Teucrium scordium L.
			Lavender	Lavandula spica L.
			Eyebright	Euphrasia officinalis L.
			Ground pine	Ajuga chamaepitys (L.) Schreber.
			St John's wort	Hypericum perforatum L.
		X	Hyssop	Hyssopus officinalis L.
			Laurel	Laurus nobilis L.
			Hart's-tongue fern	Asplenium scolopendrium Newn.
			Santa Maria, mastrante,	Lantana scabiosiflora Kunth
			Marjoram	Origanum majorana L.
	X		Sweetclover	Melilotus officinalis Medikus.
			Peppermint	Mentha piperita L.
			Oregano	Origanum vulgare L.
X			Clary sage	Salvia sclarea L.
			Pennyroyal	Mentha pulegium L.
			Rosemary	Rosemarinus officinalis L.
			Rue	Ruta graveleons L.

		Andrés Zamudio de Alfaro[a] 1590s	Hospital of Santa Ana[b] 1551	Diego de Tineo *botica*[c] 1555	Francisco de Alva *botica*[d] 1576	Bartolomé Díaz Cabeza de Vaca *botica*[e] 1608
	Sabina	X		X		X
	Salvia	X		X		
	Taray [tamariz]	X		X		
	Tomillo			X		
Total	31	24	6	15	0	5
Flowers						
	Adormidera blanca [papaveris blanci]	X	X	X		X
	Balaustrias	X	X	X		X
	Borrajas	X	X			
	Cantueso [esticados]	X	X	X		
	Epítimo	X	X	X		X
	Lengua de buey	X	X	X		
	Manzanilla	X	X	X	X	
	Nenúfar	X	X			
	Prasio			X		
	Rosas comunes [seca]	X	X		X	
	Rosas cordiales	X				
	Rosas de Toledo			X		
	Violetas	X	X	X	X	X
Total	13	11	10	9	3	4
Gums and resins						
	Alcanfor				X	X
	Almaciga	X	X	X		X
	Alquitira [blanca]	X		X		X
	Ámbar gris		X			
	Amoniaco	X	X	X	X	X
	Asafetida	X	X	X	X	
	Bálsamo		X			
	Bálsamo de romero					

Pedro de Bilbao *botica*[f] 1636	Hospital of Santa Ana[g] 1606	Fernando de Cartagena Cuzco[h] 1618	Common name	Latin name
X		X	Savin juniper, savin	Juniperus sabina L.
			Sage	Salvia officinalis L.
			French tamarisk	Tamarix gallica L.
	X	X	Thyme	Thymus sp.
3	3	5		
			White poppy	Papaver somniferum L.
		X	Flowers of pomegranate	Prunus granatum L.
			Borage	Borago officinalis L.
		X	French lavender	Lavandula stoechas L.
X	X	X	Clover dodder	Cuscuta epithymum Murray
			Bugloss	Anchusa azurea Mill.
			Chamomile	Matricaria chamomilla L.
			Water lily	Nymphaea spp.
		X	Horehound	Marrubium vulgare L.
	X	X	Common rose	Rosa spp.
			Rose for cordial	Rosa spp.
			Roses from Toledo	Rosa spp,
	X	X	Violets	Viola spp.
1	3	6		
X		X	Camphor	Cinnamonum camphora L.
			Mastic	Pistacia lentiscus L.
		X	Gum [white] tragacanth	Astragalus gummifer Labill.
			Ambergris	
X	X	X	Gum ammoniac	Dorema ammoniacum D. Don
	X	X	Asafoetida	Ferula assa-foetida L.
X			Balsam [unspecified]	
X			Rosemary balsam	Rosmarinus officinalis L.

	Andrés Zamudio de Alfaro[a] 1590s	Hospital of Santa Ana[b] 1551	Diego de Tineo *botica*[c] 1555	Francisco de Alva *botica*[d] 1576	Bartolomé Díaz Cabeza de Vaca *botica*[e] 1608
Bodelio [bedelio]		X	X		
Canfora refinada		X	X		
Caraña	X				
Carave [carabe]		X	X		X
Carpo balsamo			X		
Cerrato sandalino					
Copal de méxico					X
Córtamo [cártamo]		X		X	
Estoraque	X	X		X	
Estoraque calamita			X		
Estoraque rubio			X		
Euphorbio	X	X	X	X	X
Gálbano	X	X	X	X	
Gálbano de levante					
Goma aráviga		X	X	X	
Goma de hiedra [yedra]	X	X	X	X	X
Goma de hinojo			X		
Goma elemí			X		
Goma molle				X	
Gomalaca [laca]		X	X		X
Incenso	X	X	X		X
Junípero	X				
Láudano [and purificado]	X	X	X		X
Liquidámbar					X
Menjuy		X	X		
Miera				X	
Mirra	X	X	X	X	X
Opio [fino]	X	X		X	X

Pedro de Bilbao *botica*[f] 1636	Hospital of Santa Ana[g] 1606	Fernando de Cartagena Cuzco[h] 1618	Common name	Latin name
		X	Bdellium	Commiphora sp.
			Refined camphor	Cinnamonum camphora L.
			Gumbo limbo	Bursera simaruba (L.) Sarg.
	X	X	Amber – fossilised resin	
			Torchwood	Amyris balsamifera L.
		X	Sandalwood	Santalum spp.
		X	Copal	Bursera spp.
			Safflower	Carthamus tinctorius L.
X			Storax, Liquidambar	Styrax officnalis L
		X	Calamite storax	Styrax sp.
			Red storax	Styrax sp.
X	X	X	Spurge resin	Euphorbia esula L.
		X	Galbanum	Ferula gummosa Bioss. (F. galbaniflua Bioss & Buhse.)
		X	Galbanum from the Levant	Ferula gummosa Bioss. (F. galbaniflua Bioss & Buhse.)
		X	Gum arabic	Acacia spp.
X	X	X	Gum ivy	Hedera helix L.
			Fennel resin	Foeniculum vulgare Mill.
		X	Gum resin [unspecified]	
			Peruvian peppertree	Schinus molle L.
X	X	X	Goma laca, shellac	Laccifer lacca (Lac beetle)
		X	Frankincense	Boswellia spp.
X			Juniper gum	Ziziphus jujuba Mill.
X		X	Laudanum, tincture of opium	Papaver somniferum L.
			Liquidambar	Liquidambar styraciflua L.
			Benzoin, benjui	Lindera benzoin (L.) Blume
			Juniper oil	Juniperus oxycedrus L.
X	X	X	Myrrh	Commiphora myrrha (Nees) Engl.
X		X	Opium	Papaver somniferum L.

		Andrés Zamudio de Alfaro[a] 1590s	Hospital of Santa Ana[b] 1551	Diego de Tineo *botica*[c] 1555	Francisco de Alva *botica*[d] 1576	Bartolomé Díaz Cabeza de Vaca *botica*[e] 1608
	Opoponaco	X	X	X	X	X
	Resina blanca				X	
	Sangre de drago	X	X	X	X	
	Sangre de drago de gota			X		
	Sarcocola [sarcacola]	X	X	X	X	X
	Serapino	X	X	X	X	
	Tacamaca [tacamahaca]	X				X
	Trementina		X		X	
	Unción					X
	X [j]ilo bálsamo			X		
Total	46	19	25	27	19	18
Animal parts						
	Almizcle	X	X	X		
	Antales		X			
	Blata visansis			X		X
	Canas de vaca	X				X
	Cantárida	X				
	Castóreo	X		X	X	X
	Cebo de mano				X	
	Cuerno de ciervo	X		X		X
	Dentales		X			X
	Enjundia de caballo				X	X
	Enjundia de gallina	X				
	Estiércol de lagarto [hienda]	X				
	Estinco	X				
	Grana chermes [kermes]			X		

Pedro de Bilbao *botica*[f] 1636	Hospital of Santa Ana[g] 1606	Fernando de Cartagena Cuzco[h] 1618	Common name	Latin name
		X	Sweet myrrh, opoponax	Commiphora erythraea (Ehrenb.) Engl.
			White resin	Unspecified
X		X	Dragon's blood	Dracaena spp.
			Drops of dragon's blood	Dracaena spp.
		X	Sarcocolla	Astragalus fasciculifolius Boiss
		X	Gum of giant fennel	Ferula communis L.
		X	Balsam poplar, tacamahaca	Populus balsamifera L.
		X	Turpentine	Pistacia terebinthus L.
		X	Unction [unspecified]	
			Balsam of Gilead	Amyris opobalsamum Forsk.
13	7	27		
			Musk	
			Antlers?	
			Eye powder	Listed as an animal product
			Marrow of beef	
X			Spanish fly, blister bee	
	X	X	Castoreum	
			Animal fat or grease	
			Deer horn	
			Teeth unspecified	
			Horse fat	
			Chicken fat	
X		X	Lizard dung	
			Skink	
			Kermes insect	Coccus ilicis L.

		Andrés Zamudio de Alfaro[a] 1590s	Hospital of Santa Ana[b] 1551	Diego de Tineo *botica*[c] 1555	Francisco de Alva *botica*[d] 1576	Bartolomé Díaz Cabeza de Vaca *botica*[e] 1608
	Grasa		X		X	
	Huesos de corazón de ciervo		X	X		
	Piedra bezar	X				
	Rasuras de marfil [eboris]	X		X	X	
	Sebo de cabrón	X				
Total	19	11	5	7	5	6
Non-organic simples						
Metals and minerals						
	Acero quemado				X	
	Albayalde	X	X	X	X	X
	Alhadida		X			
	Almártaga				X	
	Alquitrán					X
	Alumbre		X	X	X	
	Alumbre de pluma			X	X	
	Antimonio			X		
	Arsénico		X			
	Atutía preparada	X	X		X	X
	Atutía [alejandrina]					
	Azarcón				X	
	Azogue					X
	Azufre			X		X
	Azufre amarillo y verde		X			
	Bol arménico		X	X		
	Bol arménico oriental preparado					X
	Caparrosa de castilla			X		X
	Cardenillo		X	X		

Pedro de Bilbao *botica*[f] 1636	Hospital of Santa Ana[g] 1606	Fernando de Cartagena Cuzco[h] 1618	Common name	Latin name
X	X	X	Grease	
			Deer bones	
			Bezoar stone	
X		X	Ivory shavings	
			Goat's grease	
4	2	4		
			Steel	
		X	White lead, lead carbonate	
			Burnt copper	
		X	Litharge, lead oxide	
		X	Tar, pitch	
		X	Alum, potassium aluminium sulphate	
			Alum in the form of filaments	
X			Antimony	
			Arsenic	
		X	Tutty, zinc oxide, chimney soot from smelting of a mineral	
		X	Tutty, zinc oxide, chimney soot from smelting of a mineral	
		X	Lead oxide	
		X	Mercury	
			Sulphur	
			Yellow and green sulphur	
		X	Armenian bole, aluminium silicate (colored with ferrous oxide to give red earth)	
X		X	Armenian bole from the Levant	
			Copper sulphate, vitriol from Castile	
			Verdigris, acetate of copper	

		Andrés Zamudio de Alfaro[a] 1590s	Hospital of Santa Ana[b] 1551	Diego de Tineo *botica*[c] 1555	Francisco de Alva *botica*[d] 1576	Bartolomé Díaz Cabeza de Vaca *botica*[e] 1608
	Escoria ferri	X			X	
	Espodio	X		X	X	X
	Espodio preparado					
	Estibio de matriolo	X				
	Estibio preparado	X				X
	Lapis hematicus preparado					X
	Lapis judaicus			X		X
	Limadura de cascitis	X				
	Litargirio	X		X		
	Litium			X		
	Minio			X		X
	Mumia			X	X	X
	Oropimente	X	X		X	X
	Panes de oro			X		
	Pez de ávila					
	Piedra lipes					X
	Plomo quemado	X				X
	Rejalgar			X		X
	Sal gema [sar jema]		X	X		
	Solimán crudo		X	X		
	Solimán labrado		X			
	Vermellon			X		X
Total	41	10	12	19	11	18
Stones						
	Albín		X			
	Coral blanco	X		X		X
	Coral blanco preparado					X
	Coral preparado				X	
	Coral rubio [colorado]					X

Pedro de Bilbao *botica*[f] 1636	Hospital of Santa Ana[g] 1606	Fernando de Cartagena Cuzco[h] 1618	Common name	Latin name
			Iron slag	
	X	X	Mineral ash	
		X	Porcessed mineral ash	
X			Antimony	
		X	Processed antimony	
			Bloodstone	
			Jews' stone	
			Filings of copper sulphate or oxide	
			Litharge, almartaga, lead oxide	
			Lithium	
		X	Minium, red lead oxide	
		X	Mummy	
		X	Oropiment, sulphide of arsenic	
			Gold leaf	
		X	Pitch from Avila	
		X	Copper sulphate	
			Burnt lead	
			ReJalgar, arsenic sulphide	
			Halite	
			Crude mercury chloride	
			Processed mercury chloride	
X			Cinnabar	
4	1	18		
			Bloodstone	
		X	White coral	
X			Ground white coral	
			Ground coral	
		X	Coloured coral	

		Andrés Zamudio de Alfaro[a] 1590s	Hospital of Santa Ana[b] 1551	Diego de Tineo *botica*[c] 1555	Francisco de Alva *botica*[d] 1576	Bartolomé Díaz Cabeza de Vaca *botica*[e] 1608
	Coral rubio preparado			X		X
	Cristal	X		X	X	X
	Cristal preparado				X	
	Esmeralda	X	X			X
	Esmeraldas preparados					X
	Granato	X	X	X	X	X
	Hematites	X		X	X	X
	Jacinto [zircón]	X	X	X		
	Jacintos preparados				X	
	Lapis					
	Lapis antalis			X		
	Lapis dentalis			X		
	Lapis lazuli	X	X	X		X
	Lapis lincis			X		X
	Perlas preparados				X	
	Piedra imán	X	X	X	X	
	Rubí	X				
	Sanguinaria			X		
	Topacio	X				
	Zafiro	X				
Total	25	11	6	12	8	11
Chemicals						
	Alcohol					X
	Atincar fino		X	X		
	Sal armoniaco		X			
	Sali[s]tre		X			
	Vitriol		X			X
Total	5	0	4	1	0	2

Pedro de Bilbao *botica*[f] 1636	Hospital of Santa Ana[g] 1606	Fernando de Cartagena Cuzco[h] 1618	Common name	Latin name
X			Ground light coloured coral	
			Crystal	
			Ground crystal	
X			Emerald	
			Ground emeralds	
			Garnet	
		X	Haematite, bloodstone	
			Zircon	
			Ground zircon	
X			Stone [unspecified]	
			Unidentified	
			Tooth tartar	
X		X	Lapis lazuli	
			Lynx stone	
			Ground pearls	
			Magnetite	
			Ruby	
			Bloodstone	
			Topaz	
			Sapphire	
5	0	4		
		X	Alcohol	
		X	Refined tincal, or borax of commerce	
			Salt of ammonium chloride	
			Saltpetre	
		X	Vitriol, sulphuric acid	
0	0	3		

	Andrés Zamudio de Alfaro[a] 1590s	Hospital of Santa Ana[b] 1551	Diego de Tineo *botica*[c] 1555	Francisco de Alva *botica*[d] 1576	Bartolomé Díaz Cabeza de Vaca *botica*[e] 1608
Oils, syrups, juices and waters					
Oils					
[A]beto					X
Absintios [asensios, ajenjos]	X		X		
Adormideras	X	X	X	X	
Alacranes	X	X	X		X
Alcaparras	X	X	X	X	X
Alegría [ajonjolí]	X		X	X	X
Almastiga [almaciga]	X	X	X	X	X
Almendras amargas	X	X	X	X	
Almendras dulces	X	X	X	X	
Amargo					
Aparicio	X				
Aragón					X
Arrayán		X	X	X	
Azafrán	X				
Azahar		X	X		
Azucenas	X				
Bálsamo	X		X		
Bayas			X		X
C[h]ipre			X		
Castóreo	X	X			
Costo [sic]		X	X		
Dialtea					
Dulce					X
Enebro			X		X
Eneldo	X	X	X	X	X
Estoraque		X	X		
Euphorbio	X	X	X		
Higuerilla					X

Pedro de Bilbao *botica*[f] 1636	Hospital of Santa Ana[g] 1606	Fernando de Cartagena Cuzco[h] 1618	Common name	Latin name
			Spruce?	Picea abies (L.) H. Karst.
		X	Absinthe, wormwood	Artemisa absinthium L.
X			Poppy	Papaver somniferum L.
		X	Scorpion	
X		X	Caper	Capparis spinosa L.
		X	Sesame	Sesamum indicum L.
	X	X	Mastic	Pistacia lentiscus L.
	X	X	Biiter almonds	Amygdalus amara Hayne
	X	X	Sweet almonds	Amygdalus dulcis Mill.
X			Bitter oil [unspecified]	
		X	Aparicio oil [of olives]. Oil prepared by Aparicio de Zubia in fifteenth century	
			Recipe prepared with variety of ingredients	
		X	Myrtle	Myrtus communis L.
			Safrron	Crocus sativus
			Orange blossom	Citrus sinensis L.
			Lily	Lilium candidium L.
			Oil of balsam [unspecified]	
		X	Bay laurel	Laurus nobilis L.
			Cypress tree	Cupressus sempervirens L.
			Castoreum	
			Costus	Costus arabicus L.
		X	Dialthea, marshmallow	Althaea officinalis L.
X			Sweet oil	
			Juniper	Juniperus communis L.
X	X	X	Dill	Enethum graveolens L.
			Storax, Liquidambar	Styrax officnalis L
X			Spurge	Euphorbia esula L.
			Castor oil	Ricinis communis L.

	Andrés Zamudio de Alfaro[a] 1590s	Hospital of Santa Ana[b] 1551	Diego de Tineo *botica*[c] 1555	Francisco de Alva *botica*[d] 1576	Bartolomé Díaz Cabeza de Vaca *botica*[e] 1608
Hipérico	X				
Jazmín			X		
Ladrillo			X		
Laurel	X			X	
Linaza [lino]	X	X			X
Liquidámbar				X	
Lirio	X	X	X		
Lombrices	X		X	X	
Manzanilla	X	X	X	X	
Marciatón					X
María					
Mata		X		X	X
Matiolo	X				
Membrillos	X	X	X	X	
Menjuy		X	X		
Menta	X		X		
Mirto	X				
Mostaza [mosta?]		X	X	X	X
Nardino [despique]	X	X	X	X	
Nenúfar			X		
Nueces			X		
Onphancino			X		
Pepitas de calabaza	X		X	X	
Pirión				X	
Rábano			X	X	
Romero			X		
Rosado	X	X	X	X	X
Rosado onfancino	X			X	X
Ruda	X	X	X		X
Sauco	X		X		
Solato	X				

Pedro de Bilbao *botica*[f] 1636	Hospital of Santa Ana[g] 1606	Fernando de Cartagena Cuzco[h] 1618	Common name	Latin name
			St John's wort	Hypericum perforatum L.
			Jasmine	Jasminum sp.
			Brick pieces	
			Laurel	Laurus nobilis L.
	X	X	Flax/Linseed	Linum usitatissimum L.
		X	Liquidambar	Liquidambar styraciflua L.
	X	X	Lily	Lilium candidum L.
	X	X	Worms [unpecified]	
	X	X	Camomile	Matricaria chamomilla L.
		X	Oil invented by Marcianus	Complex
		X	Oil of Santa Maria	Calophyllum mariae Planch et Triana
		X	Bush [unspecified]	
X			Scorpion oil recipe of Andrea Matiolli	
	X	X	Quince	Cydonia maliformis Mill.
			Benzoin, benjui	Lindera benzoin (L.) Blume
			Peppermint	Mentha piperita L.
			Myrtle	Myrtus communis L.
			White mustard	Sinapis alba L.
	X		Spikenard	Spikenard
			Water lily	Nymphaea spp.
			Nuts unspecified	
			Unripe olive	
			Calabash seeds	Cucurbita sp. or Lagenaria sp.
			Piriona insect?	Piriona fasciculata Aldrich 1928
			Radish	Raphanus L.
			Rosemary	Rosemarinus officinalis L.
	X	X	Rose	Rosa spp.
			Rose in oil of unripe olive	
		X	Rue	Ruta graveleons L.
			Elderberry	Sambucus negra L.
			Unidentified	

		Andrés Zamudio de Alfaro[a] 1590s	Hospital of Santa Ana[b] 1551	Diego de Tineo *botica*[c] 1555	Francisco de Alva *botica*[d] 1576	Bartolomé Díaz Cabeza de Vaca *botica*[e] 1608
	Sopilativo [opilativo?]			X		
	Trementina		X	X		
	Unguento agripa		X	X		
	Víboras	X				
	Violado	X	X	X		
	Vulpino [raposo]	X	X	X		
	Yerba buena					X
	Ynsensias [incensio]				X	
	Yrino [irino]		X			
	Zorra		X			
Total	69	35	29	42	22	19
Syrups (*jarabe, lamedor*)						
	[Al]miba[r] aromática					X
	Absintios [ajenjo, asensios]	X		X		X
	Acederas	X				X
	Acetoso	X			X	X
	Acetoso y diarrodón				X	X
	Achichoria con ruibarbo	X				X
	Ácido de cidras	X				
	Adormideras	X				X
	Agraz	X			X	X
	Agrega					
	Almíbar de membrillos	X		X	X	
	Apestado					
	Arrayán			X	X	X
	Arrope		X			
	Arrope de moras		X		X	X

Pedro de Bilbao *botica*[f] 1636	Hospital of Santa Ana[g] 1606	Fernando de Cartagena Cuzco[h] 1618	Common name	Latin name
X			Unspecified	To remove obstruction
			Turpentine	Pistacia terebinthus L.
			Agrippa ointment	Based on bryonia (Bryonia sp.) and invented by the King of Agrippa
			Snake oil	
	X		Violet	Viola spp.
			Fox oil	
		X	Mint	Mentha spp.
			Frankincense	Hyssopus officinalis L.
			Iris	Iris germanica L.
		X	Fox oil	
8	12	25		
			Aromatic syrup, possibly of quince	
X			Absinthe, wormwood	Artemisa absinthium L.
			Sorrel	Rumex acetosa L.
			Vinegar and sugar	
			Vinegar, sugar and roses	
			Chicory and rhubarb	Cichorium endivia L. and Reum rharbarbarum L.
	X		Citron	Citrus medica L.
	X	X	Poppy	Papaver somniferum L.
		X	Verjuice [unripe grapes]	Vitis vinifera L.
	X		Unidentified	
	X		Quince syrup	Cydonia maliformis Mill.
	X		Unidentified	
X		X	Myrtle	Myrtus communis L.
			Syrup [unspecified]	
		X	Blackberry syrup	Rubus sp or Morus sp.

	Andrés Zamudio de Alfaro[a] 1590s	Hospital of Santa Ana[b] 1551	Diego de Tineo *botica*[c] 1555	Francisco de Alva *botica*[d] 1576	Bartolomé Díaz Cabeza de Vaca *botica*[e] 1608
Arrope de nueces	X	X			
Artemisa	X				
Bisancis	X		X		
Borrajas	X			X	X
Calabaca[s]					X
Camuesas				X	X
Cantueso [Estecados]	X				X
Cinco raíces [con vinagre]	X				X
Cortezas de cidras	X				X
Culantrillo	X			X	
Culantrillo de conserva	X				
Diamorón	X		X		
Dos raíces				X	X
Endivia [Chicoria]	X		X	X	
Epítimo	X		X		
Escoria ferri	X				
Eupatorio	X		X		
Fumaria compuesto	X				
Fumaria simple (Funisterra)	X	X	X		X
Granadas	X				
Guayacán					X
Hisopo	X		X		
Incensios				X	
Infusión de rosas	X		X		
Infusión violar	X				

Pedro de Bilbao *botica*[f] 1636	Hospital of Santa Ana[g] 1606	Fernando de Cartagena Cuzco[h] 1618	Common name	Latin name
			Nut syrup [unspecified]	
			Wormwood	Artemisa vulgare L.
			Endive, borage, celery and hops	Hops are Humerus lupulus L.
	X	X	Borage	Borago officinalis L.
			Calabash	Cucurbita or Lagenaria sp.
			Apples	Pyrus malus L.
	X		French lavender	Lavandula stoechas L.
			Five roots [parsley, celery, asparagus and butcher's broom	
	X	X	Rind of Citron	Citrus medica L.
	X	X	Maidenhair fern	Adiantrum capillus-veneris L.
			Maidenhair fern conserve	Adiantrum capillus-veneris L.
			Blackberry syrup	Rubus sp or Morus sp.
			Two roots [unspecified]	
	X	X	Endive, chicory	Cichorium endivia L.
			Clover dodder	Cuscuta epithymum Murray
			Iron slag	
X			Agrimony	Eupatorium sp.
			Complex fumitory syrup	Fumaria officinalis L.
		X	Simple fumitory syrup with fumaria, canafistula and tamarind	Fumaria officinalis L.
			Pomegranate	Prunus granatum L.
			Lignum vitae	Guaiacum officinale L
			Hyssop	Hyssopus officinalis L.
			Frankincense	Boswellia spp.
			Rose	Rosa spp.
			Violet	Viola spp.

	Andrés Zamudio de Alfaro[a] 1590s	Hospital of Santa Ana[b] 1551	Diego de Tineo *botica*[c] 1555	Francisco de Alva *botica*[d] 1576	Bartolomé Díaz Cabeza de Vaca *botica*[e] 1608
Jujubas [Azofeifas]	X		X		X
Laca	X				
Limones	X				X
Liquiricia [orosus]	X				X
Llantén			X		X
Membrillos	X			X	X
Menta [compuesto]	X		X		X
Miel de abejas					X
Miel de sentaura [centaurea?]					
Miel rosada	X		X	X	
Miel rosada colada			X		
Miel rosado pérsica	X				
Mirto y tusilago	X				
Nueve infusiones				X	
Oxicara			X		
Oximel	X				X
Oximel esquélitico	X				X
Oxizacara			X		
Polipodio					
Prasio			X		
Raíces				X	
Raíces con vinagre	X				
Raíces sin vinagre	X				
Rosado	X		X	X	X
Rosas secas	X			X	X
Sabor regis [rey sabor]					X
Sen					

Pedro de Bilbao *botica*[f] 1636	Hospital of Santa Ana[g] 1606	Fernando de Cartagena Cuzco[h] 1618	Common name	Latin name
			Jujube, Chinese or Red Date	Ziziphus zizyphus
			Goma laca, shellac	Laccifer lacca (Lac beetle)
	X		Lemon	Citrus limon (L.) Burm. f.
			Liquorice	Glycyrrhiza glabra L.
	X		Plantain	Plantago major L.
	X	X	Quince	Cydonia maliformis Mill.
			Peppermint	Mentha piperita L.
		X	Bee honey	
X			Centuary	Centaurea centaurium L.
	X	X	Rose honey	
			Strained rose honey	
			Rosa persica honey	Rosa persica J.F. Gmel.
			Myrtle and coltsfoot	Myrthus communis L. And Tussilago farfara L.
			Nine infusions [based on rose]	Rosa spp.
			Type of insect, unidentified	
		X	Honey in white wine vinegar	
	X		Honey in vinegar of squill	Scilla maritima L.
			Citric acid (agrio) with milk, honey and sugar	
	X		Polypody	Polypodium vulgare L.
			Horehound	Marrubium vulgare L.
	X	X	Roots [unspecified]	
			Roots with vinegar	
			Roots without vinegar	
	X	X	Rose	Rosa spp.
		X	Dried roses	Rosa spp.
			King Sabor syrup	Syrup of apples, borage, bugloss, aniseed, saffron
	X		Senna – leaves	Cassia senna L., Cassia acutifolia Delile

		Andrés Zamudio de Alfaro[a] 1590s	Hospital of Santa Ana[b] 1551	Diego de Tineo *botica*[c] 1555	Francisco de Alva *botica*[d] 1576	Bartolomé Díaz Cabeza de Vaca *botica*[e] 1608
	Tamarindos				X	
	Verdolagas	X				
	Vino					X
	Violado	X		X		X
	Violado de conserva	X				
	Vizenzis				X	
	Yerba buena				X	
	Zarzaparilla				X	
	Zumo de limón	X				
Total	76	47	4	21	22	33
Juices and infusions (zumo, infusion)						
	Acat[c]ia			X		
	Acederas					X
	Agraz	X			X	X
	Ajenjos [abcensios, absintios]	X		X	X	X
	Almirones				X	
	Borrajas	X			X	X
	Culantrillo				X	
	Endivia	X				X
	Eupatorio	X		X	X	X
	Fumaria [Funisterra]	X		X		
	Granadas	X	X	X		
	Hipoquistidos	X		X		
	Limones	X				
	Liquiricia	X				
	Lúpulos	X				
	Manzanas					X
	Membrillos	X	X	X	X	X
	Miel de centaurea	X				X
	Miel rosada colada				X	
	Opio	X		X		

Pedro de Bilbao *botica*[f] 1636	Hospital of Santa Ana[g] 1606	Fernando de Cartagena Cuzco[h] 1618	Common name	Latin name
			Tamarind	Tamrindus indica L.
			Purslane	Portulaca oleracea L.
			Wine	
	X	X	Violet	Viola spp.
			Violet conserve	Viola spp.
			Unidentified [bisantis?]	
			Mint	Mentha spp.
	X	X	Sarsaparilla	Smilax spp.
			Lemon	Citrus limon (L.) Burm. f.
4	21	18		
			Acacia	Acacia sp.
			Sorrel	Rumex acetosa L.
			Verjuice [unripe grapes]	Vitis vinifera L.
		X	Absinthe wormwood	Artemisia absinthium L.
			Chicory, endive	Cichorium intybus L.
			Borage	Borago officinalis L.
			Maiden hair fern	Adiantrum capillus-veneris L.
			Endive, chicory	Cichorium endivia L.
		X	Agrimony	Eupatorium sp.
			Fumitory	Fumaria officinalis L.
			Pomegranate	Prunus granatum L.
		X	Daisy	Bellis sp.
			Lemon	Citrus limon (L.) Burm.f.
			Liquorice	Glycyrrhiza glabra L
			Hops	Humerus lupulus L.
			Apple	Pyrus malus L.
		X	Quince	Cydonia maliformis Mill.
			Honey of centaury	Centaurea centaurium L.
			Strained rose honey	
			Opium	Papaver somniferum L.

		Andrés Zamudio de Alfaro[a] 1590s	Hospital of Santa Ana[b] 1551	Diego de Tineo *botica*[c] 1555	Francisco de Alva *botica*[d] 1576	Bartolomé Díaz Cabeza de Vaca *botica*[e] 1608
	Orosus		X			
	Planpina				X	
	Regaliz			X		
	Rosa seca					X
	Rosas		X	X	X	X
Total	25	14	4	10	10	11
Waters						
	[En]senape		X			
	Agua ardiente de cabeza		X		X	
	Alidonia [alidona]		X			
	Almirones		X			
	Apio		X			
	Arrayán	X				
	Asensios			X		
	Azaderas [acederas]				X	X
	Azahar		X		X	X
	Borrajas		X			
	Cabezas de rosas	X			X	X
	Celidonia	X		X		
	Cerrajes					
	Colirio					X
	Culantrillo				X	
	Endivia		X		X	X
	Eufrasia	X	X	X		
	Fumisterra			X		
	Galia[sic] moscata			X		
	Guindas	X				
	Hinojo	X	X		X	X
	Incensio		X			
	Lanfranco	X				

Pedro de Bilbao *botica*[f] 1636	Hospital of Santa Ana[g] 1606	Fernando de Cartagena Cuzco[h] 1618	Common name	Latin name
		X	Liquorice	Glycyrrhiza glabra L
			Duckweed, chickweed, pimpernel	
			Licorice extract	Glycyrrhiza glabra L.
			Dried rose	Rosa spp.
		X	Rose	Rosa spp,
0	0	6		
			White mustard seed	Sinapis alba L.
			Water for head fever	
			Stone in the intestines of a swalllow	
			Chicory, endive	Cichorium intybus L.
			Celery	Apium graveolens L.
			Myrtle	Myrtus communis L.
			Absinthe, wormwood	Artemisa absinthium L.
			Sorrel	Rumex acetosa L.
	X		Orange blossom	Citrus sinensis L.
	X	X	Borage	Borago officinalis L.
			Rose buds	Rosa spp.
			Celandine	Chelidonium majus L.
	X		Sow thistle	Sonchus oleraceus L.
			Eye wash	
			Maiden hair fern	Adiantrum capillus-veneris L.
			Endive	Cichorium endivia L.
			Eyebright	Euphrasia officinalis L.
			Fumitory	Fumaria officinalis L.
			Muscat wine	
			Sour cherry	Prunus cerasus L.
	X	X	Sweet fennel	Foeniculum vulgare Mill.
			Frankincense	Boswellia spp.
			Eye wash composed of wine, rose water, plantain water, oropiment and cardenillo	

		Andrés Zamudio de Alfaro[a] 1590s	Hospital of Santa Ana[b] 1551	Diego de Tineo *botica*[c] 1555	Francisco de Alva *botica*[d] 1576	Bartolomé Díaz Cabeza de Vaca *botica*[e] 1608
	Lengua de buey		X			
	Llantén	X	X		X	
	Luminosa	X				
	Malvas					X
	Manzanilla			X	X	
	Membrillos			X	X	X
	Panplina		X			
	Rosada	X		X	X	X
	Trébol				X	
	Vinagre esquilítico	X				X
	Vinagre rosado	X		X	X	X
	Yerba buena [hierba buena]		X			
Total	35	12	15	9	13	11
Complex medicines						
Electuaries						
	Atriaca de valencia		X	X		
	Benedicta	X		X	X	
	Confección [h]amech simple y compuesto	X	X	X	X	X
	Confección de genys			X		
	Diacatolicón	X		X	X	X
	Diafinicón	X		X		X
	Diaprunus laxativo	X		X	X	
	Diaprunus simple	X		X		X
	Diasén	X		X		
	Diaseterón			X		
	Hiera logodion	X	X	X		
	Hiera picra simple (gerapliega)	X		X		X
	Indio menor (mayor)	X		X	X	X

Pedro de Bilbao *botica*[f] 1636	Hospital of Santa Ana[g] 1606	Fernando de Cartagena Cuzco[h] 1618	Common name	Latin name
			Bugloss	Anchusa azurea Mill.
	X		Plantain	Plantago major L.
			Syrup of plantain and purslane mixed with alum and beaten egg white.	
			Mallow	Malva spp.
			Chamomile	Matricaria chamomilla L.
			Quince	Cydonia maliformis Mill.
			Chickweed	Stellaria media (L.) Vill.
	X	X	Rose	Rosa spp.
			Trefoil, clover	Trefolium L.
			Vinegar with squill	Scilla maritima L.
			Rose vinegar	
			Mint	Mentha spp.
0	6	3		
			Complex	
		X	Complex	
	X	X	Complex	
			Unidentified	
	X	X	Complex	
	X	X	Date and vinegar	
		X	Plum and white wine	
	X	X	Complex	
			Burnt silk, pearls, coral and lapis lazuli	
			Complex of satiron (ragwort)	Senecio jacobaea L.
			Complex	
	X	X	Complex	
		X	Complex	

		Andrés Zamudio de Alfaro[a] 1590s	Hospital of Santa Ana[b] 1551	Diego de Tineo *botica*[c] 1555	Francisco de Alva *botica*[d] 1576	Bartolomé Díaz Cabeza de Vaca *botica*[e] 1608
	Magistral de tor[r]es				X	
	Psilio			X		
	Rosado mesue	X		X	X	
	Tamaros yndos [tamarindos?]					X
	Triaca de diaseteron			X		
	Zumo de rosas [de nicolao]	X		X		
Total	19	12	3	17	7	7
Opiates and others						
	Antídoto emagogo	X				
	Confección de galanga					X
	Diacurcuma magna	X		X		
	Dialaca magna	X		X		
	Filonio pérsico	X	X		X	
	Filonio romano	X	X		X	X
	Llitontripon	X		X		X
	Micleta	X		X	X	X
	Mitridato	X	X			
	Requies magna	X				
	Triaca de toledo					X
	Triaca magna (con y sin opio)	X	X	X		
	Trifera magna	X				
Total	13	11	4	5	3	5
Cordials and others						
	Confección alquermes	X		X		
	Confección de jacintos	X				X
	Diambra	X				

Pedro de Bilbao *botica*[f] 1636	Hospital of Santa Ana[g] 1606	Fernando de Cartagena Cuzco[h] 1618	Common name	Latin name
	X		Complex	
			Psyllium	Plantago pysllium L.
	X	X	Complex	
		X	Based on tamarind	
			Complex based on satiron (ragwort)	Senecio jacobaea L.
	X		Complex	
0	8	10		
			Emmenagogue of Nicolau	
			Based on galangal	
			Based on saffron	
			Based on laca	
		X	Complex of Mesue	
		X	Complex	
			Complex	
			Based on myrobolans	
			Complex	
			Based on opium and mandrake	
		X	Complex	
X			Complex	
			Complex	
1	0	3		
	X	X	Dried bodies of the insect Kermes ilicis	
	X	X	Complex	
			Based on ambar	

		Andrés Zamudio de Alfaro[a] 1590s	Hospital of Santa Ana[b] 1551	Diego de Tineo *botica*[c] 1555	Francisco de Alva *botica*[d] 1576	Bartolomé Díaz Cabeza de Vaca *botica*[e] 1608
	Diamusco dulce	X				
	Diarrodón abatis	X			X	X
	Aromático rosado	X			X	X
	Polvos cetrinos			X		
	Polvos de bizma	X		X		X
	Polvos de diacimino	X			X	X
	Polvos de diadraganto	X				X
	Polvos de diairis	X				
	Polvos de diamargariton frío	X			X	X
	Polvos de diamusco	X				X
	Polvos de diatrion piperon	X				
	Polvos de gera				X	
	Polvos de vigo [de juanes]	X	X	X		X
	Polvos restrictivos				X	
	Triaca de esmeraldas	X				X
	Triasándalos	X				
Total	19	16	1	4	6	10
Pills						
	Agárico	X				X
	Agregativos	X			X	
	Alefanginas	X		X		
	Artéticas	X				
	Asaharayet [asegered]	X				X
	Aúreas	X				X
	Coquias (Cochias)	X				X
	Fétidas	X				
	Fumaria [fumisterra]	X		X		X

Pedro de Bilbao *botica*[f] 1636	Hospital of Santa Ana[g] 1606	Fernando de Cartagena Cuzco[h] 1618	Common name	Latin name
		X	Based on musk	
		X	Based on rose	
		X	Based on rose	
X			Citrinous powder	
			Complex	
		X	Based on cumin	
		X	Based on gum tragacanth	
			Based on iris or lily	
	X	X	Complex	
		X	Complex	
			Based on pepper	
		X	Complex	
X		X	Red precipitate of mercury	
		X	Restrigent	
	X		Anti-venom of emeralds	
		X	Three types of sandalwood	
2	4	13		
	X	X	Agaric, mushroom	Polyporus sp.
	X	X	Complex	
			Complex	
			Complex	
			Complex	
	X	X	Complex	
	X	X	Colocynth, scammony and hiera picra	
	X		Leek	Alium porrum L.
	X	X	Fumitory	Fumaria officinalis L.

		Andrés Zamudio de Alfaro[a] 1590s	Hospital of Santa Ana[b] 1551	Diego de Tineo *botica*[c] 1555	Francisco de Alva *botica*[d] 1576	Bartolomé Díaz Cabeza de Vaca *botica*[e] 1608
	Hermodactiles	X				X
	Hiera simple	X				
	Indias	X				
	Lapis lazuli	X				
	Lucis	X				
	Mastique	X				X
	Nitro					
	Ruibarbo	X				X
	Sine quibus	X				
	Stomaticas					X
Total	19	17	0	2	1	9
Powders						
	Hiera simple de Galeno					
	Empello de bacas lauri (Baccis lauri]					X
	Alexandria					X
	Almaciga					X
	Almártaga					X
	Arrayán					X
	Coloquíntida					X
	Lirios					X
	Margaritas					
	Minio					X
	Mirra					X
	Oropimente					X
	Rosa					X
	Sándalos citrinos					X
	Sándalos colorados					X

Pedro de Bilbao *botica*[f] 1636	Hospital of Santa Ana[g] 1606	Fernando de Cartagena Cuzco[h] 1618	Common name	Latin name
			Hermodactyl, autumn crocus	Colchicum autumnale L.
			Complex	
			Prepared with salt from India	
		X	Lapis lazuli	
		X	Complex	
			Mastic	Pistacia lentiscus L.
	X		Potassium nitrate	
		X	Rhubarb	Rheum rhabarbarum L.
		X	Aloe	Aloe vera
			Stomach pills	
0	7	9		
	X		Complex Galenic mixture	
			Complex	
			Alexandrian senna	Senna alexandrina Mill.
			Mastic	Pistacia lentiscus L.
			Litharge, lead oxide	
			Myrtle	Myrtus communis L.
	X		Colocynth, bitter apple	Citrullus colocynthis (L.) Schrad.
			Lily	Lilium candidum L.
		X	Pearls	
		X	Minium, red lead oxide	Commiphora myrrha (Nees) Engl.
			Myrrh	
			Oropiment, sulphide of arsenic	
			Rose	Rosa spp.
			Yellow sandalwood	Santalum rubrum L.
			Red sandalwood	Symphytum officinale L.

		Andrés Zamudio de Alfaro[a] 1590s	Hospital of Santa Ana[b] 1551	Diego de Tineo *botica*[c] 1555	Francisco de Alva *botica*[d] 1576	Bartolomé Díaz Cabeza de Vaca *botica*[e] 1608
	Suelda					X
	Unguento citrino			X		
Total	17	0	0	1	0	14
Plasters						
	Apostilcón cirújico					X
	Cartagines					X
	Centaurea	X		X		X
	Ceroto de hisopo	X		X		X
	Ceroto confortativo de vigo	X	X	X		X
	Contrarotura	X	X	X		
	De pele arietine					X
	Diafinicón	X	X	X		
	Dialthea					
	Diapalma	X	X	X	X	
	Diaquilón armóniaco			X		
	Diaquilón mayor y menor	X	X	X	X	X
	Empello ceroneo			X		X
	Empello relato					X
	Estomaticón	X				
	Filhi zacahariae [empello zacharias?]	X				X
	Geminis	X		X		X
	Gracia dei		X	X		X
	Griego			X		
	Guillen serven	X	X	X		
	Gumado	X				

Pedro de Bilbao *botica*[f] 1636	Hospital of Santa Ana[g] 1606	Fernando de Cartagena Cuzco[h] 1618	Common name	Latin name
			Suelda con suelda, comfrey	Santalum citrimum J.B.
			Ointment of citron peel	
0	2	2		
			Surgical ointment	
	X	X	Complex	
		X	Centaury	Centaurea centaurium L.
			Hyssop	Hyssopus officinalis L.
		X	Complex	
		X	Complex	
			Sheep's skin	
		X	Based on dates	
		X	Dialthea, marshmallow	Althaea officinalis L.
	X	X	Based on litharge and palm oil	
			With ammonium chloride	
	X	X	Complex	
			Complex	
			Unidentified	
	X	X	Complex	
		X	Complex	
			Based on albayalde	
		X	Based on turpentine and incense	
	X		Based on verdgris and litharge	
		X	Complex	
	X		Based on resins of bdellio, fennel and gum amoniaco	

		Andrés Zamudio de Alfaro[a] 1590s	Hospital of Santa Ana[b] 1551	Diego de Tineo *botica*[c] 1555	Francisco de Alva *botica*[d] 1576	Bartolomé Díaz Cabeza de Vaca *botica*[e] 1608
	Isi aedigonis (Isis epigoni)	X				
	Madre					
	Meliloto	X	X	X		X
	Mundificativo de nido de golondrina	X				
	Negro					
	Nicolao massa					X
	Oxicrocio	X		X		X
	Pro matrice [Admatorcini]	X		X		
	Raíz					
	Ranas	X				X
	Sándalos colorados					X
	Triafármaco	X				X
Total	33	19	8	16	2	18
Ointments						
	Agripa	X	X	X	X	
	Almártaga					X
	Apostolorum	X	X		X	X
	Aragón	X		X	X	
	Armoniaco				X	
	Atutía			X	X	X
	Aúreo	X			X	
	Azahar					X
	Basilicón	X				X
	Blanco	X			X	X
	Castaña	X				
	Ceroto litargirio	X				

Pedro de Bilbao *botica*[f] 1636	Hospital of Santa Ana[g] 1606	Fernando de Cartagena Cuzco[h] 1618	Common name	Latin name
			Complex	
		X	No fixed ingredients, chosen by particular apothecaries	
		X	Sweetclover	Melilotus officinalis Medikus.
			Based on swallow's nests	
		X	Based on albayalde	
	X	X	Recipe of Nicolao Massa	
	X	X	Complex	
			Complex	
		X	Unspecified	
	X	X	Frogs and other ingredients	
			Red sandalwood	Santalum rubrum L.
	X		Based on litharge	
0	10	19		
	X	X	Complex	
	X	X	Litharge, lead oxide	
	X	X	Complex	
		X	Complex	
		X	Ammonium chloride	
	X		Tutty, zinc oxide, chimney soot from smelting of a mineral	
	X		Slag from gold smelting	
	X	X	Orange blossom	Citrus sinensis L.
			Complex	
	X	X	Based on white lead	
			Chestnut	Castanea sativa L.
			Litharge, almartaga	

	Andrés Zamudio de Alfaro[a] 1590s	Hospital of Santa Ana[b] 1551	Diego de Tineo *botica*[c] 1555	Francisco de Alva *botica*[d] 1576	Bartolomé Díaz Cabeza de Vaca *botica*[e] 1608
Cetrino		X			
Comitisse			X		
Condesa				X	
Confortativo [de vigo?]		X			
Cordial	X				
Deopilativo del bazo	X			X	
Deopilativo del higado	X				X
Dialthea	X	X	X		
Dolor de costado					
Egipciaco	X				
Estoraque			X		
Goma elemí	X				
Hiera logadion					X
Hisopo húmedo	X		X		
León					X
Litargirio	X				
Marciatón	X	X	X	X	
Media confección	X				X
Minio	X				X
Plomo	X				
Pompholigos [de tutía]	X		X		
Populeón	X	X	X	X	X
Pro morbo					
Resuntivo		X			
Rosado	X	X	X	X	X
Rubio de vigo	X				
Sandalino	X		X	X	X

Pedro de Bilbao *botica*[f] 1636	Hospital of Santa Ana[g] 1606	Fernando de Cartagena Cuzco[h] 1618	Common name	Latin name
			Yellow ointment unspecified	
			Complex	
	X	X	Complex	
		X	Red precipitate of mercury	
			Complex	
			Ointment to remove obstructions to spleen	
			Ointment to remove obstructions to liver	
			Dialthea, marshmallow	Althaea officinalis L.
		X	Ointment for pain in the side	
	X	X	Alum, verdigris	
			Storax, Liquidambar	Styrax officnalis L
			Diverse resins	
			Complex	
			Complex	
			Unidentified	Recipe of Doctor Leon?
			Litharge, almartaga	
	X	X	Complex ointment invented by Marcianus	
			Mastic, frankincense	
	X	X	Minium, red lead oxide	
		X	Lead	
			Tutty, lead, frankincense	
X			Complex	
	X		Ointment to treat syphilis	
			Restorative	
	X	X	Rose	Rosa spp.
		X	Red precipitate of mercury	
	X		Sandalwood	Santalum spp.

		Andrés Zamudio de Alfaro[a] 1590s	Hospital of Santa Ana[b] 1551	Diego de Tineo *botica*[c] 1555	Francisco de Alva *botica*[d] 1576	Bartolomé Díaz Cabeza de Vaca *botica*[e] 1608
	Sopilativo [desopilativo]		X	X		
	Terbenta			X		
	Trementina de abeto	X		X		
	Unción					X
	Verde			X	X	
Total	44	26	10	16	14	15
Other						
Conserves						
	Borrajas	X	X	X		
	Cantueso	X	X	X		
	Carne de ciruela	X	X		X	
	Carne de ciruelas silvestres	X				
	Carne de dátiles			X		X
	Culantrillo de pozo	X				
	Hierbabuena	X				
	Higos negros			X		
	Lengua de buey	X		X		
	Mirabolanos			X		
	Nueces de nogal			X		
	Rosada	X		X		
	Rosada de pérsica de miel y azúcar	X		X		
	Servas			X		
	Violada	X	X	X		
Total	15	10	4	11	1	1

Pedro de Bilbao *botica*[f] 1636	Hospital of Santa Ana[g] 1606	Fernando de Cartagena Cuzco[h] 1618	Common name	Latin name
	X	X	Unspecified	To remove obstruction
			Turpentine, terebinth?	Pistacia terebinthus L
			Turpentine from spruce	Picea abies (L.) H. Karst.
			Unction unspecified	
			Green	
1	15	17		
			Borage	Borago officinalis L.
			French lavendar, spike	Lavandula stoechas L.
			Plum	Prunus spp.
			Wild plums	Prunus spp.
X		X	Date palm	Phoenix dactylifera L.
			Fern from the well	Adiantum capillus-veneris L.
			Mint	Mentha spp.
			Black fig	Ficus carica L.
			Bugloss	Anchusa azurea Mill.
			Myrobalan	Terminalia chebula L.
			Walnut	Junglans regia L.
			Rose	Rosa spp.
			Rosa persica	Rosa persica J.F. Gmel.
			Fruits of serbal (rowan)	Scorbus sp.
		X	Violet	Viola spp.
1	0	2		

		Andrés Zamudio de Alfaro[a] 1590s	Hospital of Santa Ana[b] 1551	Diego de Tineo *botica*[c] 1555	Francisco de Alva *botica*[d] 1576	Bartolomé Díaz Cabeza de Vaca *botica*[e] 1608
Medicinal pieces and ingredients (trociscos)						
	Absintios	X				X
	Alcaparras	X				
	Alquequenje	X				
	Bérbero	X	X			
	Blancos de rasis	X				
	Carabe	X				X
	Diarrodon	X	X			X
	Espodio con simientes	X	X			X
	Eupatorio	X				
	Galia alefangina	X				X
	Galia moscata de mesue	X	X			
	Galia moscata de nicolao	X				
	Mirica [mirra]	X	X			
	Ramich	X				
	Rasis con/sin opio					X
	Ruibarbo	X				X
	Terra sigillata	X		X		
Total	17	16	5	1	0	7
Flour						
	Abenate [avenate]			X		
	Almidón		X			
	Almortas	X				

Pedro de Bilbao *botica*[f] 1636	Hospital of Santa Ana[g] 1606	Fernando de Cartagena Cuzco[h] 1618	Common name	Latin name
			Absinthe, wormwood	Artemisa absinthium L.
			Caper	Capparis spinosa L.
	X		Bladder cherry, winter cherry, Chinese lantern	Physalis alkekengi L.
		X	Barberry	Berberis vulgaris L.
			Lead carbonate prepared after Rasis	
			Amber – fossilised resin	
			Rose	Rosa spp.
			Mineral ash with seeds	
			Agrimony	Eupatorium sp.
			Complex based on myrobolanos	
			Complex based on nutmeg after Mesue	
			Complex based on nutmeg after Nicolao	
			Myrrh	Commiphora myrrha (Nees) Engl.
			Sorrel and others	Rumex acetosa L.
	X	X	Resin with/out opium	
			Rhubarb	Rheum rhabarbarum L.
	X		A clay mineral	
0	3	2		
			Oatmeal	
			Flour unspecified	
			White pea	Lathyrus sativus L.

		Andrés Zamudio de Alfaro[a] 1590s	Hospital of Santa Ana[b] 1551	Diego de Tineo *botica*[c] 1555	Francisco de Alva *botica*[d] 1576	Bartolomé Díaz Cabeza de Vaca *botica*[e] 1608
	Bizcocho blanco [del coro]		X			
	Cebada	X				
	Habas	X				
	Lenteja	X				
	Tortas de rosas			X		
	8	4	2	2	0	0
Miscellaneous						
	Aguardiente		X			
	Amento dulce			X		
	Aureo pigmento			X		
	Azúcar cande		X		X	
	Azúcar quebrados				X	
	Azúcar rosado		X			X
	Capullos de seda					X
	Cera amarilla					
	Rasuras blancas [de vino]			X		
	Rasuras tintas			X		
	Seda cruda			X		
	Sumu terra				X	
	Taco almagra					
Total	13	0	3	5	3	2
Total	863	429	251	391	214	325

Sources:

a Recommendations of the Spanish *protomédico* Andrés Zamudio de Alfaro in the 1590s: *Catálogo de las cosas que los boticarios han de tener en sus boticas* in Davis and López Terrada, "Protomedicato y farmacia," 591–612.

b First pharmacy of the hospital of Santa Ana: AGNP Protocolos Siglo XVI 63 Diego Gutiérrez fols. 177v.-178 Quenta de la botica 20 Oct. 1551. See also: Miguel Rabí Chara, "La primera botica de los hospitales de la ciudad de Lima en el siglo XVI," *Asclepio* 52 (2000): 276–77.

Pedro de Bilbao *botica*[f] 1636	Hospital of Santa Ana[g] 1606	Fernando de Cartagena Cuzco[h] 1618	Common name	Latin name
			White biscuit	
			Barley flour	Hordeum vulgare L.
			Bean flour	Vicia faba L.
			Lentils	Lens culinaris Medikus
			Rose cakes	
0	0	0		
			Aguardiente	
			Nut wine	
			Gold pigment	
			Sugar candy	
			Brown sugar	
	X	X	Rose sugar	
X			Silkworm cacoon	
		X	Yellow wax	
		X	White wine residue dye	
			Red wine residue dye	
			Raw silk	
			Unidentified	
		X	Unidentified	
1	1	4		
94	136	290		

c *Botica* imported by Diego de Tineo: JCB Mss codex Sp 136 fols. 861–868 Francisco Martínez y compañía obligación – Diego de Tineo y consortes 1555.

d Inventory of the *botica* of Francisco de Alva: AGNP Real Audiencia. Causas Civiles legajo 16 cuad. 81 fols. 10v-15 Inventario de los bienes del bachiller Alba [sic], después de fallecido. 1576.

e Inventory of the *botica* of Bartolomé Diaz Cabeza de Vaca: AAL Testamentos 5 exp. 1 fols. 44–49v. Memoria de drogas y medicinas … 6 Sep. 1608.

f Inventory of the *botica* of Pedro de Bilbao: AGNP Protocolos Siglo XVII 1789 Sánchez Vadillo fols. 1372v.-1378v. Inventario de bienes de Pedro de Bilbao 25 Aug. 1636.
g Inspection of the pharmacy of the hospital of Santa Ana in 1606. ABPL 9086 fols. 118–123 Visita of the *botica* of the hospital of Santa Ana by Doctor Melchor Romero 9 and 10. Oct. 1606.
h Inventory of the *botica* of Fernando de Cartagena, Cuzco 1618: "Una botica colonial," *Revista del archivo histórico del Cuzco* 4 (1953): 263–82.

The total number of plant species found in pharmacy inventories is slightly smaller than is indicated in the above table. Pharmacies often listed different parts of the same plant separately, for example, as simples, seeds and roots, so that some are counted more than once. Since duplicate species entries accounted for less than ten per cent of the total number of botanical products listed above and since pharmacists distinguished the different forms in which they were stored, they are retained here as separate entries.

While every attempt has been made to identify the *materia medica* and provide English names, it is recognised that this is not always an easy task. This applies also to the Latin names for plants, some of which have changed over time. The names of products in different pharmacies varied slightly. A large number of texts and databases were used, among the most useful were:

The Plant List. http://www.theplantlist.org/ [Accessed 1 Nov. 2015]; USDA United States Department of Agriculture, Plants Database. http://plants.usda.gov/java/ [Accessed 1 Nov. 2015]; International Plant Names Index. http://www.ipni.org/[Accessed 1 Nov. 2015]; Lily Y. Beck (trans.) *Pedanius Dioscorides of Anazarbus: De Materia Medica* (Hildesheim, Zurich, New York: Olms-Weidmann); Hermilio Valdizán and Angel Maldonado, *La medicina popular peruana* (Lima: Torres Aguirre, 1922); Enrique Laval, *Botica de los Jesuitas de Santiago* (Santiago: Asociación Chilena de Asistencia Social, 1953); Paula De Vos, "European Materia Medica in Historical Texts: Longevity of a Tradition and Implications for Future Use," *Journal of Ethnopharmacology*, 132 (2010): 31–39; Charles Davis, and María Luz López Terrada, "Protomedicato y farmacia en Castilla a finales del siglo XVI: edición crítica del *Catálogo de las cosas que los boticarios han de tener en sus boticas*, de Zamudio de Alfaro, protomédico general (1592–1599)," *Asclepio* 62 (2) (2010): 579–626.

For the composition of complex medicines the following were used: Enrique Laval, *Botica de los Jesuitas de Santiago* (Santiago: Asociación Chilena de Asistencia Social, 1953) and Luis de Oviedo, *Methodo de la colección, y reposición de las medicinas simples, de su corrección, y preparación* (Madrid: Luis Sánchez, 1622). Although first published in 1581, the 1622 edition of Oviedo is used since it is more comprehensive. The second edition in 1595 added electuaries, syrups, pills, oils and medicinal ingredients, and the third edition in 1609 included ointments and plasters.

APPENDIX C

Books Shipped from Spain to Doctor Melchor de Amusco in Nombre de Dios, 1584

Title in manuscript	Author	First date and place of publication	Suggested title of book
un gualtero bruelo practico	Gualtherus Bruele	Antwerp, 1579	Praxis medicinae theorica et empirica familiarissima
controversias de valles dos pares	Francisco Valles	Alcalá, 1556	Controversiarum medicarum et philosophicarum libri decem.
los galenos en quatro cuerpos	Galen		Four unspecified volumes of Galen
un avicena	Avicenna		Unspecified
tabula consilios joanes mater y de gracia			Unidentified
donato antonio	Donato Antonio Altomare	Venice, 1562	De medendis febribus ars medica
las obras de vega	Cristóbal de Vega	Lyon, 1564	Opera, nempe, Liber de arte medendi
paulo	Paulus Aegineta	Basel, 1556	Medicinæ totius enchiridion
gordonio	Bernardo de Gordonio	Sevilla, 1495	Lilio de medicina
savanorola y herculano			Unidentified
arnaldo	Arnau de Vilanova		Unspecified
consiliador	Petrus de Abano	Venice, 1565	Conciliator differentiarum quae inter philosophos et medicos versantur
practica de savanarola	Joannes Michael (Savonarola)	Venice, 1497	Practica maior
una biblia			A bible

 | DOI 10.1163/9789004351271_012

Title in manuscript	Author	First date and place of publication	Suggested title of book
laguna	Andrés Laguna		Unspecified
galeno de sanitate tuenda y otros cuerpos chicos todos de galeno	Galen	Paris, 1517	De sanitate tuenda: libri sex
sensuras de mesue	Bartolomaeus Angelo Paglia	Venice, 1543	In Antidotarium Joannis filii Mesue censura cum declaratione simplicium medicinarum
bernaldo de senio	Bernardino de Siena		Sermons, unspecified
tagavejo	Vittore Trincavelli	Basel, 1571	De compositione et usu medicinalium
articela	Franciscus Argilagnes	Venice, 1483	Articella, seu, Opus artis medicinae.
nicolao proposito	Nicole Prévost	Lyon, 1491	Dispensarium magistri Nicolai prepositi ad aromatarios
estilo descrevir cartas	Juan de Icíar	Zaragoza, 1552	Nuevo estilo descreuir cartas mensageras sobre diversas materias
mercado de tavardillo	Luis Mercado	Valladolid, 1574	De essentia causis signis et curatione febris malignae in qua maculae rubentes similes morsibus pulicum per cutem erumpunt.
valerio	Valerio Cordo	Venice, 1546	Dispensatorium, hoc est Pharmacorum conficiendorum ratio
guerra de portugal	Francisco Díaz de Vargas	Zaragoza, 1581	Discurso y summario de la guerra de Portugal y successos della

Title in manuscript	Author	First date and place of publication	Suggested title of book
coloquio sobre la dentadura	Francisco Martínez de Castrillo	Valladolid, 1557	Coloquio breve y compendioso sobre la materia de la dentadura y maravillosa obra de la boca
aviso de sanidad	Francisco Núñez de Oria	Madrid, 1572	Auiso de sanidad que trata de todos los géneros de alimentos y del regimiento de la sanidad co[m]prouado por los más insignes y graues doctore
libro de cozina	Ruberto de Nola	Toledo, 1525	Libro de cozina [sic]
el parto de la virgen	Jacopo Sannazaro	Salamanca, 1569	El parto de la virgen
metafora de medicina	Bernardino de Laredo	Sevilla, 1522	Metaphora medicine
modus facendi	Bernardino de Laredo	Sevilla, 1527	Modus faciendi cum ordine medicandi
batalla de roncesvalles	Francisco Garrido de Villena	Valencia, 1555	El verdadero sucesso de la famosa batalla de Roncesualles
problemas de aristoteles	Aristotle	Unspecified	Problemata Aristotelis ac philosophorum medicorumque complurium
franco de peste	Francisco Franco	Sevilla, 1569	Libro de enfermedades contagiosas y de la preservación dellas
Manuscript torn			
Manuscript torn			
marco aurelio	Antonio de Guevara	Sevilla, 1528	Libro aureo de Marco Aurelio
lucio apuleyo	Lucio Apuleyo	Sevilla, 1520	Libro del Lucio Apuleyo del asno de oro
galateo espanol	Lucas Gracián Dantisco	Madrid, ca 1582	Galateo español
luzman	Jerónimo de Contreras	Barcelona, 1565	Selva de aventuras

Title in manuscript	Author	First date and place of publication	Suggested title of book
viaje y nufragio del macedonico	Juan Bautista de Loyola	Salamanca, 1578	Viaje y naufragio del macedonio de Loyola
filonio de taranta	Valesco de Tarenta	Barcelona, 1484	Practica, quae alias Philonium dicitur
collado de indicatione	Luis Collado	Valencia, 1572	De indicationibus librum unum
joanes de vigo	Giovanni de Vigo	Rome, 1514	Practica in arte chirurgica continens novem libros infra scriptos
virgilio en romance			Virgil in Spanish
pedraza de conciencia	Iuan de Pedraza	Barcelona, 1566	Summa de casos de conciencia
valles de revictus in acutis	Francisco Valles	Alcalá, 1569	Commentaria in libros Hippocratis de Ratione victus in morbis acutis
antonio musa de morbo galico	Antonio Musa Brasavola	Venice, 1553	De morbo gallico
farmacopaea de silvio	Jacobo Sylvio	Lyons, 1548	Pharmacopoeae Iacobi Sylvii
rondoleto	Guillaume Rondelet	Paris, ca 1545	Methodus curandorum omnium morborum corporis humani in tres libros distincta
dino florentino sobre avicena	Dino del Garbo	Ferrara, 1489	Expositio super tertia, quarta, et parte quintae fen IV: Canonis Avicennae.
anatomia de montanana [sic]	Bernardino Montana de Monserrate	Valladolid, 1551	Libro de la anothomía del hombre
lemos sobre el metodo	Luis de Lemos	Salamanca, 1581	Methodus medendi
araucana primera y segunda parte	Alonso de Ercilla y Çuñiga	Madrid, 1569	La Araucana, primera y segunda partes
silvio sobre me [mesue]	Jacobo Sylvio	Paris, 1542	Joannis Mesuae Damaceni ... Jacobo Sylvio

Title in manuscript	Author	First date and place of publication	Suggested title of book
fragoso de botica	Juan Fragoso	Alcalá, 1566	Catalogus simplicium medicamentorum
actuario de urinis	Joannes Actuarius	Venice, 1519	De urinis
antonio musa de simples	Antonio Musa Brasavola	Lyon, 1537	Examen omnium simplicium medicamentorum
valdes de la utilidad de la sangria	Fernando de Valdés	Sevilla, 1582	Tratado de la utilidad de la sangría en las viruelas y otras enfermedades de los muchachos
reportorio de chaves	Jerónimo de Chaves	Sevilla, 1548	Chronographía o repertorio de los tiempos
gaynero	Antonio Guainerio	Pavia, 1481	Opera medica
petro bacitano			Unidentified
joanes bacanelus	Giovanni Battista Baccanelli	Venice, 1553	De consensu medicorum in curandis morbis libri quatuor
epidemias	Hippocrates	Paris, 1546	Libri epidemiorum Hippocratis primus, tertius, et sextus, cum Galeni in eos commentariis
daca de urinis	Alfonso Díez Daza	Sevilla, 1577	Artium et medicinae doctoris
aphorisimos de hipocrates 2 pares	Hippocrates	Unspecified	Aphorismi
cornelio celso	Aulus Cornelius Celso	Florence, 1478	De medicina
inquiridion	Alfonso López Corella	Zaragoza, 1549	Enchiridion medicinae
antonio musa de simples	Antonio Musa Brasavola	Rome, 1536	Examen omnium simplicium medicamentorum
pronosticos			Unspecified almanac
la conquista del penon	Baltasar de Collazos	Valencia, 1566	Commentarios de la fundación y conquistas y toma del Peñon

Title in manuscript	Author	First date and place of publication	Suggested title of book
pronosticos de valles	Francisco Valles	Alcalá, 1567	Commentaria in prognosticum Hippocratis
carta pacio para confesar			Copy book for confession
merino practico	Diego Merino	Burgos, 1574	De morbis internis libri sex

SOURCE: AGI CONTRATACIÓN 1081 N2 R1 FOLS. 67–68 REGISTRO DEL NAVÍO NUESTRA SEÑORA DE CANDELARIA 1584.

Note: The titles are the author's interpretation and are not definitive. The dates and place of publication are those of known first editions and not necessarily the editions contained in the shipment.

Glossary

For the names of all plants and minerals see Appendix B

Alambique	Glass still used for distillation
Alcalde	Magistrate
Almojarifazgo	Ad valorem tax
Alquitara	Metal still used for distillation
Aposento	Room or store
Arancel	Schedule or list, here of prices
Armada	Armed fleet
Avería	Tax levied to support the *armadas*
Bachiller	Holder of bachelor's degree
Barbero	Barber
Barrio	Neighbourhood or suburb
Beata	Nun who lived outside a convent
Beaterio	Residence for *beatas*
Benemérito	Worthy citizen
Botica	Pharmacy
Boticario	Apothecary
Cabildo	Town council
Capellanía	Chaplaincy
Casa de Contratación	Board of Trade
Casta	Person of mixed ethnic background
Cédula	Decree
Censo	Agreement for the payment of an annuity
Cerero	Wax worker or trader
Chácara	Small holding or farm
Cirujano	Surgeon
Cofradía	Brotherhood
Colegio	College, school
Collación	District of a parish
Compañía	Company or co-partnership
Compuesto	Compound medicine
Confitero	Confectioner
Converso	New Christian or Crypto-Jew
Corregidor	Provincial magistrate
Droguero	Druggist
Ducado	Ducat, 375 maravedís

Educanda	Secular student at convent school
Encomienda	An allocation of Indians to an individual who was entitled to exact tribute and labour from them in return for Christian instruction.
Escribano	Notary, scribe
Especiero	Spicer, grocer
Espumadera	Skimmer
Estanco	Monopoly
Estanco de naipes	Monopoly for the sale of playing cards
Familiar	Officer of the Inquisition
Fiel ejecutor	Public inspector, often of weights and measures
Gremio	Guild
Hábito	Dress of a person belonging to a military order
Hechicera/o	Witch, sorcerer
Hermano venticuatro	Original founder of a cofradía
Hidalgo	Noble
Jarabe	Syrup
Lamedor	Thick medicinal syrup
Licenciado	Title given to a person with a bachelor's degree or licenciate.
Licenciatura	Degree, licence
Limpieza de sangre	Purity of blood
Maestro	Master tradesman/artisan
Maravedí	Small Spanish coin
Mayordomo	Steward, administrator
Médico	Physician
Mercader	Merchant
Mita de plaza	Forced labour system
Morisco	Person of Moorish heritage
Obraje	Textile workshop
Oficial	Skilled journeyman
Perulero	Peruvian born trader with Spain
Peso corriente	272 maravedís
Peso de oro	450 maravedís
Peso de plata de a nueve	450 maravedís, used in business accounts
Peso de plata ensayada	307 maravedís, used in business accounts
Protomedicato	Medical board, entrusted with examining and licensing medical practitioners.
Protomédico	Chief medical officer and examiner
Pulpero	Owner of a general store

Rebotica	Back room of a pharmacy used for preparing medicines
Quarterón	A quarter, in this case of indigenous of African ancestry
Receta	Recipe, prescription
Recetario	Book of récipes or prescriptions
Regidor	Councillor, member of the town council
Sarna	Scabies
Seglar	Secular student at convent school
Tienda	Shop
Trastienda	Back room of a shop
Unguento	Ointment
Vale	Promissary note
Vecino	Citizen
Velero	Tallow chandler
Vestido	Suit
Visita	Inspection
Yanacona	Indigenous person who had left his/her community and worked for a Spaniard
Zambo/a (also Zambahigo)	Person of mixed African and Indian descent

Bibliography

Primary Sources

Acosta, José de. *Historia natural y moral de las Indias in Obras del Padre José de Acosta*. Biblioteca de autores españoles 73, 1–247. Madrid: Ediciones Atlas, 1954 [1590].

Agricola, Georgius. *De re metallica*. Translated by Herbert Clark Hoover and Lou Henry Hoover. New York: Dover Publications, Inc., 1950 [1556].

Aguilera, Antonio de. *Exposición sobre las preparaciones de Mesue*. Alcalá: Juan de Villanueva, 1569.

Angulo, Domingo. "Las ordenanzas del hospital de Santa Ana." *Revista del Archivo Nacional del Perú* 11 (1938): 131–56.

Antúnez y Acevedo, Rafael. *Memorias históricas sobre la legislación y gobierno del comercio de los españoles con sus colonias en las Indias occidentales*. Madrid: Imprenta de Sancha, 1797.

Arriaga, Pablo Joseph de. *La extirpación de idolatrías del Perú*. Lima: Geronymo de Contreras, 1621.

Azpilcueta, Martín de. *Compendio y summario de confessores y penitentes*. Alcalá: Casa de Iuan Iñíguez de Lequerica, 1580.

Blair, Emma H. and James A. Robertson. *The Philippine Islands, 1493–1803*. 55 vols. Cleveland, OH: A.H. Clark, 1903–1909.

Calancha, Antonio de la. *Corónica moralizada del orden de San Agustín en el Perú*. Barcelona: Pedro Lacavalleria, 1639.

Cardano, Girolamo. *Book of My Life*. Translated by Jean Stoner. New York: New York Review Books, 2002 [1575].

Castillo, Juan de. *Pharmacopoea, universa medicamenta in officinis pharmaceuticis usitata complectens, et explicans*. Cádiz Gadibus: Apud Joannem de Borja, 1622.

Cieza de León, Pedro. *Obras completas*. 2 vols. Madrid: Consejo Superior de Investigaciones Científicas, 1984.

Cobo, Bernabé. *Obras*, 2 vols. Biblioteca de autores españoles 91–92. Madrid: Ediciones Atlas, 1956 [1653].

Colmenero, José. *Reprobación de pernicioso abuso de los polvos de la corteza de quarango o China china*. Salamanca, Eugenio Antonio Garcia, 1697.

Constituciones y ordenanças de la Vniuersidad, y Studio General de la Ciudad de los Reyes del Piru, 1602. https://archive.org/details/constitucionesy001univ [Accessed 28 Oct. 2015].

Constituciones y ordenanzas del hospital real de Santa Ana, Lima, 28 Feb. 1609. https://archive.org/details/constitucionesy000hosp [Accessed 2 Jul. 2016].

Córdoba y Salinas, Diego de. *Crónica franciscana de las provincias del Perú*. Edited by Lino Gómez Canedo. Washington, DC: American Academy of Franciscan History, 1957.

Figueroa, Juan de. *Opusculo de astrología en medicina*. Lima. No publisher. 1660.

Fragoso, Juan. *Discursos de las cosas aromáticas, árboles y frutales, y de otras muchas medicinas simples, que se traen de la India oriental y sirven al uso de medicina*. Madrid: Francisco Sánchez, 1572.

Garzoni, Tomaso. *La piazza universale di tutte le professioni del mondo*, 2 vols. Edited by Paolo Cherchi and Beatrice Collina. Turin: Giulio Einaldi, 1996 [1587].

Hernández, Francisco. *Quatro libros: De la natvraleza, y virtvdes de las plantas, y animales que estan receuidos en el vso de medicina en la Nueua España, y la methodo, y correcion, y preparacion, que para administrallas se requiere con lo que el Doctor Francisco Hernandez escriuio en lengua Latina*. Mexico: Casa de la viuda de Diego Lopez Daualos, 1615.

Hidalgo Agüero, Bartolomé de. *Thesoro de la verdadera cirugía*. Seville: En Casa de Francisco Pérez, 1604.

Index et catalogus librorum prohibitorum. Madrid: Apud Alphonsum Gomezium, 1583.

Laredo, Bernardino de. *Modus faciendi: cum ordine medicandi: a médicos y boticarios muy común y necesario*. Seville: Jacobo Cromberger, 1527.

León, Andrés de. *Tratados de medicina, cirugía, y anatomía*. Valladolid: Luis Sánchez, 1605.

Lewin, Boleslao, ed. *Descripción del Virreinato del Perú; crónica inédita de comienzos del siglo XVII* [Pedro de León Portocarrero]. Rosario: Universidad Nacional del Litoral, Facultad de Filosofía, Letras y Ciencias de la Educación, 1958.

Libros de cabildos de Lima. Edited by Bertram T. Lee, Juan Bromley, Sophy E. Schofield, and Emilio Harth-Terré. 23 vols. Lima: Torres Aguirre, 1935–1963.

Lizárraga, Reginaldo de. *Descripción breve de toda la tierra del Perú, Tucumán, Río de la Plata y Chile*. Biblioteca de autores españoles 216. Madrid: Ediciones Atlas, 1968.

López de León, Pedro. *Prática [sic] y teórica de las apostemas en general y particular: questiones y prácticas de cirugía, de heridas, llagas, y otras cosas nuevas, y particulares*. Seville: Oficina de Luys Estupiñan, 1628.

Méndez Nieto, Juan. *Discursos medicinales*. Edited by Luis S. Granjel. Salamanca: Universidad de Salamanca, 1989.

Mendiburu, Manuel de. *Diccionario histórico-biográfico del Perú*. 2 vols. Lima: Imprenta de J. Francisco Solís, 1876.

Monardes, Nicólas. *Primera y segunda y tercera parte de la historia medicinal de las cosas que se traen de nuestras Indias Occidentales, que sirven en medicina*. Seville: Alonso Escriuano, 1574.

Monardes, Nicolas. *Joyfull Newes Out of the Newfound World*. Edited by John Frampton. London: William Norton, 1580.

Mugaburu, Josephe and Francisco. *Chronicle of Colonial Lima: The Diary of Josephe and Francisco Mugaburu, 1640–1697*. Translated and edited by Robert R. Miller. Norman: University of Oklahoma Press, 1975.

Muñoz, Miguel Eugenio. *Recopilación de las leyes, pragmáticas, reales decretos, y acuerdos del Proto-medicato*. Valencia: Imprenta de la Viuda de Antonio Bordazar, 1751.

Murillo, Jerónimo. *Therapeutica, método de Galeno en lo que toca a cirugía*. Zaragoza, 1561.

Murúa, Martín de. *Historia general del Perú*. 2 vols. Madrid: Instituto Gonzalo Fernández de Oviedo, 1962–1964 [1616].

Novísima recopilación de las leyes de España. Madrid, 1805.

Ocaña, Fray Diego de. Memoria viva de una tierra de olvido: Relación del viaje al Nuevo Mundo de 1599 a 1607. Barcelona: CECAL, 2013.

Ortega, Francisco de. *Directorio de boticarios, en que se ponen ocho advertencias*. Seville: Matías Clavijo, 1625.

Oviedo, Luis de. *Methodo de la colección, y reposición de las medicinas simples, de su corrección, y preparación*. Madrid: Luis Sánchez, 1622 [1581].

Padrón de los Indios de Lima en 1613. Comp. Miguel de Contreras. Edited by Noble David Cook. transcribed by Mauro Escobar Gamboa. Lima: Universidad Nacional de San Marcos, 1968.

Palacios, Felix. *Palestra pharmaceutica chymico-galenica*. Madrid: Juan Garcia Infançon, 1706.

Porres, Matías de. *Breve advertencias para beber frío con nieve*. Lima, 1621.

Recopilación de las leyes de los reynos de las indias. 3 vols. Madrid: Gráficas Ultra, 1943 [1681].

Recopilación de las leyes destos reynos, hecha por mandado de la Magestad Catholica del Rey don Philippe Segundo nuestro Señor. Alcalá de Henares: Casa de Iuan Iñiguez de Lequerica, 1598.

Recopilación de las leyes, pragmáticas reales, decretos, y acuerdos del Real Proto-Medicato: hecha por encargo y dirección del mismo Real Tribunal por Don Miguel Eugenio Muñoz. Valencia: Imprenta de la Viuda de Antonio Bordazar, 1751.

Recopilación de los ordenamientos de Sevilla. Seville: Juan Varela de Salamanca, 1527. http://bvpb.mcu.es/es/consulta/registro.cmd?id=451883 [Accessed 17 Mar. 2016].

Relaciones geográficas de Indias: Peru. Biblioteca de autores españoles 183–185. Edited by Marcos Jiménez de Espada. Madrid: Ediciones Atlas, 1965.

Saladino de Ascoli. *Compendio de los boticarios*. Translated by Alonso Rodríguez Tudela. Valladolid: Arnao Guillén de Brocar, 1515.

Salinas y Córdova, Fray Buenaventura de. *Memorial de las historias del Nuevo Mundo Pirú* [sic]. Vol. 1. Lima: Universidad Nacional Mayor de San Marcos, 1957 [1630].

Suardo, José Antonio. *Diario de Lima (1629–1639)*. Edited by Rubén Vargas Ugarte. 2 vols. Lima: "Lumen", 1936.

Suárez de Figueroa, Christóval. *Plaza universal de todas ciencias y artes*. Madrid: Luis Sánchez, 1615.
Susio, Giovanni Battista. *Libro del conoscere la pestilenza*. Mantova: Giacomo Ruffinello, 1576.
"Una botica colonial." *Revista del archivo histórico del Cuzco* 4 (1953): 263–82.
Valles, Francisco de. *Tratado de las aguas destiladas, pesos y medidas de que los boticarios deuen vsar por nueva ordenança y mandato de su Magestad y su Real Consejo*. Madrid: Luis Sánchez, 1592.
Vargas Machuca, Captain Bernardo de. *Milicia y descripción de las Indias*. Madrid: Librería de Victoriano Suárez, 1892.
Vargas Machuca, Captain Bernardo de. *The Indian Militia and Description of the Indies*. Edited by Kris Lane, translated by Timothy F. Johnson. Durham, NC: Duke University Press, 2008.
Vargas Ugarte, Rubén. *Concilios limenses (1551–1772)*. 3 vols. Lima: Imp. Juan Cardenal Guevara, 1951–1954.
Vázquez de Espinosa, Antonio. *Compendio y descripción de las Indias occidentales*. Biblioteca de autores españoles 231. Madrid: Ediciones Atlas, 1969.
Veitia Linage, Joseph de. *Norte de la contratación de las Indias Occidentales*. Seville: Imprenta de Francisco de Blas, 1672.

Secondary Sources

Aguirre Beltrán, Gonzalo. *Medicina y magia: el proceso de aculturación en la estructura colonial*. México: Universidad Veracruzana, Instituto Nacional Indigenista, 1992 [1963].
Alberro, Solange. *De gachupín al criollo: o de cómo los españoles* de *México dejaron de serlo*. México: El Colegio *de* México, 1992.
Alchon, Suzanne A. *Native Society and Disease in Colonial Ecuador*. Cambridge: Cambridge University Press, 1991.
Alperrine-Bouyer, Monica. *La educación de las élites indígenas en el Perú colonial*. Lima: Instituto de Estudios Peruanos, 2007.
Altman, Ida. *Emigrants and Society: Extremadura and Spanish America in the Sixteenth Century*. Berkeley: University of California, 1989.
Anagnostou, Sabine. "Jesuits in Spanish America: Contributions to the Exploration of the American Materia Medica." *Pharmacy in History* 47 (1) (2005): 5–10.
Anderson, Ruth Matilda. *Hispanic Costume 1480–1530*. New York: Hispanic Society of America, 1979.
Aragón Noriega, Ilana. "El pan, el vino, y otros negocios: aspectos laborales de la mujer." In *Lima en el siglo XIV*, edited by Laura Gutiérrez Arbulú, 257–85. Lima: Pontificia Universidad Católica del Perú, Instituto Riva-Agüero, 2005.

Armus, Diego, ed. *Disease in the History of Modern Latin America: From Malaria to Aids*. Duke University Press: Durham, 2003.

Armus, Diego. "Legados y tendencias en la historiografía sobre la enfermedad en América Latina moderna." In *Higienizar, medicar, gobernar: historia, medicina y sociedad en Colombia*, edited by Jorge Márquez, Álvaro Casa, and Victoria Estrada, 13–39. Medellín: Universidad Nacional de Colombia, 2004.

Arnold, David, ed. *Imperial Medicine and Indigenous Societies*. Manchester: Manchester University Press, 1988.

Arrebola Nacle, Pilar and Guillermina López Andújar. "Suministro de medicamentos a la gente de mar y guerra de la Real Armada (1636)." *Boletin de la sociedad española de historia de la farmacia* 27 (1986): 43–52.

Arrow, Kenneth J. "Uncertainty and the Welfare Economics of Medical Care." *The American Economic Review* 53 (5) (1963): 941–73.

Ashworth, William B. "Catholicism and Early Modern Science." In *God and Nature: Historical Essays on the Encounter between Christianity and Religion*, edited by David C. Lindberg and Ronald L. Numbers, 136–66. Berkeley: University of California Press, 1986.

Attfield, Robin. "Christian Attitudes to Nature." *Journal of the History of Ideas* 44 (3) (1983): 369–86.

Aulet Sastre, Guillermo. "Precios autorizados de libros españoles en Indias." *Revista de Indias* 24 (1946): 311–12.

Azzolini, Monica. "Reading Health in the Stars: Politics and Medical Astrology in Renaissance Milan." In *Horoscopes and Public Spheres: Essays in the History of Astrology*, edited by Günter Oestmann, H. Dareel Rytkin, and Kocku von Stukrad, 183–205. Berlin: Walter de Gruyter, 2005.

Barrera-Osorio, Antonio. *Experiencing Nature: The Spanish American Empire and the Early Scientific Revolution*. Austin: University of Texas Press, 2006.

Barrera-Osorio, Antonio. "Knowledge and Empiricism in the Sixteenth-century Spanish Atlantic World." In *Science in the Spanish and Portuguese Empires: 1560–1800*, edited by Daniela Bleichmar, Paula De Vos, Kristin Huffine, and Kevin Sheehan, 219–32. Stanford: Stanford University Press, 2009.

Basalla, George. "The Spread of Western Science." *Science* 156 (1967): 611–22.

Bastien, Joseph W. *Healers of the Andes: Kallawaya Herbalists and Their Medicinal Plants*. University of Utah Press: Salt Lake City, 1987.

Bauer, Arnold J. "The Church in the Economy of Spanish America: Censos and Depósitos in the Eighteenth and Nineteenth Centuries." *Hispanic American Historical Review* 63 (4) (1983): 707–33.

Binde, Per. "Nature in Roman Catholic Tradition." *Anthropological Quarterly* 74 (1) (2001): 15–27.

Bleichmar, Daniela. "Books, Bodies, and Fields: Sixteenth-Century Transatlantic Encounters with New World *Materia Medica*." In *Colonial Botany: Science, Commerce, and Politics in the Early Modern World*, edited by Londa Schiebinger and Claudia Swan, 83–99. Philadelphia, PA: University of Pennsylvania Press, 2004.

Bleichmar, Daniela, Paula De Vos, Kristine Hufine and Kevin Sheehan, eds. *Science in the Spanish and Portuguese Empires, 1500–1800*. Stanford: Stanford University Press, 2009.

Borah, Woodrow. Early *Trade and Navigation between Mexico and Peru, Ibero-Americana* 38. Berkeley and Los Angeles: University of California, 1954.

Boumediene, Samir. *La colonisation du savoir: une histoire des plantes medicinales de "Nuevo Mundo"* (1492–1750). Vaulx-en-Velin: Éditions des mondes à faire, 2016.

Bowers, Kristy Wilson "Tradition and Innovation in Spanish Medicine: Bartolomé Hidalgo de Agüero and the Vía Particular." *The Sixteenth Century Journal* 41(1) (2010): 29–47.

Bowers, Kristy Wilson. *Plague and Public Health in Early Modern Seville*. Rochester: University of Rochester Press, 2013.

Bowser, Frederick P. *The African Slave in Colonial Peru, 1524–1650*. Stanford CA: Stanford University Press, 1974.

Boyd-Bowman, Peter. "Spanish Patterns of Emigration to the Indies to 1600." *Hispanic American Historical Review* 56 (4) (1976): 580–604.

Brockliss, Lawrence and Colin Jones. *The Medical World of Early Modern France*. Oxford: Clarendon Press, 1997.

Bromley, Juan. "La ciudad de Lima en el año 1630." *Revista histórica* 24 (1959): 268–317.

Bromley, Juan. *Las Viejas Calles de Lima*. Lima: Municipalidad Metropolitana de Lima, 2005. At: http://www.limateca.org/uploads/tinymce/2005_Las_viejas_calles_de_Lima_-_Juan_Bromley.pdf [Accessed 28 Feb. 2017].

Bronner, Fred. "Elite Formation in Seventeenth-Century Peru." *Boletín de estudios latinoamericanos y del caribe* 24 (1978): 3–26.

Brosseder, Claudia. "Astrology in Seventeenth-century Peru." *Studies in History and Philosophy of Biological and Biomedical Sciences* 41 (2) (2010): 146–57.

Brosseder, Claudia. *The Power of Huacas: Change and Resistance in the Andean World of Colonial Peru*. Austin: University of Texas Press, 2014.

Browman, David L. and James N. Gundersen. "Altiplano Comestible Earths: Prehistoric and Historic Geophagy of Highland Peru and Bolivia." *Geoarchaeology* 8 (5) (1993): 413–25.

Burns, Kathryn. *Colonial Habits: Convents and the Spiritual Economy of Cuzco, Peru*. Durham: Duke University Press, 1999.

Burns, Kathryn. "Andean Women in Religion: *Beatas*, 'Decency', and the Defence of Honour in Colonial Cuzco." In *Gender, Race and Religion in the Colonization of the Americas*, edited by Nora E. Jaffary, 81–91. Aldershot: Ashgate, 2007.

Cabanillas Cárdenas, Carlos F., ed. *Juan del Valle y Caviedes: guerras físicas, proezas medicales, hazañas de la ignorancia*. Madrid: Iberoamericana Editorial Vervuert, 2013.

Cagle, Hugh. "Cultures of Inquiry, Myths of Empire: Natural History in Colonial Goa." In *Medicine, Trade and Empire: Garcia de Orta's Colloquies on the Simples and*

Drugs of India (1563), edited by Palmira Fontes da Costa, 107–28. Farnham: Ashgate, 2015.

Cahill, David. "Financing Health Care in the Viceroyalty of Peru: The Hospitals of Lima in the Late Colonial Period." *The Americas* 52 (2) (1995): 123–54.

Cañizares-Esguerra, Jorge. "New World, New Stars: Patriotic Astrology and the Invention of Indian and Creole Bodies in Colonial Spanish America, 1600–1650." *The American Historical Review* 104 (1) (1999): 33–68.

Cañizares-Esguerra, Jorge. "Spanish America: From Baroque to Modern Colonial Science." In *The Cambridge History of Science*, vol. 4, edited by Roy Porter, 718–38. Cambridge: Cambridge University Press, 2003.

Cañizares-Esguerra, Jorge. "Iberian Science in the Renaissance: Ignored How Much Longer?." *Perspectives on Science* 12 (1) (2004): 86–124. Also, modified in *Nature, Empire, and Nation: Explorations of the History of Science in the Iberian World* as "The Colonial Iberian Roots of the Scientific Revolution," 14–45. Stanford: University of Stanford Press, 2006.

Cañizares-Esguerra, Jorge. "Iberian Colonial Science." *Isis* 96 (2005): 64–70.

Cantuarias Acosta, Ricardo. "Las modas limeñas." In *Lima en el siglo XIV*, edited by Laura Gutiérrez Arbulú, 287–307. Lima: Pontificia Universidad Católica del Perú, Instituto Riva-Agüero, 2005.

Celestino, Olinda and Albert Meyers. *Las cofradías en el Perú: región central*. Frankfurt: Editorial Klaus d. Vervuert, 1981.

Chakrabarti, Pratik. *Medicine and Empire 1600–1960*. Basingstoke: Palgrave Macmillan, 2014.

Chance, John K. and William B. Taylor. "Estate and Class in a Colonial City: Oaxaca in 1792." *Comparative Studies in Society and History* 19 (4) (1977): 451–87

Charles, John. "Trained by Jesuits: Indigenous *Letrados* in Sixteenth-Century Peru." In *Indigenous Intellectuals: Knowledge, Power and Colonial Culture in Mexico and the Andes*, edited by Gabriela Ramos and Yanna Yannakakis, 60–78. Durham: Duke University Press, 2014.

Charney, Paul. "A Sense of Belonging: Colonial Indian Cofradías and Ethnicity in the Valley of Lima, Peru." *The Americas* 54 (3) (1998): 379–407.

Charney, Paul. *Indian Society in the Valley of Lima, Peru, 1532–1765*. Lanham, MD: University Press of America, 2001.

Charney, Paul. "'For My Necessities': The Wills of Andean Commoners and Nobles in the Valley of Lima, 1597–1607." *Ethnohistory* 59 (2) (2012): 323–51.

Chaunu, Huguette and Pierre Chaunu. *Séville et l'Atlantique, 1450–1650*. 8 vols. Paris: A. Colin, 1955–1959.

Chirlone, Quintín and Carlos Mallaína. "Biografía Antonio Robles Cornejo." *El restaurador farmacéutico*, año 11 no. 8 20 Mar. 1855. On line at: http://bibliotecavirtual.ranf.com/es/catalogo_imagenes/grupo.cmd?path=2010122 [Accessed 11 May 2016].

Chirlone, Quintín and Carlos Mallaína. *Historia de la farmacia*. 2nd ed. Madrid: Imp. José M. Ducazcal, 1865.

Clayton, Lawrence A. "Trade and Navigation in the Seventeenth-Century Viceroyalty of Peru." *Journal of Latin American Studies* 7 (1) (1975): 1–21.

Clouse, Michele. *Medicine, Government and Public Health in Philip II's Spain: Shared Interests, Competing Authorities*. Farnham: Ashgate, 2011.

Colmeiro, Miguel. "Noticias acerca de un manuscrito perteneciente al licenciado Antonio Robles Cornejo." *Revista de los progresos de las ciencias exactas, físicas y naturales* (Madrid) 9 (7) (1859): 440–44.

Cook, Harold J. "Medicine." In *The Cambridge History of Science*, vol. 3, edited by Katherine Park and Loraine Daston, 407–34. Cambridge: Cambridge University Press, 2006.

Cook, Harold J. *Matters of Exchange: Commerce, Medicine, and Science in the Dutch Golden Age*. New Haven: Yale University Press, 2007.

Cook, Harold J. "Physicians and Natural History." In *Cultures of Natural History*, edited by Nicholas Jardine, James A. Secord and Emma C. Spary, 91–105. Cambridge: Cambridge University Press, 2008.

Cook, Harold J. "Markets and Cultures: Medical Specifics and the Reconfiguration of the Body in Early Modern Europe." *Transactions of the Royal Historical Society* 21 (2011): 123–45.

Cook, Harold J. and Timothy D. Walker. "Circulation of Medicine in the Early Modern Atlantic World." *Social History of Medicine* 26 (3) (2013): 337–51.

Cook, Noble David. *Demographic Collapse: Indian Peru 1520–1620*. Cambridge: Cambridge University Press, 1981.

Cook, Noble David. *Born to Die: Disease and New World Conquests, 1492–1650*. Cambridge University Press: Cambridge, 1998.

Cooper, Alix. *Inventing the Indigenous: Local Knowledge and Natural History in Early Modern Europe*. Cambridge: Cambridge University Press, 2007.

Cope R. Douglas. *The Limits of Racial Domination: Plebeian Society in Colonial Mexico City, 1660–1720*. Madison: The University of Wisconsin Press, 1994.

Cordero del Campillo, Miguel. *Crónicas de Indias: ganadería, medicina y veterinaria*. Salamanca: Junta de Castilla y León, 2001.

Corfield, Penelope J. "From Poison Peddlers to Civic Worthies: The Reputation of the Apothecaries in Georgian England." *Social History of Medicine* 22 (1) (2009): 1–21.

Crawford, Matthew J. *The Andean Wonder Drug: Cinchona Bark and Imperial Science in the Spanish Atlantic, 1630–1800*. Pittsburgh, PA: University of Pittsburgh Press, 2016.

Crespo Rodríguez, María Dolores. *Arquitectura doméstica de la Ciudad de los Reyes (1535–1750)*. Seville: Consejo Superior de Investigaciones Científicas, 2006.

Crosby, Alfred W. "Virgin Soil Epidemics as a Factor in the Aboriginal Depopulation in America." *The William and Mary Quarterly* 33 (2) (1976): 289–99.

Cueto, Marcus and Stephen Palmer. *Medicine and Public Health in Latin America: A History*. New York: Cambridge University Press, 2015.

Cunningham, Andrew. "The Culture of Gardens." In *Cultures of Natural History*, edited by Nicholas Jardine, James A. Secord and Emma C. Spary, 38–56. Cambridge: Cambridge University Press, 2008.

Cunningham, Andrew and Bridie Andrews, eds. *Western Medicine as Contested Knowledge*. Manchester University Press: Manchester, 1997.

Curtin, Philip D. *Death by Migration: Europe's Encounter with the Tropical World in the Nineteenth Century*. Cambridge University Press: Cambridge, 1989.

Cushner, Nicholas P. *Lords of the Land: Sugar, Wine, and Jesuit Estates of Coastal Peru, 1600–1767*. New York: SUNY Press, 1980.

Cussen, Celia. *Black Saint of the Americas: The Life and Afterlife of Martín de Porres*. Cambridge: Cambridge University Press, 2014.

Daston, Lorraine and Katharine Park. *Wonders and the Order of Nature 1150–1750*. New York: Zone Books, 1998.

Davis, Charles and María Luz López Terrada. "Protomedicato y farmacia en Castilla a finales del siglo XVI: edición crítica del *Catálogo de las cosas que los boticarios han de tener en sus boticas*, de Zamudio de Alfaro, protomédico general (1592–1599)." *Asclepio* 62 (2) (2010): 579–626.

De Vos, Paula S. "The Art of Pharmacy in Seventeenth- and Eighteenth-Century Mexico." PhD diss., University of California, Berkeley, 2001.

De Vos, Paula S. "The Apothecary in Seventeenth- and Eighteenth-Century Mexico: Historiography and Case Studies in Medical Regulation, Charity, and Science." *Colonial Latin American Historical Review* 13 (3) (2004): 249–83.

De Vos, Paula S. "The Science of Spices: Empiricism and Economic Botany in the Early Spanish Empire." *Journal of World History* 17 (4) (2006): 399–427.

De Vos, Paula S. "From Herbs to Alchemy: the Introduction of Chemical Medicine to Mexican Pharmacies in the Seventeenth and Eighteenth Centuries." *Journal of Spanish Cultural Studies* 8 (2) (2007): 135–68.

De Vos, Paula S. "European Materia Medica in Historical Texts: Longevity of a Tradition and Implications for Future Use." *Journal of Ethnopharmacology*, 132 (1) (2010): 28–47.

De Vos, Paula S. "The 'Prince of Medicine', Yūhannā ibn Māsawayh and the Foundations of the Western Pharmaceutical Tradition." *Isis* 104 (4) (2013): 667–712.

De Vos, Paula S. "Apothecaries, Artists and Artisans: Early Industrial Material Culture in the Biological Old Regime." *Journal of Interdisciplinary History*, 45 (3) (2015): 277–336.

Debus, Allen G. *The English Paracelsians*. New York: Franklin Watts, Inc., 1968.

Debus, Allen G. *The Chemical Philosophy: Paracelsian Science and Medicine in the Sixteenth and Seventeenth Centuries* vol. 1. New York: Science History Publication, 1977.
Debus, Allen G. "Chemists, Physicians and the Changing Perspectives on the Scientific Revolution." *Isis* 89 (1) (1998): 66–81.
Debus, Allen G. "Paracelsus and the Delayed Scientific Revolution in Spain: A Legacy of Philip II." In *Reading the Book of Nature: the Other Side of the Scientific Revolution*, edited by Allen G. Debus and Michael T. Walton, 147–61. Kirksville-Missouri: Sixteenth Century Journal Publishers, Inc., 1998.
Delgado Matallana, Gustavo. *Historia de la enseñanza médica en el Perú*. Lima: Universidad Nacional Mayor de San Marcos, Fondo Editorial, 2008.
Deza Bringas, Luis. *Testimonios del linaje médico peruano en los libros del cabildo de Lima, siglo XVI*. Lima: Universidad de San Martín de Porres, Facultad de Medicina Humana, 2004.
Díaz Piedrahita, Santiago and Luis Carlos Mantilla. *La terapéutica en el Nuevo Reino de Granada: un recetario franciscano del siglo XVIII*. Bogotá: Academia Colombiana de Ciencias Exactas, Físicas y Naturales, 2002.
Drayton, Richard. *Nature's Government: Science, Imperial Britain and the "Improvement" of the World*. New Haven: Yale University Press, 2000.
Dueñas, Alcira. *Indians and Mestizos in the "Lettered City"*. Boulder: University Press of Colorado, 2010.
Durán Montero, María Antonia. *Lima en el siglo XVII: arquitectura, urbanismo y vida cotidiana*. Seville: Diputación Provincial, 1994.
Duviols, Pierre. *La lutte contre les religions autochtones dans le Pérou colonial: l'extirpation de l'idolâtrie, entre 1532 et 1660*. Lima: Institut Français d'Études Andines, 1971.
Eamon, William. *Science and the Secrets of Nature: Books of Secrets in Medieval and Early Modern Culture*. Princeton: Princeton University Press, 1994.
Earle, Rebecca. "'Two Pairs of Pink Satin Shoes!!' Race, Clothing and Identity in the Americas (17th–19th centuries)." *History Workshop Journal* 52 (2001): 175–95.
Earle, Rebecca. "'If You Eat Their Food': Diets and Bodies in Early Colonial Spanish America." *American Historical Review* 115 (3) (2010): 688–713.
Earle, Rebecca. *The Body of the Conquistador: Food, Race and the Colonial Experience in Spanish America, 1492–1700*. Cambridge: Cambridge University Press, 2012.
Eguiguren, Luis Antonio. *Alma mater: origines de la Universidad de San Marcos (1551–1579)*. Lima 1939. No publisher.
Eguiguren, Luis Antonio. *Diccionario histórico cronológico de la universidad real y pontificia de San Marcos y sus colegios*, Tomo I. Lima 1940. Printer: Torres Aguirre.
Elliott, John H. *Empires of the Atlantic World: Britain and Spain in America 1492–1830*. New Haven: Yale University Press, 2006.

Estes, J. Worth. "The European Reception of the First Drugs from the New World." *Pharmacy in History* 37 (1) (1995): 3–23.

Fernández-Carrión, Mercedes and David Martín-Castilla. "Francisco de Ortega y la profesión farmacéutica sevillana en el primer tercio del siglo XVII." *Boletín de la sociedad española de historia de la farmacia* 37 (148) (1986): 289–301.

Fernández-Carrión, Mercedes and José Antonio Pérez Romero. "Visitas de inspección sanitaria en especierías, confiterías, cererías y aguardenterías de Antequera durante el siglo XVII." *Baética* 8 (1985): 435–46.

Fernández-Carrión, Mercedes and José Luis Valverde. *Farmacia y sociedad en Sevilla en el siglo XVI*. Seville: Servicio de Publicaciones del Ayuntamiento de Sevilla. 1985.

Fernández-Carrión, Mercedes and José Luis Valverde. "Research Notes on Spanish American Drug Trade." *Pharmacy in History* 30–31 (1988–1989): 27–32.

Few, Martha. *Women Who Live Evil Lives: Gender, Religion and the Politics of Power in Colonial Guatemala.* Austin: University of Texas Press, 2002.

Few, Martha. "Medical *Mestizaje* and the Politics *of Pregnancy in Colonial Guatemala, 1660–1730*" In *Science in the Spanish and Portuguese Empires, 1500–1800*, edited by Daniela Bleichmar, Paula De Vos, Kristine Hufine and Kevin Sheehan, 132–46. Stanford: Stanford University Press, 2009.

Findlen, Paula. *Possessing Nature: Museums, Collecting, and Scientific Culture in Early Modern Italy*. Berkeley and Los Angeles: University of California Press, 1994.

Findlen, Paula. "Courting Nature." In *Cultures of Natural History*, edited by Nicholas Jardine, James A. Secord and Emma C. Spary, 57–74. Cambridge: Cambridge University Press, 2008.

Fisher, Andrew B. and Matthew D. O'Hara, "Introduction: Racial Identities and Their Interpreters in Colonial Latin America." In *Imperial Subjects: Race and Identity in Colonial Latin America*, edited by Andrew B. Fisher and Matthew D. O'Hara, 15–23. Durham: Duke University Press, 2009.

Flores Hernández, Benjamín. "Medicina de los conquistadores, en la *Milicia Indiana* de Bernardo de Vargas Machuca." *Boletín mexicano de historia y filosofía de la médicina* 6 (1) (2003): 5–10.

Forbes, Robert J. *A History of the Art of Distillation*. Leiden: E.J. Brill, 1970.

Ford, Thayne R. "Stranger in a Foreign Land: José de Acosta's Scientific Realisations in Sixteenth-century Peru." *Sixteenth Century Journal* 29 (1) (1998): 19–33.

Foster, George M. "Relationship between Spanish and Spanish-American Folk Medicine." *Journal of American Folklore* 66 (1953): 201–17.

Foster, George M. "On the Origin of Humoral Medicine in Latin America." *Medical Anthropology Quarterly* NS 1 (4) (1987): 355–93.

Foster, George M. *Hippocrates' Latin American Legacy: Humoral Medicine in the New World.* Langhorne, PA: Gordon and Breach, 1994.

Foust, Clifford M. *Rhubarb: The Wondrous Drug*. Princeton: Princeton University Press, 1992.

Freedman, Paul. *Out of the East: Spices and the Medieval Imagination*. New Haven: Yale University Press, 2008.

Fresquet Febrer, José Luis. "Uso de productos del reino mineral en la terapéutica del siglo XVI: el libro de los *Medicamentos Simples* de Juan Fragoso (1581) and el *Antidotario* de Juan Calvo (1580)." *Asclepio* 51 (1999): 55–92.

Fresquet Febrer, José Luis. "*Juan Fragoso y los 'Discursos de las cosas aromáticas, árboles y frutales…'(1572)*." Valencia: CISC – Universitat de Valencia, 2001.

Friedenwald, Harry. *The Jews and Medicine*, vol. 2. Baltimore: The John Hopkins Press, 1944.

Garcia Ballester, Luis. "The Inquisition and Minority Medical Practitioners in Counter Reformation Spain: Judaizing and Morisco Practitioners, 1560–1610." In *Medicine and the Reformation*, edited by Ole Peter Grell and Andrew Cunningham, 156–91. London: Routledge, 1993.

García Ballester, Luis. *La búsqueda de la salud: sanadores y enfermos en la España medieval*. Barcelona: Ediciones Peninsula, 2001.

García Cáceres, Uriel. *Juan del Valle y Caviedes: Cronista de la medicina*. Lima: Banco Central de Reserva del Perú and Universidad Peruana Cayetano Heredia, 1999.

García Fuentes, Lutgardo. *Los peruleros y el comercio de Sevilla con las Indias, 1580–1630*. Seville: Universidad de Sevilla, 1997.

Garland Ponce, Beatriz. "Las cofradías en Lima durante la colonia: una primera aproximación." In *La venida del reino. Religión, evangelización y cultura en América, Siglos XVI–XX*, edited by Gabriela Ramos, 199–228. Cuzco: Centro de Estudios Regionales Andinos "Bartolomé de las Casas", 1994.

Garofalo, Leo. "Conjuring with Coca and the Inca: the Andeanisation of Lima's Afro-Peruvian Ritual Specialists, 1580–1690." *The Americas* 63 (1) (2006): 53–80.

Gentilcore, David. *Healers and Healing in Early Modern Italy*. Manchester: Manchester University Press, 1998.

Giglioni, Guido. "Girolamo Cardano: University Student and Professor." *Renaissance Studies* 27 (4) (2013): 517–32.

Gómez, Pablo F. "The Circulation of Bodily Knowledge in the Seventeenth-century Black Spanish Caribbean," *Social History of Medicine* 26 (3) (2013): 383–402.

González Arce, José Damián. "Los proyectos de ordenanzas generales de médicos, cirujanos y boticarios de Castilla (ca.1491–1513)." *Dynamis* 31 (1) (2011): 207–26.

González Sánchez, Carlos A. "El libro y la carrera de Indias: registro de ida de navíos." *Archivo hispalense* 72 (1989): 93–103.

González Sánchez, Carlos A. "Los libros de los españoles el virreinato del Perú. Siglos XVI y XVI." *Revista de Indias*, 56 (no. 206) (1996): 7–47.

González Sánchez, Carlos A. "Consideraciones sobre el comercio de libros en Lima a principios del siglo XVII." *Anuario de Estudios Americanos* 54 (2) (1997): 665–92.

González Sánchez, Carlos A. *New World Literacy: Written Culture across the Atlantic, 1500–1700*. Translated by Tristan Platt, 46–54. Lewisburg: Bucknell University Press, 2011.

Goodman, David C. "Philip II's Patronage of Science and Engineering." *The British Journal for the History of Science* 16 (1) (1983): 49–66.

Goodman, David C. *Power and Penury: Government, Technology and Science in Philip II's Spain*. Cambridge: Cambridge University Press, 1988.

Grafton, Anthony and Nancy Siraisi. "Between the Election and My Hopes: Girolamo Cardano and Medical Astrology." In *Secrets of Nature: Astrology and Alchemy in Early Modern Europe,* edited by William R. Newman and Anthony Grafton, 69–132. Cambridge, MA: MIT Press, 2001.

Granjel, Luis S. *La medicina española renacentista*. Salamanca: Ediciones Universidad de Salamanca, 1980.

Granjel, Luis S. *Discursos medicinales: vida y obra de Juan Méndez Nieto*. Salamanca: Universidad de Salamanca y Junta de Castilla y León, 1989.

Grant, Edward. "When Did Modern Science Begin?" *The American Scholar* 66 (1997): 105–13.

Graubert, Karen B. *With Our Labor and Our Sweat: Indigenous Women and the Formation of Colonial Society in Peru, 1550–1700*. Stanford: Stanford University Press. 2007.

Green, Monica. "Bodies, Gender, Health, Disease: Recent Work on Medieval Women's Medicine." *Studies in Medieval and Renaissance History*, 3rd ser. 5 (2005): 1–46.

Griffiths, Nicholas. *The Cross and the Serpent. Religious Repression and Resurgence in Colonial Peru*. Norman: University of Oklahoma Press, 1995.

Griffiths, Nicholas. "Andean *Curanderos* and their Repressors: The Persecution of Native Healing in Late Seventeenth- and Early Eighteenth-Century Peru." In *Spiritual Encounters: Interactions between Christianity and Native Religions in Colonial America*, edited by Nicholas Griffiths and Fernando Cervantes, 185–97. Birmingham: University of Birmingham Press, 1999.

Guerra, Francisco. *Nicolás Bautista Monardes: su vida y su obra (ca. 1493–1588)*. Mexico City: Compañía Fundidora de Cierro y Acero de Monterrey, 1961.

Guerra, Francisco. "Drugs from the Indies and the Political Economy of the Sixteenth Century." In *Analecta medico-historica: materia medica in the XVIth century*, edited by M. Florkin, 29–55. Oxford: Pergamon, 1966.

Guerra, Francisco. "The Role of Religion in Spanish American Medicine." In *Medicine and Culture*, edited by F.N.L. Poynter, 179–88. Wellcome Institute: London, 1969.

Guerra, Francisco. *El Tesoro de Medicinas de Gregorio López* 1542–1596. Madrid: Ediciones Cultura Hispánica del Instituto de Cooperación Iberoamericana, 1982.

Guerra, Francisco. *El hospital en hispanoamérica y filipinas 1492–1898*. Madrid: Ministerio de Sandiad y Consumo, 1994.

Guibovich, Pedro. Autores, censores y producción de libros en el virreinato peruano (2015). Unpublished manuscript at: https://escriturasvirreinales.files.wordpress.com/2015/05/guibovich-autores_y_censores-u-c.pdf [Accessed 2 Jul. 2016].

Guibovich Pérez, Pedro M. "Printing Press in Colonial Peru: Production Process and Literary Categories in Lima." *Colonial Latin American Review* 10 (2) (2001): 167–88.

Guibovich Pérez, Pedro M. "Bibliotecas de médicos en Lima colonial." In *Del autor al lector*, edited by Carmen Castañeda, 293–304. Mexico City: CIESAS, 2002.

Guibovich Pérez, Pedro M. *Censura, libros e inquisición en el Perú colonial, 1570–1754*. Seville: Universidad de Sevilla, Secretariado de publicaciones, 2004.

Gunther, Robert T. *The Greek Herbal of Disocorides*. New York: Harner Publishing Co., 1959.

Gutiérrez de Pineda, Virginia. *Medicina tradicional de Colombia: el triple llegado*. 2 vols. Bogotá: Universidad Nacional de Colombia, 1985.

Halikowski-Smith, Stefan. "'Profits Sprout Like Tropical Plants': A Fresh Look at What Went Wrong with the Eurasian Spice Trade c. 1550–1800." *Journal of Global History* 3 (2008): 389–418.

Hamnett, Brian R. "Church Wealth in Peru: Estates and Loans in the Archdiocese of Lima in the Seventeenth Century." *Jahrbuch für Geschichte Lateinamerikas* 10 (1973): 113–32.

Hampe-Martínez, Teodoro. "The Diffusion of Books and Ideas in Colonial Peru: A Study of Private Libraries in the Sixteenth and Seventeenth Centuries." *Hispanic American Historical Review* 273 (2) (1993): 211–33.

Haring, Clarence H. *Trade and Navigation between Spain and the Indies*. Cambridge, Mass.: Harvard University Press, 1918.

Harris, Steven J. "Jesuit Scientific Activity in the Overseas Missions, 1540–1773." *Isis* 96 (1) (2005): 71–79.

Hellmer, Marie. "Le Callao (1617–1815)" *Jahrbuch für Geschicte von Staat, Wirtschaft und Gesellschaft Lateinamerikas* 2 (1965): 145–95.

Henry, John. "Doctors and Healers: Popular Culture and the Medical Profession." In *Science, Culture and Popular Belief in Renaissance Europe*, edited by Stephen Pumfrey, Paolo L. Rossi and Maurice Slawinski, 191–221. Manchester: Manchester University Press, 1991.

Hernández Sáenz, Luz María. *Learning to Heal: The Medical Profession in Colonial Mexico, 1767–1831*. New York: Peter Lang, 1997.

Hernández Sáenz, Luz María and Foster, George M. "Curers and Their Cures in Colonial New Spain and Guatemala: The Spanish Component." In *Mesoamerican Healers*, edited by Brad R. Huber and Alan R. Sandstrom, 19–46. Austin: University of Texas Press, 2001.

Herrera Dávila, Joaquín. "Las boticas sevillanas de 1631." *Boletín de la sociedad española de historia de la farmacia* 39 (nos. 154–155) (1988): 63–81.

Herrera Hinojo, Pilar. "Contratos de aprendizaje en farmacia en el siglo XVI. Notas para la historia de la farmacia española." *Boletín de la sociedad española de historia de farmacia* 19 (no. 73) (1968): 73–79.

Herzog, Tamar. *Upholding Justice: Society, State, and the Penal System in Quito (1650–1750)*. Ann Arbor: University of Michigan Press, 2004.

Horkheimer, Max and Theodor Adorno. *Dialectic of Enlightenment*. New York: Herder and Herder, 1972.

Israel, Jonathan I. "Counter Reformation, Economic Decline, and the Delayed Impact of the Medical Revolution in Catholic Europe." In *Health Care and Relief in Counter-Reformation Europe*, edited by Ole Peter Grell, Andrew Cunningham, with Jon Arrizabalaga, 40–55. London: Routledge, 1999.

Jarcho, Saul. "Medicine in Sixteenth Century New Spain as Illustrated by the Writings of Bravo, Farfán and Vargas Machuca." *Bulletin of the History of Medicine* 31 (1957): 425–41.

Jarcho, Saul. "Origin of the American Indian as Suggested by Fray Joseph de Acosta (1589)." *Isis* 50 (4) (1959): 430–38.

Jenner, S.R. Mark and Patrick Wallis. *The Medical Marketplace*. In *Medicine and the Market in England and its Colonies, c.1450–1850*, 1–23. Basingstoke: Palgrave Macmillan, 2007.

Jouve Martín, José R. *The Black Doctors of Colonial Lima: Science, Race, and Writing in Colonial and Early Republican Peru*. Montreal: McGill-Queen's University Press, 2014.

Kamen, Henry. *The Spanish Inquisition: An Historical Revision*. London: Weidenfeld and Nicolson, 1997.

Katzew, Ilona. *Casta Painting: Images of Race in Eighteenth-Century Mexico*. New Haven: Yale University Press, 2004.

Kremers and Urdang's History of Pharmacy. Revised by Glenn Sonnedecker. 4th ed. Madison, WI: American Institute for the History of Pharmacy, 1986.

Lafuente, Antonio. "Enlightenment Science in an Imperial Context: Local Science in the Late-Eighteenth-Century Hispanic World." *Osiris* 15 (2001): 155–73.

Lanning, John Tate. *Academic Culture in the Spanish Colonies*. Oxford: Oxford University Press, 1940.

Lanning, John Tate. "Legitimacy and *Limpieza de Sangre* in the Practice of Medicine in the Spanish Empire." *Jahrbuch für Geschicte von Staat, Wirtschaft und Gesellschaft Lateinamerikas* 4 (1967): 37–60.

Lanning, John Tate. *The Royal Protomedicato: The Regulation of the Medical Professions in the Spanish Empire*, edited by John Jay TePaske. Durham: Duke University Press, 1985.

Lastres, Juan B. *Historia de la medicina peruana*. 3 vols. Lima: Impr. Santa María, 1951.

Laval, Enrique. *Botica de los Jesuitas de Santiago*. Santiago: Asociación Chilena de Asistencia Social, 1953.

Lavrin, Asunción. "Female Religious." In *Cities and Society in Colonial Latin America*, edited by Louisa Schell Hoberman and Susan Migden Socolow, 165–95. Albuquerque: University of New Mexico Press, 1986.

Leonard, Irving A. "Best Sellers of the Lima Book Trade." *Hispanic American Historical Review* 22 (1) (1942): 5–33.

Leonard, Irving A. "On the Lima Book Trade, 1591." *Hispanic American Historical Review* 33 (4) (1953): 511–25.

Leonard, Irving A. *Books of the Brave: Being an Account of Books and of Men in the Spanish Conquest and Settlement of the Sixteenth-century New World*. Berkeley and Los Angeles: University of California, 1992 [1949].

Leonard, Irving A. *Los libros del conquistador*. Mexico: Fondo de Cultura Económica, 1996 [1949].

Leong, Elaine. "Making Medicines in the Early Modern Household." *Bulletin of the History of Medicine* 82 (2008): 145–68.

Lévano Medina, Diego. "La administración de los bienes temporales en las cofradías limeñas del siglo XVII." In *Corporaciones religiosas y evangelización en Iberoamérica, siglos XVI–XVII*, edited by Diego Lévano Medina and Kelly Montoya Estrada, 111–42. Lima: Universidad Nacional Mayor San Marcos, 2010.

Lewis, Laura A. *Hall of Mirrors: Power, Witchcraft, and Caste in Colonial Mexico*. Durham, NC: Duke University Press, 2003.

Lindemann, Mary. *Medicine and Society in Early Modern Europe*. Cambridge: Cambridge University Press, 1999.

Lissón Chaves, Emilio, ed. *La iglesia de España en el Perú: colección de documentos para la historia de la iglesia en el Perú, que se encuentran en varios archivos*. 5 vols. Seville: Editorial Católica Española, 1943–1947.

Livingstone, David N. *Putting Science in its Place: Geographies of Scientific Knowledge*. Chicago: The University of Chicago Press, 2003.

Livingstone, David N. and Charles W.J. Withers, eds. *Geography and Enlightenment*. Chicago: The University of Chicago Press, 1999.

Lockhart, James. *Spanish Peru 1532–1560: A Colonial Society*. Madison: The University of Wisconsin Press, 1968.

Lohmann Villena, Guillermo. *Plata del Perú, riqueza de Europa: los mercaderes peruanos y el comercio con la metropolí en el siglo XVII*. Lima: Fondo Editorial del Congreso del Perú, 2004.

Lohmann Villena, Guillermo and María Justina Sarabia Viejo. *Francisco de Toledo: disposiciones gubernativas para el virreinato del Perú 1575–1580*. Seville: Escuela de Estudios Hispano-Americanos, 1989.

López Pérez, Miguel. "La influencia de la alquimia medieval hispana en la Europa moderna." *Asclepio* 54 (2002): 211–29.

López Pérez, Miguel. "Spanish Paracelsus Revisited and Decontaminated." *Azogue* 7 (2013) URL: http://www.revistaazogue.com. [Accessed 23 May 2016].

López Pérez, Miguel and Mar Rey Bueno. "Simón de Tovar (1528–1596): redes familiares, naturaleza Americana y comercio de maravillas en la Sevilla del XVI." *Dynamis* 26 (2006): 69–91.

López Piñero, José María. "Paracelsus and His Work in Sixteenth and Seventeenth Century Spain." *Clio Medica* 8 (2) (1973): 113–41.

López Piñero, José María. *Ciencia y técnica en la sociedad española de los siglos XVI y XVII*. Barcelona: Labor Universitaria 1979.

López Piñero, José María. "Las 'nuevas medicinas' americanos en la obra (1565–1574) de Nicolás Monardes." *Asclepio* 42 (1) (1990): 3–67.

López Piñero, José María and Francesc Bujosa Homar. "Tradición y renovación en la medicina española del siglo XVI." *Asclepio* 30–31 (1978–1979): 285–306.

López Piñero, José María and Francisco Calero, eds. *De pulvere febrifugo Occidentalis Indiae (1663) de Gaspar Caldera de Heredia y la introducción de la quina en Europa*. Valencia: Instituto de Estudios Documentales e Históricos sobre la Ciencia, Universidad de Valencia-CSIC, 1992.

López Terrada, María Luz. "Medical Pluralism in the Iberian Kingdoms: The Control of Extra-academic Practitioners in Valencia." *Medical History Supplement* 29 (2009): 7–25.

Lowry, Lyn Brandon. "Forging an Indian Nation: Urban Indians under Spanish Colonial Control, Peru 1535–1765." PhD diss. University of California, Berkeley, 1991.

Lozoya, Xavier. "Natural History and Herbal Medicine in Sixteenth-century America." In *Science in Latin America: A History*, edited by Juan José Saldaña, 29–50. Austin: University of Texas Press, 2006.

Lynch, John. *Spain under the Habsburgs*. 2 vols. Oxford: Blackwell, 1965.

MacCormack, Sabine. *On the Wings of Time: Rome, the Incas, Spain and Peru*. Princeton: Princeton University Press, 2009.

MacLeod, Roy. "Introduction." In *Disease, Medicine and Empire: Perspectives on Western Medicine and the Experience of European Expansion*, edited by Roy MacLeod and Milton Lewis, 1–18. London: Routledge, 1988.

MacLeod, Roy. "Nature and Empire: Science and the Colonial Enterprise." *Osiris* 15 (2000): 1–13.

Mangan, Jane E. *Trading Roles: Gender, Ethnicity and the Urban Economy in Colonial Potosí*. Durham, NC: Duke University Press, 2005.

Mangan, Jane E. *Transatlantic Obligations: Creating the Bonds of Family in Conquest-Era Peru and Spain*. New York: Oxford University Press, 2015.

Mannarelli, María Emma. "Inquisición y mujeres: las hechiceras en el Perú durante el siglo XVII." *Revista Andina* 3 (1) 1985: 141–56.

Mannarelli, María Emma. *Hechiceras, beatas y expósitas: mujeres y poder inquisitorial en Lima*. Lima: Fondo Editorial del Congreso del Perú, 1998.

Mannarelli, María Emma. *Private Passions and Public Sins: Men and Women in Seventeenth Century Lima*. Translated by Sidney Evans and Meredith D. Dodge. Albuquerque: University of New Mexico Press, 2007.

Mansilla Justo, Judith. "Poder y prestigio social en las cofradías españolas, siglo XVII y XVIII." In *Corporaciones religiosas y evangelización en Iberoamérica, siglos XVI–XVII*, edited by Diego Lévano Medina and Kelly Montoya Estrada, 229–72. Lima: Universidad Nacional Mayor San Marcos, 2010.

Marañón, Pedro Manuel Alonso. "Constituciones del colegio real de San Felipe y San Marcos de Lima." *Estudios de historia social y económica de América* 15 (1997): 419–34.

Martín, Luis. *The Intellectual Conquest of Peru: The Jesuit College of San Pablo, 1568–1767*. New York: Fordham University Press, 1968.

Martín, Luis. *Daughters of the Conquistadores: Women of the Viceroyalty of Peru*. Albuquerque: University of New Mexico Press, 1983.

Martín, Luís. "La biblioteca del Colegio de San Pablo (1568–1767), antecedente de la Biblioteca Nacional." In *La Biblioteca Nacional: aportes para su historia*, 25–37. http://www.comunidadandina.org/bda/docs/PE-CA-0015.pdf [Accessed 2 Jul. 2016].

Martín, Luis and Jo Ann Geurin Pettas. *Scholars and Schools in Colonial Peru*. Dallas, TX: Southern Methodist University, 1973.

Martínez, María Elena. *Genealogical Fictions: Limpieza de Sangre, Religion, and Gender in Colonial Mexico*. Stanford: Stanford University Press, 2008.

Martínez Bermejo, Saúl. "Beyond Luxury: Sumptuary Legislation in 17th-Century Castile." In *Making, Using and Resisting the Law in European History*, edited by Günther Lottes, Eero Medijainen, and Jon Vidar Sigurdsson, 94–108. Pisa: Plus-Pisa University Press, 2008.

Matallana, Gustavo Delgado. *Historia de la enseñanza médica en el Perú*. Lima: Universidad Nacional Mayor de San Marcos, 2008.

Maya Restrepo, Luz Adriana. "Botánica y medicina africanas en la Nueva Granada, siglo XVII." *Historia crítica* 19 (2000): 27–47.

McAlister, Lyle N. "Social Structure and Social Change in New Spain." *Hispanic American Historical Review* 43 (3) (1963): 349–70.

McClellan, James E. III. *Colonialism and Science: Saint Domingue in the Old Regime*. Baltimore: Johns Hopkins University Press, 1992.

McNeill, J.R. *Mosquito Empires: Ecology and War in the Greater Caribbean, 1620–1914*. Cambridge: Cambridge University Press, 2010.

McVaugh, Michael R. *Medicine before the Plague: Practitioners and their Patients in the Crown of Aragon 1285–1345*. Cambridge, Cambridge University Press, 1993.

McVaugh, Michael R. "Chemical Medicine in the Medical Writings of Arnau de Vilanova." II *Trobada Internacional d'Estudis sobre Arnau de Vilanova. Arxiu de Textos Catalans Antics* 23–24: (2004–2005): 239–64.

McVaugh, Michael R. "The 'Experience-Based Medicine' of the Thirteenth Century." *Early Science and Medicine* 14 (2009): 105–30.

Mecham, John Lloyd. *Church and State in Latin America: A History of Politico-Ecclesiastical Relations*. Chapel Hill: University of North Carolina Press, 1934.

Medina, José Toribio de. *La imprenta en Lima,* 1584–1824. 4 vols. Santiago: Impreso y grabado en casa del autor, 1904–1907.

Medina, José Toribio de. *Historia del tribunal de Santo Oficio de la Inquisición de Lima (1569–1820)*. Vol. 1. Santiago, Chile: Fondo Histórico y Bibliográfico J.T. Medina, 1956.

Meléndez, Mariselle and Karen Stolley. "Introduction: Enlightenments in Ibero-America." *Colonial Latin American Review* 24 (1) (2015): 1–16.

Memarzadeh, Maher. "Medical Practitioners in Early Colonial Mexico." PhD diss., UCLA, 2005.

Millones-Figueroa, Luis. "La historia natural del padre Bernabé Cobo: algunas claves para su lectura." *Colonial Latin American Review* 12 (1) (2003): 85–97.

Mills, Kenneth. *Idolatry and Its Enemies: Colonial and Andean Religion and* Extirpation, 1640–1750. Princeton: Princeton University Press, 1997.

Montes, Javier Alejo and María Carmen Rodríguez García. "Los estudios de la facultad de medicina en la Universidad de Salamanca de finales del siglo XVI." In *Espacio, tiempo y forma, Serie IV, Historia moderna* 7 (1994): 37–56.

Montoya Estrada, Kelly. "Una procesión de Viernes Santo en Lima XVII." In *Corporaciones religiosas y evangelización en Iberoamérica, siglos XVI–XVII*, edited by Diego Lévano Medina and Kelly Montoya Estrada, 143–60. Lima: Universidad Nacional Mayor San Marcos, 2010.

Moran, Bruce T. "The *Herbarius* of Paracelsus." *Pharmacy in History*, 35 (3) (1993): 99–127.

Moran, Bruce T. *Distilling Knowledge. Alchemy, Chemistry, and the Scientific Revolution*. Cambridge, Mass.: Harvard University Press, 2005.

Moreau, Emmanuelle. "Le Père Bernabé Cobo et la nomenclature botanique." *Caravelle* 82 (2004): 195–204.

Mörner, Magnus. *Race Mixture in the History of Latin America*. Boston: Little, Brown & Company, 1967.

Mörner, Magnus. "Economic Factors and Stratification in Colonial Spanish America with Special Regard to Elites." *Hispanic American Historical Review* 63 (2) (1978): 335–69.

Muldrew, Craig. *The Economy of Obligation: The Culture of Credit and Social Relations in Early Modern England.* Basingstoke: Macmillan, 1998.

Multhauf, Robert P. "John of Rupescissa and the Origin of Medical Chemistry." *Isis* 45 (2) (1954): 359–67.

Multhauf, Robert P. "Medical Chemistry and 'the Paracelsians'." *Bulletin of the History of Medicine* 28 (1954): 101–26.

Multhauf, Robert P. "The Significance of Distillation in Renaissance Medical Chemistry." *Bulletin of the History of Medicine* 30 (1956): 329–46.

Navarro Brotóns, Victor and William Eamon. "Spain and the Scientific Revolution: Questions and Conjectures." In *Más allá de la Leyenda Negra: España y la revolución científica*, edited by Victor Navarro Brotons and William Eamon, 27–38. Valencia: Instituto de Historia de la Ciencia y Documentación López Piñero, 2007.

Newman, William R. "From Alchemy to 'Chymistry'." *The Cambridge History of Science.* Vol. 3, edited by Katharine Park and Lorraine Daston, 497–517. Cambridge: Cambridge University Press, 2006.

Newson, Linda A. "A Historical-Ecological Perspective on Epidemic Disease." In *Advances in Historical Ecology*, edited by William Balée, 42–63. New York: Columbia University Press, 1998.

Newson, Linda A. *Life and Death in Early Colonial Ecuador*. Norman: University of Oklahoma Press, 1995.

Newson, Linda A. and Minchin, Susie. *From Capture to Sale: The Portuguese Slave Trade to Spanish South America in the Early Seventeenth Century.* Leiden: Brill, 2007.

Nieto Olarte, Mauricio. "Scientific Practices in the Sixteenth-Century Iberian Atlantic." In *Theorising the Ibero-American Atlantic*, edited by Harald E. Braun and Lisa Vollendorf, 141–58. Leiden: Brill, 2013.

Numbers, Ronald L. "Conclusion." In *Medicine in the New World: New Spain, New France, and New England*, edited by Ronald L. Numbers, 154–58. Knoxville: University of Tennessee Press, 1987.

O'Phelan, Scarlett. "El vestido como identidad étnica e indicador social de una cultura material." In *El barroco peruano* II, edited by Ramón Mujica Pinilla, 99–133. Lima: Banco de Crédito del Perú, 2003.

O'Toole, Rachel S. *Bound Lives: Africans, Indians, and the Making of Race in Colonial Peru.* Pittsburgh: University of Pittsburgh Press, 2012.

Osorio, Alejandra B. "'El callejón de la soledad': Vectors of Cultural Hybridity in Seventeenth-century Lima." In *Spiritual Encounters: Interactions between Christianity and Native Religions in Colonial America*, edited by Nicholas Griffiths and Fernando Cervantes, 198–229. Birmingham: University of Birmingham Press, 1999.

Osorio, Alejandra B. "Hechicerías y curanderías en la Lima del siglo XVII: Formas femeninas de control y acción social." In *Mujeres y género en la historia del Perú*, edited by Margarita Zegarra, 59–75. Lima: CENDOC-Mujer, 1999.

Osorio, Alejandra B. *Inventing Lima: Baroque Modernity in Peru's South Sea Metropolis*. New York: Palgrave Macmillan, 2008.

Otte, Enrique. *Cartas privadas de emigrantes a Indias, 1540–1616*. Mexico: Fondo de Cultura Económica, 1993.

Palmer, Richard. "Pharmacy in the Republic of Venice in the Sixteenth Century." In *The Medical Renaissance of the Sixteenth Century*, edited by A. Wear, R.K. French and I.M. Lonie, 100–17. Cambridge: Cambridge University Press, 1985.

Pardo-Tomás, José. "Autores médicos en los índices inquisitoriales españoles del siglo XVI." *Dynamis* 5–6 (1985–1986): 201–14.

Pardo-Tomás, José. *Ciencia y censura: la inquisición española y los libros científicos en los siglos* XVI *y* XVII. Madrid: CSIC, 1991.

Pardo-Tomás, José. "Two Glimpses of America from a Distance: Carolus Clusius and Nicolás Monardes." In *Carolus Clusius: Towards a Cultural History of a Renaissance Naturalist*, edited by Florike Egmond, Paul Hoftijzer and Robert Visser, 173–93. Amsterdam: Edita, 2007.

Pardo-Tomás, José. "'Antiguamente vivían más sanos que ahora': Explanations of Native Mortality in the *Relaciones Geográficas de Indias*." In *Medical Cultures of the Early Modern Spanish Empire*, edited by John Slater, Mariluz López Terrada and José Pardo-Tomás, 41–65. Farnham: Ashgate, 2014.

Pastor Frechoso, Feliz Francisco. *Boticas, boticarios y materia médica en Valladolid: siglos XVI y XVI*. Salamanca: Junta de Castilla y León. Salamanca: Consejería de Cultura y Turismo, 1993.

Perdiguero, Enrique. "Protomedicato y curanderismo." *Dynamis* 16 (1996): 91–108.

Peterson, Georg G. *Mining and Metallurgy in Ancient Perú. The Geological Society of America Special Paper 467*. Translation of *Minería y metalurgia en el antiguo Perú*, by William E. Brooks. Boulder, Col.: The Geological Society of America, Inc., 2010.

Pike, Ruth. *Aristocrats and Traders: Sevillian Society in the Sixteenth Century*. Ithaca, NY: Cornell University Press, 1972.

Pomata, Gianna. "Practising between Heaven and Earth: Women Healers in Seventeenth-century Bologna." *Dynamis* 19 (1999): 119–43.

Poole, Stafford. "The Politics of *Limpieza de Sangre*: Juan de Ovando and His Circle in the Reign of Philip ll." *The Americas* 55 (1999): 359–89.

Porter, Roy. "The Patient's View." *Theory and Society* 14 (2) (1985): 175–98.

Porter, Roy. *The Greatest Benefit to Mankind: The Medical History of Humanity from Antiquity to the Present*. London: HarperCollins, 1997.

Portuondo, María M. *Secret Science: Spanish Cosmography and the New World*. Chicago: University of Chicago Press, 2009.

Premo, Bianca. *Children of the Father King: Youth, Authority and Legal Minority in Colonial Lima*. Chapel Hill: University of North Carolina, 2005.

Prieto, Andrés I. *Missionary Scientists: Jesuit Science in Spanish South America, 1570–1810*. Nashville, TN: Vanderbilt University Press, 2011.

Puerto Sarmiento, Francisco Javier. "La farmacia renacentista española y la botica de El Escorial." In *La ciencia en el Monasterio del Escorial*, edited by Francisco Javier Campos and Fernández de Sevilla, 73–132. Madrid: Real Universitario Escorial-María Cristina, 1993.

Puerto Sarmiento, Francisco Javier. "Alquimia áurea: alquimia y destilación en la corte de Felipe II (1527–1598)." *Dynamis* 17 (1997): 107–40.

Quezada, Noemí. *Enfermedad y maleficio: el curandero en el México colonia*, 2nd ed. México: Universidad Nacional Autónoma de México, Instituto de Investigaciones Antropológicas, 2000.

Quiroz Chueca, Francisco. *Gremios coloniales peruanos*. Lima: Universidad Nacional Mayor de San Marcos, Seminario de Historia Rural Andina, 1991.

Quiroz Chueca, Francisco. "Compañías y gremios artesanales, Lima siglo XVI." *Sequilao* 1 (1992): 1–14.

Quiroz Chueca, Francisco. *Gremios, razas y libertad de industria: Lima colonial*. Lima: Universidad Nacional Mayor de San Marcos, 1995.

Quiroz Chueca, Francisco. "Gremios en Lima." In *Lima en el siglo XVI*, edited by Laura Gutiérrez Arbulú, 493–525. Lima: Instituto Riva-Agüero, 2005.

Quiroz Chueca, Francisco. *Artesanos y manufactureros*. Lima: Instituto de Estudios Peruanas, 2008.

Quiroz Chueca, Francisco. Artisans and Journeymen in Colonial Lima. Unpublished manuscript, 2014.

Rabí Chara, Miguel. *Del hospital de Santa Ana (1549 a 1924) al hospital nacional Arzobispo Loayza (1925 a 1999): 450 años de protección de la salud de las personas*. Lima: INDECOPI, 1999.

Rabí Chara, Miguel. "La primera botica de los hospitales de la ciudad de Lima en el siglo XVI." *Asclepio* 52 (2000): 269–80.

Rabí Chara, Miguel. *El hospital de San Bartolomé de Lima : la protección y asistencia de la gente de color de 1646 a 1821 y evolución posterior. Examen de un proceso histórico, social y sanitaria*. Lima: Grahuer, 2001.

Raina, Dhruv. "From West to Non-West Basalla's Three-stage Model Revisited." *Science as Culture* 8 (4) (1999): 497–516.

Ralley, Robert. "Medical Economies in Fifteenth Century England." In *Medicine and the Market in England and its Colonies, c.1450–1850*, edited by Mark Jenner and Patrick Wallis, 24–46. New York: Palgrave Macmillan, 2007.

Ramírez, Susan E. *Provincial Patriarchs: Land Tenure and the Economics of Power in Colonial Peru*. Albuquerque: University of New Mexico Press, 1985.

Ramos, Gabriela. *Death and Conversion in the Andes: Lima and Cuzco, 1532–1670*. Notre Dame, Indiana: University of Indiana Press, 2010.

Ramos, Gabriela. "Indian Hospitals and Government in the Colonial Andes." *Medical History* 57 (2) (2013): 186–205.

Reedy, Daniel R. *The Poetic Art of Juan Valle Caviedes*. Chapel Hill: University of North Carolina, 1964.

Reff, Daniel T. *Disease, Depopulation, and Culture Change in Northwestern New Spain, 1518–1764*. Salt Lake City: University of Utah Press, 1991.

Rey Bueno, Mar. "El informe Valles: los desdibujados límites del arte de boticarios a finales del siglo XVI (1589–1594)." *Asclepio* 56 (2) (2004): 243–68.

Rey Bueno, Mar. "Concordias medicinales de entrambos mundos." *Revista de Indias* 66 (no. 237) (2006): 347–62.

Rey Bueno, Mar. "Los proyectos sobre materia medica peruana de Antonio de Robles Cornejo y Matías de Porres (1617–1621)." In *Rotas da Natureza: cientistas, vaigens, expedições, instituições*, edited by Ana Leonor Pereira and João Pita, 27–31. Coimbra: Universidade de Coimbra, 2006.

Rey Bueno, Mar. "Los paracelsistas españoles: medicina química en la España moderna." In *Más allá de la Leyenda Negra: España y la revolución científica*, edited by Victor Navarro Brotons and William Eamon, 41–55. Valencia: Instituto de Historia de la Ciencia y Documentación López Piñero, 2007.

Rey Bueno, Mar and María Esther Alegre Pérez. "Los distiladores de Su Magestad: distilación, espagiria y paracelsismo en la corte de Felipe II." *Dynamis* 21 (2001): 323–50.

Rey Bueno, Mar and María Esther Alegre Pérez. "Renovación en la terapéutica real: los destiladores de su magestad, maestros simplicistas y médicos herbolarios de Felipe II." *Asclepio* 53 (1) (2001): 27–55.

Ribeiro, Aileen. "Dress in the Early Modern Period, c.1500–1780." In *The Cambridge History of Western Textiles*, edited by David Jenkins, 659–90. Cambridge: Cambridge University Press, 2003.

Ricardo, Antonio. "Testamento de don Antonio Ricardo." *Revista del Archivo Nacional del Perú* 19 (1955): 293–305.

Risse, Guenter B. "Transcending Cultural Barriers: the European Reception of Medicinal Plants from the Americas." In *Botanical Drugs of the Americas in the Old and New Worlds: Invitational Symposium at the Washington-Congress, 1983*, edited by Wolfgang-Hagen Hein, 31–42. (Amerikanische pflanzliche Arzneien in der Alten und Neuen Welt : Einladungs-Symposium anlässlich des Kongresses in Washington, 1983). Stuttgart: Wissenschaftliche Verlagsgesellschaft, 1984.

Risse, Guenter B. "Medicine in New Spain." In *Medicine in the New World: New Spain, New France, and New England*, edited by Ronald L. Numbers, 12–63. Knoxville: University of Tennessee Press, 1987.

Risse, Guenter B. *Mending Bodies, Saving Souls: A History of Hospitals*. New York: Oxford University Press, 1999.

Rizo-Patrón, Paul. "La nobleza de Lima en tiempos de los Borbones." *Bulletin de l'Institut Français d'Études Andines* 19 (1) (1990): 129–63.

Rocco, Fiammetta. *Miraculous Fever-Tree: Malaria, Medicine and the Cure That Changed the World*. New York: HarperCollins, 2003.

Rodríguez Guerrero, José. "Censura y Paracelsismo durante el reinado de Felipe II."*Azogue* 4 (2001) URL: http://www.revistaazogue.com [Accessed 2 Jul. 2016].

Rodríguez-Buckingham, Antonio. "Change and the Printing Press in Sixteenth-Century Spanish America." In *Agent of Change: Print Culture Studies after Elizabeth L. Einstein*, edited by Sabrina Alcorn Baron, Eric N. Lindquist and Eleanor F. Shevlin, 216–37. Amherst and Boston: University of Massachusetts Press, 2007.

Rojo Vega, Anastasio. *Enfermos y sanadores en la Castilla del siglo XVI*. Valladolid: Universidad De Valladolid, 1993.

Romero Romero, Catalina. "Libros, cultura y sociedad en el Perú Virreinal: Las bibliotecas de los colegios jesuitas de Arequipa, Potosí y las misiones de Moxos en el siglo XVIII." 2 vols. PhD diss., Universidad de Complutense de Madrid, 1992.

Ronderos, Paula. "El arte de boticarios durante la primera mitad del siglo XVII en el Nuevo Reino de Granada." *Fronteras de historia* 12 (2007): 175–96.

Ronderos, Paula. *El dilema de los rótulos: Lectura del inventario de una botica santafereña de comienzos del siglo XVII*. Bogotá: Pontificia Universidad Javeriana, 2007.

Sabater, Pilar Gardeta. "El nuevo modelo del Real Tribunal del Protomedicato en la América española: trasformaciones sufridas ante las Leyes de Indias y el cuerpo legislativo posterior." *Dynamis* 16 (1996): 237–59.

Saldaña, Juan José. "Science and Public Happiness during the Latin American Enlightenment." In *Science in Latin America: A History*, edited by Juan José Saldaña, 50–92. Austin: University of Texas Press, 2006.

Sánchez, Ana. *Amancebados, hechicheros y rebeldes (Chancay, siglo XVII)*. Cuzco: Centro de estudios regionales andinos, Bartolomé de las Casas, 1991.

Sánchez-Menchero, Mauricio. "'From Where They Are Now to Whence They Came From': News About Health and Disease in New Spain (1550–1615)." In *Medical Cultures of the Early Modern Spanish Empire*, edited by John Slater, Mariluz López Terrada and José Pardo-Tomás, 91–105. Farnham: Ashgate, 2014.

Sayre, Meredith Beck. "Cultivating Soils and Souls: The Jesuit Garden in the Americas." MA diss., Simon Fraser University, 2005.

Schiebinger, Londa. *Plants and Empire: Colonial Bioprospecting in the Atlantic World*. Cambridge, Mass.: Harvard University Press, 2004.

Schiebinger, Londa and Claudia Swan, eds. *Colonial Botany: Science, Commerce, and Politics in the Early Modern World*. Philadelphia: University of Pennsylvania Press, 2005.

Schurz, William Lyle. "Mexico, Peru and the Manila Galleon." *Hispanic American Historical Review* 1 (1918): 389–402.

Scott, Heidi V. *Contested Territory: Mapping Peru in the Sixteenth and Seventeenth Centuries*. Notre Dame, IN: University of Notre Dame Press, 2009.

Shapin, Steven. *The Scientific Revolution*. Chicago: University of Chicago Press, 1996.

Silverblatt, Irene. *Modern Inquisitions: Peru and the Colonial Origins of the Civilized World*. Durham: Duke University Press, 2004.

Siraisi, Nancy G. *Medieval and Early Renaissance Medicine: An Introduction to Knowledge and Practice*. Chicago: University of Chicago, 1990.

Slater, John. "The Green Gold Fallacies: Myth and Reality in the Transatlantic Trade in Medicinal Plants (1493–1663)." In *Geografías médicas: orillas y fronteras culturales de la medicina hispanomericana (siglos XVI y XVII)*, edited by José Pardo-Tomás and Mauricio Sánchez Menchero, 99–122. Mexico DF: Universidad Autónoma de México, 2014.

Slater, John and Andrés Prieto. "Was Spanish Science Imperial?" *The Colorado Review of Hispanic Studies* 7 (2009): 3–10.

Smith, Pamela H. *The Business of Alchemy: Science and Culture in the Holy Roman Empire*. Princeton: Princeton University Press, 1997.

Smith, Pamela H. "Science on the Move: Recent Trends in the History of Early Modern Science." *Renaissance Quarterly* 62 (2) (2009): 345–75.

Smith, Pamela H., Amy R. Meyers and Harold J. Cook, eds. *Ways of Making and Knowing: The Material Culture of Empirical Knowledge*. Ann Arbor: University of Michigan Press, 2014.

Socolow, Susan Migden. *The Women of Colonial Latin America*. Cambridge: Cambridge University Press, 2000.

Solano Alonso, J. *Salud, cultura y sociedad en Cartagena de Indias siglos XVI y XVII*. Universidad del Atlántico: Barranquilla, 1998.

Solomon, Michael. *Fictions of Well-Being: Sickly Readers and Vernacular Medical Writing in Late Medieval and Early Modern Spain*. Philadelphia: University of Pennsylvania Press, 2010.

Sorní, Javier and José María Suñe. "Barcelona. Baja edad media: especieros o boticarios?" *Boletín de la sociedad española de historia de la farmacia* 34 (no. 136) (1983): 139–50.

Sorní, Javier and José María Suñe. "La farmacia en Barcelona desde *Alfons el Benigne* a *Pere el Ceremoniós* (1327–1387)." *Boletín de la sociedad española de historia de la farmacia* 36 (nos. 141–42) (1985): 67–79.

Sosadias, Inés. "El negro curandero en la Inquisición de Cartagena de Indias siglo XVII." MA diss., Universidad de los Andes, 1981.

Sowell, David. *The Tale of Healer Miguel Perdomo Neira: Medicine, Ideologies, and Power in the Nineteenth-Century Andes*. Wilmington, Delaware: Scholarly Resources Inc., 2001.

Spargo, John Webster. *Virgil the Necromancer: Studies in Virgilian Legends*. Cambridge, Mass.: Harvard University Press, 1934.

Stephenson, Marcia. "From Marvelous Antidote to the Poison of Idolatry: The Transatlantic Role of Andean Bezoar Stones during the Late Sixteenth and Early Seventeenth Centuries." *Hispanic American Historical Review* 90 (1) (2010): 3–39.

Stolberg, Michael. *Experiencing Illness and the Sick Body in Early Modern Europe*. Basingstoke: Palgrave Macmillan, 2011.

Strochhia, Sharon T. "The Nun Apothecaries of Renaissance Florence: Marketing Medicines in the Convent." *Renaissance Studies* 25 (5) (2011): 627–47.

Studnicki-Gizbert, Daviken. A Nation Upon the Ocean Sea*: Portugal's Atlantic Diaspora and the Crisis of the Spanish Empire, 1492–1640*. Oxford: Oxford University Press, 2007.

Suárez, Margarita. *Comercio y fraude el en Perú colonial: las estrategias mercantiles de un banquero*. Lima: Banco Central de Reserva del Perú, Fondo Editorial: Instituto de Estudios Peruanos, 1995.

Suárez, Margarita. *Desafíos transatlánticos: mercaderes, banqueros y el estado en el Perú virreinal, 1600–1700*. Lima: Pontificia Universidad Católica del Perú, Instituto Riva-Agüero: Fondo de Cultura Económica: Instituto Francés de Estudios Andinos, 2001.

Sweet, James H. *Domingos Álvares, African Healing, and the Intellectual History of the Atlantic World*. Chapel Hill: University of North Carolina Press, 2011.

Urdang, George. "The Development of Pharmacopoeias." *Bulletin of the World Health Organisation* 4 (4) (1951): 577–603.

Urdang, George. "How Chemicals Entered the Official Pharmacopoeias." *Archives internationales d'histoire des sciences* 7 (1954): 303–14.

Valcarcel, Carlos Daniel. *Historia de la educación colonial*. Lima: Editorial Universo, 1968.

Valdizán, Hermilio and Angel Maldonado. *La medicina popular peruana*. 3 vols. Lima: Torres Aguirre, 1922.

Valle y Caviedes, Juan del. *Obra completa*. Edited by Daniel R. Reedy. Caracas: Biblioteca Ayacucho, 1984.

Valverde, José Luis. "Regulaciones de farmacia contenidas en las ordenanzas municipales de las ciudades españolas." *Sonderdruck aus Veröffentlichungen der Internationalen Gesellschaft für Geschichte der Pharmazie* [Die Vörträge der Hauptversammlung in Luxemburg] 36: 135–57. Stuttgart, 1970.

Valverde, José Luis and Guillermina López Andújar. "Los disposiciones legales sobre el certificado de 'limpieza de sangre' en relación al ejército profesional farmacéutico." *Boletín de la sociedad española de la historia de la farmacia* 31 (1981): 259–71.

Van Deusen, Nancy E. *Between the Sacred and the Worldly: The Institutional and Cultural Practice of Recogimiento in Colonial Lima*. Stanford: Stanford University Press, 2001.

Van Deusen, Nancy E. *The Souls of Purgatory: The Spiritual Diary of a Seventeenth-Century Afro-Peruvian Mystic, Ursula de Jesús*. Albuquerque: University of New Mexico Press, 2004.

Van Deusen, Nancy E. "Circuits of Knowledge among Women in Early Seventeenth-Century Lima." In *Gender, Race and Religion in the Colonization of the Americas*, edited by Nora E. Jaffary, 137–50. Aldershot: Ashgate, 2007.

Varey, Simon, ed. *The Mexican Treasury: The Writings of Dr. Francisco Hernández*. Translated by Cynthia L. Chamberlain and Simon Varey. Stanford: Stanford University Press, 2002.

Varey, Simon, Rafael Chabrán and Dora B. Weiner, eds. *Searching for the Secrets of Nature: The Life and Works of Dr. Francisco Hernández*. Stanford: Stanford University Press, 2002.

Vargas Ugarte, Rubén. *Impresos peruanos (1584–1650)*. Lima: Universidad Nacional San Marcos, 1953.

Vargas Ugarte, Rubén. *Historia de la iglesia en el Perú*. Tomo II (1570–1640). Burgos: Imprenta de Aldecoa, 1959.

Vega Jácome, Walter. "Cofradías limeñas." In *Lima en el siglo XIV*, edited by Laura Gutiérrez Arbulú, 703–52. Lima: Pontificia Universidad Católica del Perú, Instituto Riva-Agüero, 2005.

Vergara, Teresa C. "Growing Up Indian: Migration, Labor, and Life in Lima (1570–1640)." In *Raising an Empire: Children in Early Modern Iberia and Colonial Latin America*, edited by Ondina E. González and Bianca Premo, 75–106. Albuquerque: University of New Mexico Press, 2007.

Vila Vilar, Enriqueta. *Hispanoamérica y el comercio de esclavos: los asientos Portugueses*. Seville: Escuela de Estudios Hispanoamericanos, 1977.

Vila Vilar, Enriqueta. "Las ferias de Portobelo: apariencia y realidad del comercio con Indias." *Anuario de estudios americanos* 39 (1982): 275–340.

Walker, Tamara J. "'He Outfitted his Family in Notable Decency': Slavery, Honour and Dress in Eighteenth-Century Lima." *Slavery and Abolition* 30 (3) (2009): 383–402.

Walker, Timothy D. *Doctors, Folk Medicine and the Inquisition: the Repression of Magical Healing in Portugal during the Enlightenment*. Leiden: Brill, 2005.

Wallis, Patrick. "Consumption, Retailing, and Medicine in Early-Modern London." *Economic History Review* 61 (1) (2008): 26–53.

Ward, Christopher. *Imperial Panama: Commerce and Conflict in Isthmian America, 1550–1800*. Albuquerque: University of New Mexico Press, 1994.

Warren, Adam. *Medicine and Politics in Colonial Peru: Population Growth and the Bourbon Reforms*. Pittsburgh, PA: University of Pittsburgh Press, 2010.

Wear, Andrew. "The Early Modern Debate about Foreign Drugs: Localism versus Universalism in Medicine." *The Lancet* 354 (1999): 149–51.

Wear, Andrew. *Knowledge and Practice in English Medicine, 1550–1680*. Cambridge University Press, Cambridge, 2000.

Whittet, T.D. "Pepperers, Spicers and Grocers – Forerunners of the Apothecaries." *Proceedings of the Royal Society of Medicine* 61 (1968): 25–30.

Winterbottom, Anne E. "Of the China Root: A Case study of the Early Modern Circulation of Materia Medica." *Social History of Medicine*, 28 (1) (2015): 22–44.
Withers, Charles W.J. *Placing the Enlightenment: Thinking Geographically about the Age of Reason*. Chicago: the University of Chicago Press, 2007.
Withers, Charles W.J. and David N. Livingstone. "Introduction: On Geography and Enlightenment." In *Geography and Enlightenment*, edited by David N. Livingstone and Charles W.J. Withers, 1–28. Chicago: The University of Chicago Press, 1999.
Zarate, Eduardo C. *Inicios de la escuela de medicina de Lima*. Lima: No publisher, 2005.

Index

www.ingramcontent.com/pod-product-compliance
Ingram Content Group UK Ltd.
Pitfield, Milton Keynes, MK11 3LW, UK
UKHW022224051026
15624UKWH00002B/346

9 789004 350632